Biomineralization, Environmental Microbiology and Earth science

Biomineralization, Environmental Microbiology and Earth science

Edited by **Joe Carry**

SYRAWOOD
PUBLISHING HOUSE

New York

Published by Syrawood Publishing House,
750 Third Avenue, 9th Floor,
New York, NY 10017, USA
www.syrawoodpublishinghouse.com

Biomineralization, Environmental Microbiology and Earth science
Edited by Joe Carry

International Standard Book Number: 978-1-68286-017-5 (Hardback)

Contents

Preface

Biomineralization, environmental microbiology and earth science are closely related fields with interdisciplinary concepts and applications. This book encompasses research work of internationally renowned scientists and academicians. It aims to broaden the scope of these disciplines and aid their progress. Some significant topics discussed in this text are ecohydrology, geochemistry, climate variability, etc. It will prove extremely beneficial for students, researchers and professionals engaged in these fields.

After months of intensive research and writing, this book is the end result of all who devoted their time and efforts in the initiation and progress of this book. It will surely be a source of reference in enhancing the required knowledge of the new developments in the area. During the course of developing this book, certain measures such as accuracy, authenticity and research focused analytical studies were given preference in order to produce a comprehensive book in the area of study.

This book would not have been possible without the efforts of the authors and the publisher. I extend my sincere thanks to them. Secondly, I express my gratitude to my family and well-wishers. And most importantly, I thank my students for constantly expressing their willingness and curiosity in enhancing their knowledge in the field, which encourages me to take up further research projects for the advancement of the area.

Editor

Spatiotemporal variations of nitrogen isotopic records in the Arabian Sea

S.-J. Kao[1], B.-Y. Wang[1], L.-W. Zheng[1], K. Selvaraj[1], S.-C. Hsu[2], X. H. Sean Wan[1], M. Xu[1], and C.-T. Arthur Chen[3]

[1]State Key Laboratory of Marine Environmental Science, Xiamen University, Xiamen, China
[2]Research Center for Environmental Changes, Academia Sinica, Taipei, Taiwan
[3]Department of Oceanography, National Sun Yat-sen University, Kaohsiung, Taiwan

Correspondence to: S.-J. Kao (sjkao@xmu.edu.cn)

Abstract. Available reports of dissolved oxygen, $\delta^{15}N$ of nitrate ($\delta^{15}N_{NO_3}$) and $\delta^{15}N$ of total nitrogen ($\delta^{15}N_{bulk}$) for trap material and surface/downcore sediments from the Arabian Sea (AS) were synthesized to explore the AS' past nitrogen dynamics. Based on 25 $\mu mol\,kg^{-1}$ dissolved oxygen isopleth at a depth of 150 m, we classified all reported data into northern and southern groups. By using $\delta^{15}N_{bulk}$ of the sediments, we obtained geographically distinctive bottom-depth effects for the northern and southern AS at different climate stages. After eliminating the bias caused by bottom depth, the modern-day sedimentary $\delta^{15}N_{bulk}$ values largely reflect the $\delta^{15}N_{NO_3}$ supply from the bottom of the euphotic zone. Additionally to the data set, nitrogen and carbon contents vs. their isotopic compositions of a sediment core (SK177/11) collected from the most southeastern part of the AS were measured for comparison. We found a one-step increase in $\delta^{15}N_{bulk}$ starting at the deglaciation with a corresponding decrease in $\delta^{13}C_{TOC}$ similar to reports elsewhere revealing a global coherence. By synthesizing and reanalyzing all reported down core $\delta^{15}N_{bulk}$, we derived bottom-depth correction factors at different climate stages, respectively, for the northern and southern AS. The diffusive sedimentary $\delta^{15}N_{bulk}$ values in compiled cores became confined after bias correction revealing a more consistent pattern except recent 6 ka. Such high similarity to the global temporal pattern indicates that the nitrogen cycle in the entire AS had responded to open-ocean changes until 6 ka BP. Since 6 ka BP, further enhanced denitrification (i.e., increase in $\delta^{15}N_{bulk}$) in the northern AS had occurred and was likely driven by monsoon, while, in the southern AS, we observed a synchronous reduction in $\delta^{15}N_{bulk}$, implying that nitrogen fixation was promoted correspondingly as the intensification of local denitrification at the northern AS basin.

1 Introduction

Biogeochemical processes of nitrogen in the ocean are intimately related to various elemental cycles (synergistically modulate atmospheric CO_2 and N_2O concentrations), hence the feedback on the climate being on a millennial time scale (Gruber, 2004; Falkowski and Godfrey, 2008; Altabet et al., 2002). Though oxygen deficient zones (ODZs) occupy only $\sim 4\%$ of ocean volume, the denitrification process therein contributes remarkably to the losses of nitrate, leaving excess P in the remaining water mass to stimulate N_2 fixation while entering the euphotic zone (Morrison et al., 1998; Deutsch et al., 2007) and thus controlling the budget of bioavailable nitrogen in ocean. Denitrification leaves $^{15}NO_3^-$ in residual nitrate (Sigman et al., 2001), whereas N_2 fixation introduces new bioavailable nitrogen with low $\delta^{15}N$ values (Capone et al., 1997) into ocean for compensation. The Arabian Sea (AS) – one of the three largest ODZs in the world ocean with distinctive monsoon driven upwelling – accounts for at least one-third of the loss of marine fixed nitrogen (Codispoti and Christensen, 1985) playing an important role in the past climate via regulating atmospheric N_2O concentration (Agnihotri et al., 2006) or nitrogen inventory to modulate CO_2 sequestration through a biological pump (Altabet, 2006).

Sedimentary nitrogen isotope, measured as standard δ notation, with respect to standards of atmospheric nitrogen, is an important tool to study the past marine nitrogen cycle.

Nitrogen isotope compositions of sedimentary organic matter potentially reflect biological processes in water columns, such as denitrification (Altabet et al., 1995; Ganeshram et al., 1995, 2000), nitrogen fixation (Haug et al., 1998), and the degree of nitrate utilization by algae (Altabet and Francois, 1994; Holmes et al., 1996; Robinson et al., 2004). However, alteration may occur (through various ways or processes; e.g., diagenesis) before the signal of $\delta^{15}N$ of exported production is buried.

Previous measurements of $\delta^{15}N_{bulk}$ in various cores and surface sediments in the AS showed the following points: (1) near-surface NO_3^- in the AS is completely utilized in an annual cycle, resulting in small isotopic fractionation between $\delta^{15}N$ of exported sinking particles and $\delta^{15}N$ of NO_3^- supplied to the euphotic zone (Altabet, 1988; Thunell et al., 2004); (2) monsoon-driven surface productivity and associated oxidant demand were regarded as the main control on water column denitrification in the past (Ganeshram et al., 2000; Ivanochko et al., 2005); (3) sedimentary $\delta^{15}N_{bulk}$ primarily reflects the relative intensity of water column denitrification in this area (Altabet et al., 1995, 1999); (4) oxygen supply at intermediate depth by the Antarctic intermediate waters (AAIWs) can modulate the denitrification intensity in the northern AS (Schulte et al., 1999; Schmittner et al., 2007; Pichevin et al., 2007). Among previous research, the geographical features in sedimentary $\delta^{15}N_{bulk}$ between the north and south basins of the AS have not been discussed, particularly on the basis of bottom-depth effect, which might be different during glacial and interglacial periods.

In this study, a sediment core (SK177/11) collected from the slope of the southeastern AS was measured for organic C and N contents and their stable isotopes. We synthesized previous hydrographical and isotopic data, such as dissolved oxygen (DO), N^* ($N^* = NO_3^- 16 \times PO_4^{3-} + 2.9$; Gruber and Sarmiento, 2002), and $\delta^{15}N$ of nitrate, as well as trapped material and surface/downcore sediments, among which surface and downcore sediments may have experienced more intensified diagenetic alteration. Based on the subsurface of a DO concentration of $25 \mu mol\,kg^{-1}$ isopleth at 150 m, the data sets in the AS were separated into north and south basins by time span (glacial, Holocene and modern) for comparison. We aim to (1) investigate the geographic and glacial–interglacial differences in bottom-depth effect and to (2) retrieve extra information from sedimentary $\delta^{15}N_{bulk}$ by removing basin/climate stage specific bottom-depth effects, thus better deciphering the environmental history of the Arabian Sea.

2 Study area

The Arabian Sea is characterized by seasonal reversal of monsoon winds, resulting in large seasonal physical/hydrographic/biological/chemical variations in water columns (Nair et al., 1989). Cold and dry northeasterly winds

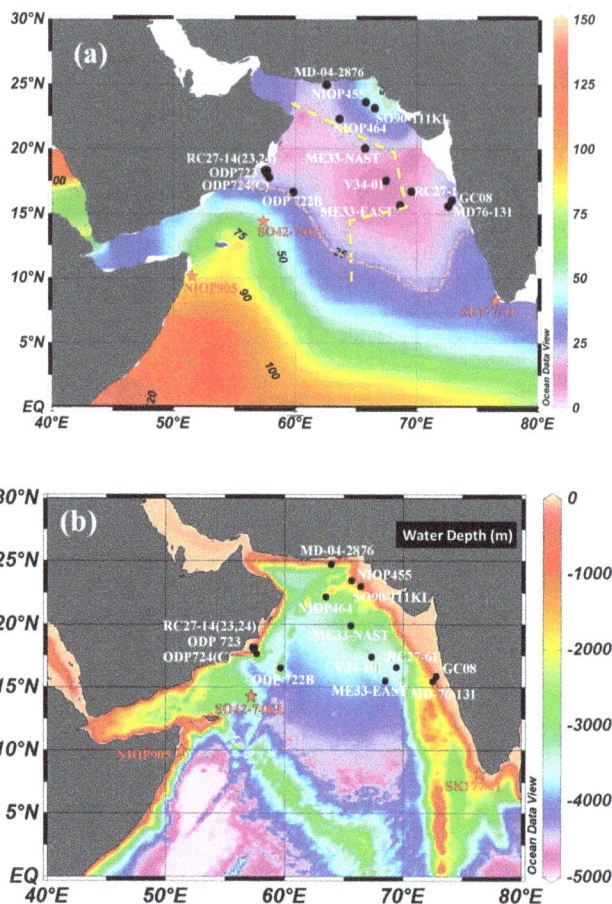

Figure 1. (a) Map of the Arabian Sea. Dissolved oxygen (DO) concentration at 150 m (World Ocean Atlas 09) was shown in color contour. Southern ($\star$) and northern ($\bullet$) categories of available cores and SK177/11, in this study, were defined by DO of $25 \mu mol\,kg^{-1}$ (see text; purple dash curve). (b) Bathymetric map superimposed by core locations; (c), (d) and (e) are DO, nitrate and N^* transects (yellow dashed line in (a), online data originated from cruises of JGOFS in 1995), respectively, for upper 2000 m. (f) N^* transect for the upper 300 m with arrows revealing the flow direction. In (a), the northern cores include core MD-04-2876 (828 m; Pichevin et al., 2007), core NIOP455 vs. NIOP464 (1002 m vs. 1470 m; Reichart et al., 1998), SO90-111KL vs. ME33-NAST (775 m vs. 3170 m; Suthof et al., 2001), ODP724C vs. ME33-EAST (603 m vs. 3820 m; Möbius et al., 2011), RC27-24 vs. RC27-61 (1416 m vs. 1893 m; Altabet et al., 1995), ODP723, ODP722(B) vs. V34-101 (808 m, 2028 m vs. 3038 m; Altabet et al., 1999), RC27-14 vs. RC27-23 (596 m vs. 820 m; Altabet et al., 2002), GC08 (2500 m; Banakar et al., 2005), MD-76-131 (1230 m; Ganeshram et al., 2000); the southern cores include core SO42-74KL (3212 m; Suthof et al., 2001), NIOP905 (1586 m; Ivanochko et al., 2005) and SK177/11 (776 m; this study).

blow during winter from a high-pressure cell of the Tibetan Plateau, whereas heating of the Tibetan Plateau in summer (June to September) reverses the pressure gradient leading to warm and moist southwesterly winds and

Table 1. Accelerator mass spectrometry (AMS) ^{14}C dates of sediment core SK177/11. Radiocarbon ages were calibrated using the CALIB 6.0 program (http://calib.qub.ac.uk/calib/calib.html, Reimer et al., 2009).

Lab code	Depth cm	Dating materials	pMC	Raw ^{14}C age (yr BP)	Calibrated age (yr BP) (1σ)	δ^{13}C (‰)
KIA24386	58	OM	65.58 ± 0.17	3390 ± 20	3186 ± 24	-18.55 ± 0.04
KIA26327	125	OM	46.65 ± 0.20	6125 ± 35	6504 ± 26	-20.02 ± 0.10
KIA24387	155	OM	31.38 ± 0.13	9310 ± 30	10054 ± 104	-19.50 ± 0.08
KIA26328	175	OM	21.96 ± 0.12	12180 ± 45	13618 ± 104	-17.71 ± 0.18
KIA24388	205	OM	13.94 ± 0.11	15830 ± 60	18646 ± 54	-21.65 ± 0.15
KIA24389	275	OM	9.81 ± 0.12	$18650 + 100(-90)$	21774 ± 194	-18.02 ± 0.10
KIA26329	355	OM	2.76 ± 0.06	28830 ± 180	32857 ± 207	-19.23 ± 0.17

OM – organic matter; pMC – percent modern.

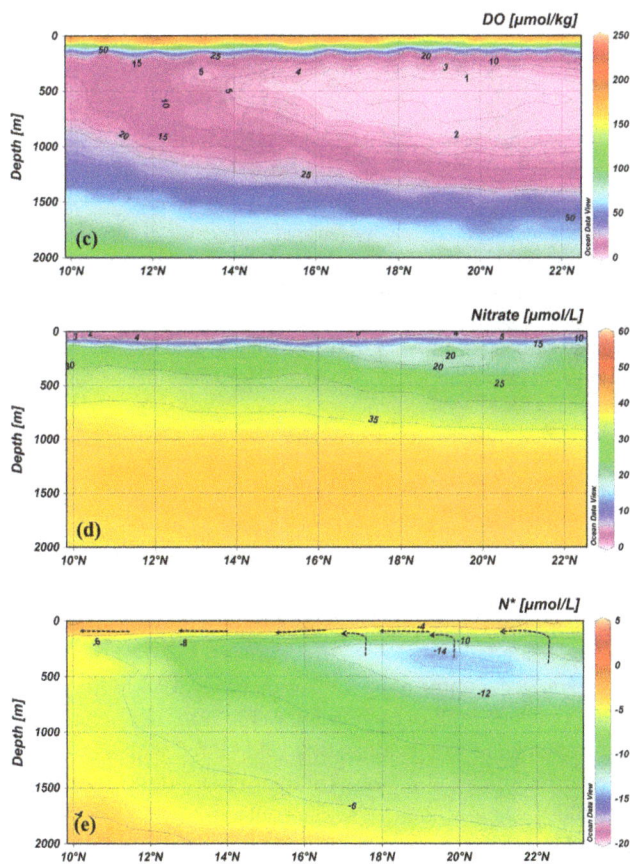

Figure 1. Continued.

Figure 1. Continued.

precipitation maximum. In the present day, the SW monsoon is much stronger than its northeastern counterpart.

The spatial distribution of DO at a depth of 150 m for the AS is shown in Fig. 1a (World Ocean Atlas 2009, http://www.nodc.noaa.gov/OC5/WOA09/woa09data.html), which shows a clear southwardly increasing pattern with DO having increased from ~ 5 to $> 100\,\mu\mathrm{mol\,kg^{-1}}$, and the lowest DO value appears northeast of the northern basin. As den-

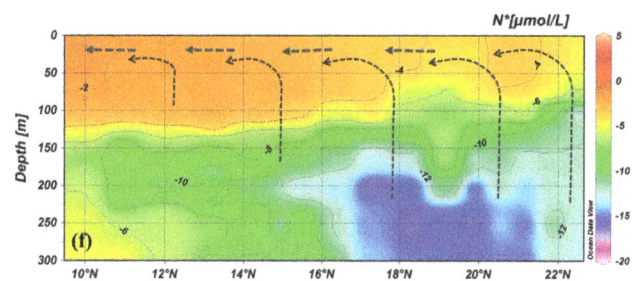

itrification, the dominant nitrate removal process generally occurs in the water column, where DO concentration ranges from 0.7 to $20\,\mu\mathrm{mol\,kg^{-1}}$ (Paulmier et al., 2009). The intensity of denitrification was reported to descend gradually, corresponding to the DO spatial pattern from the northern to the southern parts of the AS, and did not become obvious at 11 or 12° N (Naqvi et al., 1982). As indicated by the upper 2000 m N–S transect of DO (Fig. 1c), a southwardly decrease in ODZ thickness can be observed and the contour line of $5\,\mu\mathrm{mol\,kg^{-1}}$ extends to around 13° N. Since the nitrate source is mainly from the bottom of the euphotic zone at around 150 m, we postulate a geographically distinctive sedimentary δ^{15}N$_{\mathrm{bulk}}$ underneath ODZs. Thus, an isopleth of $25\,\mu\mathrm{mol\,kg^{-1}}$ DO at 150 m is applied as a geographic boundary to separate the northern from the southern part of the AS basin. The interface where DO concentration changed from 20 to $30\,\mu\mathrm{mol\,kg^{-1}}$ was such a transition zone. On the other hand, the bottom layer of the ODZ moves shallower toward the south, as shown previously by Gouretski and Koltermann (2004). Accordingly, the bottom oxygen content may also be a factor to influence the degree of alteration in sedimentary δ^{15}N$_{\mathrm{bulk}}$.

As mentioned in the introduction, nitrate is removed via denitrification in ODZs resulting in excess P to stimulate N$_2$ fixation. In Figs. 1d, e and f, we presented the N–S transect of nitrate and N^* (for both the upper 2000 m and 300 m) in January. Even though there is nitrate in the surfacewater

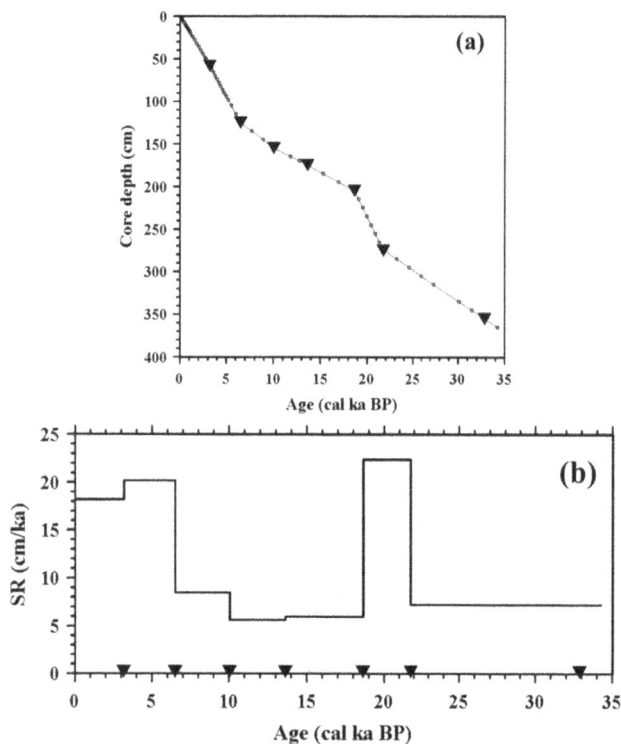

Figure 2. (a) Plot of calendar age against depth; (b) Linear sedimentation rate (indicates the ^{14}C age controlling points).

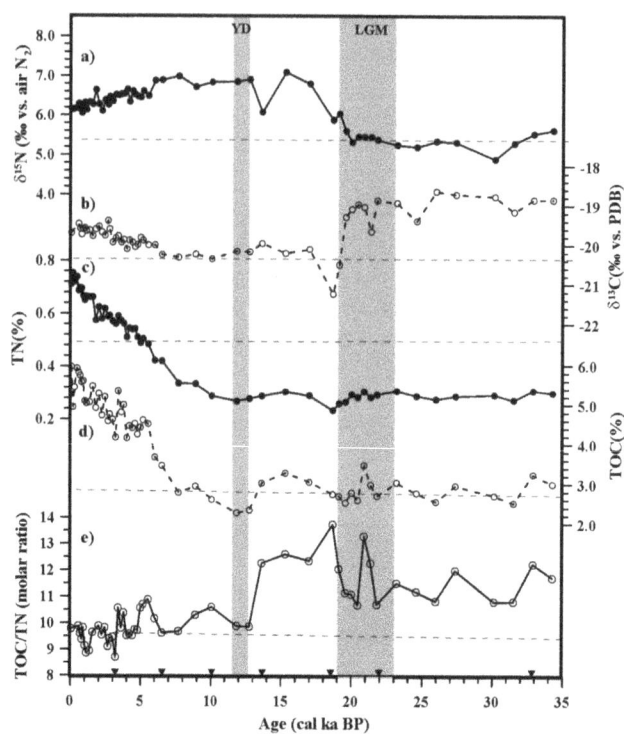

Figure 3. Temporal variations of (a) stable isotopic compositions of bulk nitrogen (δ^{15}N); (b) stable isotopic compositions of total organic carbon (TOC) (δ^{13}C); (c) contents of total nitrogen; (d) total organic carbon; (e) TOC/TN ratio. Horizontal dashed lines are references for low value periods.

(Fig. 1d), as mentioned earlier, near-surface NO_3^- in the AS is completely utilized in an annual cycle (Altabet, 1988; Thunell et al., 2004). Furthermore, negative N^* (P excess) throughout the water column represents a nitrate deficit, and the lowest N^* value appears at ~ 300 m at 18–20° N, where DO is $< 1\,\mu$mol kg^{-1}. Meanwhile, a gradual southwardly increase in N^* can be observed for upper 100 m (Fig. 1f) and the isopleth of N^* of -4 deepens southward with the highest N^* (-2) appearing at ~ 10–12° N. The volume expansion of high N^* water, as well as a simultaneous increase in N^*, strongly indicate an addition of bioavailable nitrogen when surface water is traveling southward.

3   Material and method

A sediment gravity core, SK177/11 (8.2° N and 76.47° E), was collected at water depths of 776 m on the continental slope off the southwest coast of India (Kerala) during the 177th cruise of *ORV SagarKanya* in October 2002. Although the core MD77-191 locates further south in the AS (Bassinot et al., 2012), SK177/11 is, so far, the southernmost core with reference to δ^{15}N record. The 3.65 m long core was subsampled at interval of 2 cm for upper 1 m and of 5 cm for the rest (open circles in Fig. 2a). There is a distinct boundary at ~ 1.7 m, above which the core consists mainly of brownish gray clayey sediments. Neither distinct laminations nor tur-

bidities can be observed by visual contact immediately after collection or at the time during sub-sampling (Pandarinath et al., 2007). All sub-samples were freeze-dried and ground into powder in an agate mortar with pestle. Sand was almost absent (< 1 wt %) throughout the core.

The calendar chronology for core SK177/11 was based on seven accelerator mass spectrometry (AMS) radiocarbon (^{14}C) dates of bulk organic matter (Fig. 2a). Calendar years were calculated using calibration CALIB 6.0 with a reservoir age correction of 402 years (Stuiver et al., 1998; Reimer et al., 2009). Details on the ^{14}C age controlling points were presented in Table 1. Given that the AMS ^{14}C dates of SK177/11 were obtained on total organic carbon (TOC), we may not be able to avoid the mixture of organics of different ages during transport (Mollenhauer et al., 2005) or interference by preaged organics sourced from land (Kao et al., 2008). However, besides the reservoir age correction, due to higher TOC content (range: 2.2–5.5 %) of sediments and their marine-sourced organic carbon, as confirmed by stable C isotope data and C/N ratio, shown in Figs. 3b and e, we are confident that our age model is reliable and less likely affected by age heterogeneity.

Bulk sedimentary nitrogen content and $\delta^{15}N$ analyses were carried out using a Carlo-Erba EA 2100 elemental analyzer connected to a Thermo Finnigan Delta V Advantage isotope ratio mass spectrometer (EA-IRMS). Sediments for TOC analyses were acid-treated with 1N HCl for 16 h, and then centrifuged to remove carbonate. The acid-treated sediments were further dried at 60 °C for TOC content and $\delta^{13}C$. The nitrogen isotopic compositions of acidified samples were obtained at the same time for comparison. Carbon and nitrogen isotopic data were presented by standard δ notation with respect to PDB (Pee Dee Belemnite) carbon and atmospheric nitrogen. USGS 40, which has certified $\delta^{13}C$ of -26.24 and $\delta^{15}N$ of -4.52% and acetanilide (Merck) with $\delta^{13}C$ of -29.76 and $\delta^{15}N$ of -1.52% were used as working standards. The reproducibility of carbon and nitrogen isotopic measurements is better than 0.15%. The precision of nitrogen and carbon content measurements were better than 0.02 and 0.05 %, respectively. Meanwhile, the acidified and non-acidified samples exhibited identical patterns in $\delta^{15}N$ (not shown) with mean deviation of 0.3%.

4 Results

4.1 Sedimentation rate

The age–depth curve was shown in Fig. 2a, in which age dates were evenly distributed throughout the core, although not at a high resolution. In Mollenhauer et al. (2005), the largest age offset between total organic carbon and co-occurring foraminifera is ~ 3000 years and mostly < 2000 years. Meanwhile, the offset remains more or less constant throughout past 20 ka, regardless of the deglacial transition. The youngest date in our core is 3180 cal ka BP at 58 cm. We may expect younger age on the surface. Thus, if our TOC samples contain any pre-aged organics, as indicated by Mollenhauer et al. (2005), the offset should not be too large to alter our interpretation for the comparison between glacial and Holocene periods. The linear sedimentation rates derived from seven date intervals range from 6 to 20 cm ka^{-1} (Fig. 2b), with relatively constant value (~ 6 cm ka^{-1}) prior to Holocene, except for the excursion around the last glacial maximum. The linear sedimentation rates started to increase since Holocene and reached 18–20 cm ka^{-1} when the sea level reached modern-day level.

4.2 Nitrogen and carbon contents and their isotopes

Values of $\delta^{15}N_{bulk}$ ranged from 4.7 to 7.1 ‰ with significantly lower values during the glacial period (Fig. 3a). The $\delta^{15}N$ values increased rapidly since ~ 19 ka BP, with a peak at ~ 15 ka BP, and then started to decrease gradually toward the modern day, except for the low $\delta^{15}N$ excursion at ~ 14 ka BP. Figure 3b shows that values of $\delta^{13}C_{TOC}$ (-21.5 to -18.5%) were consistent with the $\delta^{13}C$ of typical marine organic matter end member (-22 to -18%; Meyers, 1997).

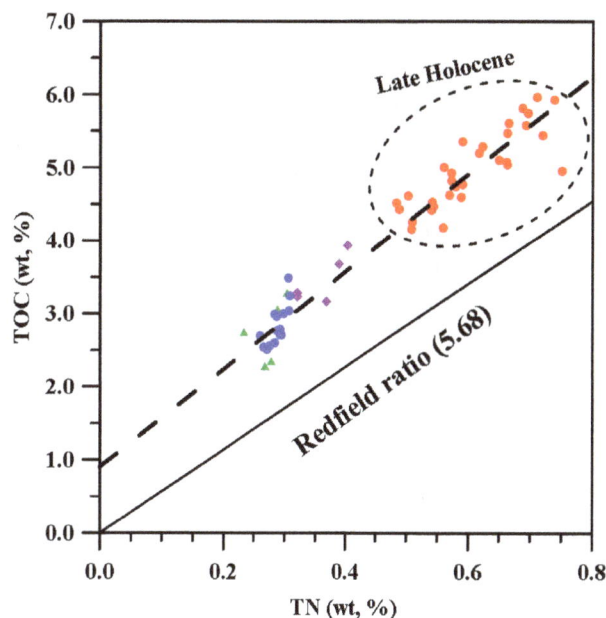

Figure 4. Scatter plot of the total organic carbon content against total nitrogen. Redfield field ratio of 5.68 is shown in line. Bold dashed line stands for regression. Red, purple, green and blue dots represent the late Holocene, early Holocene, deglacial and glacial periods, respectively.

An abrupt decrease in $\delta^{13}C$ was observed in concert with the dramatic increase in $\delta^{15}N_{bulk}$ at the start of deglaciation.

Bulk nitrogen content (TN) had a range of 0.23–0.75 % (Fig. 3c), and the TOC content ranged from 2.2 to 5.5 % (Fig. 3d). Both TN and TOC showed similar trends over the last 35 ka BP with relatively constant values prior to Holocene and an afterward elevation until the modern day. The upward increasing TOC and TN patterns since Holocene were consistent with the increasing pattern of the sedimentation rate, suggesting a higher organic burial flux induced by enhanced productivity, which had been reported elsewhere in the AS (Altabet et al., 2002).

As for the TOC/TN ratio, higher values appeared during the deglacial transition and the glacial period (Fig. 3e). The highest value coincides with the $\delta^{13}C_{TOC}$ drop, implying that there is still some influence from terrestrial organics. However, terrestrial organics contain less nitrogen (C/N of 20; Meyers, 1997), thus the $\delta^{15}N$ did not drop correspondingly. In the first meter (since ~ 5 ka), the downward decreasing pattern of TOC and TN can also be attributed to syn-sedimentary degradation; if so, a downward increasing in TOC/TN should be evident. However, TOC/TN values varied in a narrow range not revealing a significant increasing trend. Nevertheless, $\delta^{15}N$ and $\delta^{13}C$ did not show concomitant variations with C/N in the first meter or throughout the core. The influence from both organic degradation and changes in terrestrial organic input on isotopic signals is thus limited.

Figure 4 shows the scatter plot of TOC against TN. The slope of the linear regression line for TOC against TN ($TOC = (6.67 \pm 0.22) \times TN + (0.99 \pm 0.11)$, $R^2 = 0.94$, $n = 57$, $p < 0.0001$) is 6.67, again indicating that organic matter is mainly marine-sourced. Although this slope is slightly higher than the Redfield ratio of 5.68 (wt/wt), it is lower than that observed on the East China Sea shelf (7.46; Kao et al., 2003). Meanwhile, the intercept of TN is negative when TOC downs to zero, implying that inorganic nitrogen can be ignored in our core. Obviously, if we force the regression through the origin point, TOC/TN values for samples during the Holocene will have the lower ratios reflecting even less contribution from terrestrial organics.

5   Discussion

5.1   Downward transfer and transformation of N isotopic signal

As mentioned, the signal of sedimentary $\delta^{15}N$ may be altered under different burial conditions. Altabet and Francois (1994) reported little diagenetic alteration of the near-surface $\delta^{15}N$ in the equatorial Pacific, while there was an apparent $+5\,\permil$ enrichment relative to sinking particles in the Southern Ocean, south of the polar front. In the Sargasso Sea, sedimentary $\delta^{15}N$ also enriched by 3–6‰ relative to sinking particles (Altabet et al., 2002; Gruber and Galloway, 2008). The degree of alteration was attributed to particle sinking rate and organic matter (OM) preservation (Altabet, 1988). Gaye-Haake et al. (2005) also suggested that low sedimentation rates benefit organic matter decomposition, resulting in a positive shift in bulk sedimentary $\delta^{15}N$ comparing to sinking particles in the South China Sea. Finally, Robinson et al. (2012) concluded that oxygen exposure time at the seafloor is the dominant factor controlling the extent of N isotopic alteration. Thus, it is necessary to follow the track of $\delta^{15}N$ signal to clarify the occurrence of deviation during transfer.

The reported depth profiles of $\delta^{15}N_{NO_3}$ in the AS were shown in Fig. 5, in which $\delta^{15}N_{NO_3}$ values of water depth deeper than 1200 m range narrowly around 6–7‰, which is slightly higher than the global average of the deep oceans ((4.8 ± 0.2)‰ for > 2500 m, Sigman et al., 2000; (5.7 ± 0.7)‰ for > 1500 m, Liu and Kaplan, 1989). Below the euphotic layer, $\delta^{15}N_{NO_3}$ increases, rapidly peaking at around 200–400 m. The preferential removal of $^{14}NO_3$ by water column denitrification accounts for these subsurface $\delta^{15}N_{NO_3}$ highs (Brandes et al., 1998; Altabet et al., 1999; Naqvi et al., 2006). The subsurface $\delta^{15}N_{NO_3}$ maximum ranges from 10 to 18‰ for different stations, implying a great spatial heterogeneity in water columns denitrification intensity. It is worth mentioning that higher values, in general, appear in the northeastern AS (15 ∼ 18‰) (Fig. 5), highlighting that the focal area of water column denitrifi-

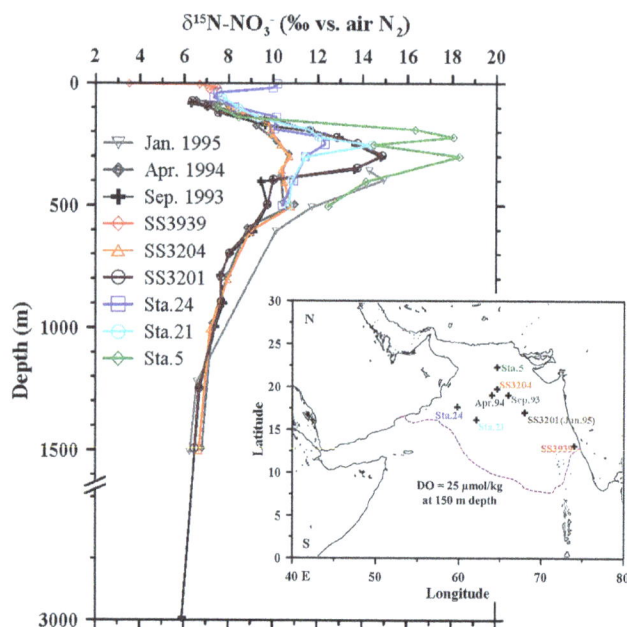

Figure 5. Depth profiles of nitrogen isotope of nitrate ($\delta^{15}N_{NO_3}$) in water columns; data not marked are all from August; the location of Station January 1995 overlaps with Station SS3201 (data digitized from Brandes et al., 1998; Altabet et al., 1999; Naqvi et al., 2006).

cation is prone to the northeastern Arabian Sea (Naqvi et al., 1994; Pichevin et al., 2007), also revealed by the DO spatial distribution (Fig. 1a). Contrary to higher denitrification in the northeastern AS, the export production is always higher in the northwestern AS throughout a year (Rixen et al., 1996). Such decoupling between productivity and denitrification was attributed to the oxygen supply by intermediate water exchange besides primary productivity oxygen demand (Pichevin et al., 2007). Note that the $\delta^{15}N_{NO_3}$ values at a water depth of 100–150 m, which correspond to the bottom depth of the euphotic zone (Olson et al., 1993), from different stations fall within a narrow range of 7–9‰ despite wide denitrification intensity underneath. The rapid addition of new nitrogen, as mentioned earlier, might account for the relatively uniform $\delta^{15}N_{NO_3}$ at the bottom of the euphotic layer. Unfortunately, there are no $\delta^{15}N_{NO_3}$ profiles or sediment trap data from the southern basin for comparison.

Interestingly, reported $\delta^{15}N$ of sinking particles ($\delta^{15}N_{SP}$) collected by five sedimentation traps deployed from 500 m throughout a depth of 3200 m ranged narrowly from 5.1 to 8.5‰ (Fig. 6), which is slightly lower but overlaps largely with $\delta^{15}N_{NO_3}$ values at 100–150 m. Such similarity in $\delta^{15}N_{NO_3}$ at 100–150 m and sinking particles strongly indicated that (1) the NO_3^- source for sinking particles was coming from a depth of around 100-150 m, instead of the ODZs (200-400 m) where the maximum $\delta^{15}N_{NO_3}$ value occurred (Schäfer and Ittekkot, 1993; Altabet et al., 1999), and (2) little alteration had occurred in $\delta^{15}N_{SP}$ during sinking in the water column, as indicated by Altabet (2006). Only these five trap stations with nitrogen isotope information were available in the AS (Gaye-Haaake et al., 2005). The trap locations were in the same area but not as far south compared to the $\delta^{15}N_{NO_3}$ stations (insert map in Fig. 6). The slightly lower $\delta^{15}N$ in sinking particles is attributable to their geographic locations (see below), since incomplete relative utilization of surface nitrate has been documented to have a very limited imprint on the $\delta^{15}N$ signal in the AS (e.g., Schäfer and Ittekkot, 1993).

The uniformly low values of $\delta^{15}N_{NO_3}$ at the bottom of the euphotic zone should be a consequence resulting from various processes in the euphotic zone, such as remineralization, nitrification and N_2 fixation. Nevertheless, the distribution pattern of N^* (Figs. 1e and f) illustrates that there must be an addition of $^{14}NO_3$ into the system to cancel out the isotopic enrichment caused by denitrification. Note that the positive offset in $\delta^{15}N_{NO_3}$ ($\Delta\delta^{15}N_{NO_3}$, 6~ 12‰) in ODZs caused by various degrees of denitrification was narrowed down significantly, while nitrate was transported upward. This implies that a certain degree of addition processes, most likely the N_2 fixation, varied in concert with the intensity of denitrification underneath. Since the upwelling zones distribute at the very north and the west of the AS and the upwelled water travels southward (or outward) on the surface, as shown in Fig. 1e, it is reasonable to see the phenomenon of denitrification-induced N_2 fixation to compensate the nitrogen deficiency. Consistent to this notion, Deutsch et al. (2007) discovered the spatial coupling between denitrification in eastern tropical Pacific (upstream) and N_2 fixation in western equatorial Pacific (downstream). Such a horizontal nitrogen addition process can also be seen clearly in our background information of N^* (Fig. 1f). In fact, fixed N had been proved to account for a significant part of surface nitrate in the modern-day AS, where denitrification is exceptionally intense (Brandes et al., 1998; Capone et al., 1998; Parab et al., 2012).

Compared with reported $\delta^{15}N$ of surface sediments retrieved from trap locations, a significant positive shift in $\delta^{15}N$ can be seen at the seafloor (Fig. 6). Such a positive deviation can be seen elsewhere in previous reports (Altabet, 1988; Brummer et al., 2002; Kienast et al., 2005) due to prolonged oxygen exposure after deposition (Robinsson et al., 2012) associated with sedimentation rate (Pichevin et al., 2007). Although Cowie et al. (2009) found an ambiguous relation between contents of sedimentary organic carbon and oxygen in deep water, they also noticed the appearance of maximum organic carbon contents at the lower boundary of ODZs, where oxygen content was relatively higher. Accordingly, they believed that other factors controlling the preservation of organic carbon existed, such as the chemical characteristics of organic matter, the interaction between organic matters and minerals, the enrichment and activity of benthic organism or the physical factor, including the screening and water dynamic effect.

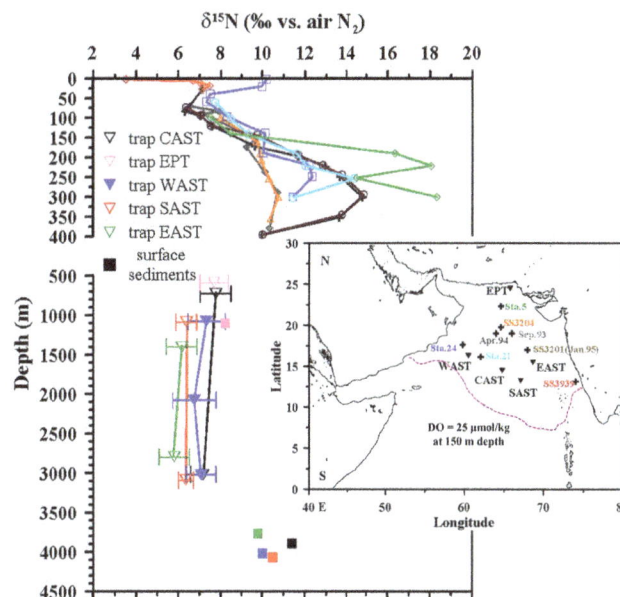

Figure 6. Vertical profiles for nitrogen isotope of nitrate (crosses in inserted map), sinking particles (inverse triangles in map) and trap-corresponding surface sediments. Data for sediment traps and surface sediments are from Gaye-Haake et al. (2005). Depth profile of $\delta^{15}N_{NO_3}$ follows that in Fig. 5.

5.2 Geographically distinctive bottom-depth effects in the modern day

As classified by oxygen content of $25\,\mu mol\,kg^{-1}$ at 150 m, the documented surface sedimentary $\delta^{15}N_{bulk}$ (Gaye-Haake et al., 2005) was separated into northern and southern groups to examine the geographic difference in bottom-depth effect. Both groups exhibit positive linear relationships between $\delta^{15}N_{bulk}$ and bottom depth (deeper than 200 m) (Fig. 7a). The regression equations were shown in Table 2. Interestingly, the regressions generally differ statistically from each other in terms of slope and intercept. The slope represents the degree of positive shift of sedimentary $\delta^{15}N$ due to bottom-depth effect. For the southern AS, the slope is (0.76 (±0.14) $\times 10^{-3}\,km^{-1}$), which is close to the correction factor (0.75 $\times 10^{-3}\,km^{-1}$) for the world ocean, proposed by Robinson et al. (2012) and further applied by Galbraith et al. (2012). By contrast, the slope for the northern AS is significantly lower (0.55 (±0.08) $\times 10^{-3}\,km^{-1}$), implying that the depth-associated alteration in the northern AS is smaller. The correction factor for bottom-depth effect was predicted to vary in different regions such as that in the South China Sea (Gaye et al., 2009). Since the magnitude of oxygen exposure is the primary control of depth effect (Gaye-Haake et al., 2005; Mobius et al., 2011; Robinson et al., 2012), we attributed this lower slope in the northern AS to relatively higher sedimentation rates (not shown) and lower

Table 2. Linear equations of bottom-depth effect during different climate stages.

Location	Northern AS	Southern AS
Modern	$\delta^{15}N = 0.55 \, (\pm 0.08) \times 10^{-3} \times depth + 8.1 \, (\pm 0.2)$ $(R^2 = 0.40, n = 78, P < 0.0001)$	$\delta^{15}N = 0.76 \, (\pm 0.14) \times 10^{-3} \times depth + 6.0 \, (\pm 0.3)$ $(R^2 = 0.66, n = 18, P < 0.0001)$
Holocene	$\delta^{15}N = 0.70 \, (\pm 0.20) \times 10^{-3} \times depth + 6.7 \, (\pm 0.3)$ $(R^2 = 0.61, n = 16, P = 0.0067)$	$\delta^{15}N = 0.93 \, (\pm 0.06) \times 10^{-3} \times depth + 5.7 \, (\pm 0.1)$ $(R^2 = 1.00, n = 3, P = 0.0152)$
Glacial	$\delta^{15}N = 0.64 \, (\pm 0.20) \times 10^{-3} \times depth + 5.2 \, (\pm 0.3)$ $(R^2 = 0.68, n = 16, P = 0.0013)$	$\delta^{15}N = 1.01 \, (\pm 0.31) \times 10^{-3} \times depth + 4.3 \, (\pm 0.7)$ $(R^2 = 0.91, n = 3, P = 0.1899^*)$

* Insignificant by P value.

oxygen contents, as indicated by previous research (Olson et al., 1993; Morrison et al., 1999; Brummer et al., 2002).

On the other hand, the intercept for the northern AS regression (8.1 ± 0.2) is significantly higher than that for the southern AS (6.0 ± 0.3). As mentioned above, $\delta^{15}N$ values of sinking particle resembled the $\delta^{15}N$ of nitrate sourced from a depth of 100–150 m. According to the depth-dependent correction factor, we may convert sedimentary $\delta^{15}N_{bulk}$ values at various water depths into their initial condition when the digenetic alteration is minimal to represent the $\delta^{15}N$ of source nitrate. Higher intercept suggests that a stronger denitrification had occurred in northern AS surface sediments. The 2.1 ‰ lower intercept in the southern AS likely reflects the addition of N_2 fixation in the upper water column while it travels southward. The progressive increase of N^* toward the southern AS supports our speculation, although no $\delta^{15}N_{NO_3}$ profiles had been published in the southern basin. Future works about $\delta^{15}N_{NO_3}$ and $\delta^{15}N_{SP}$ in the southern AS are needed.

In Fig. 7b, we presented corrected $\delta^{15}N_{bulk}$ values along with bottom depth for the northern and southern AS surface sediments for comparison. After removing site-specific bias caused by bottom-depth effect, the values and distribution ranges of $\delta^{15}N_{bulk}$ for both the northern and southern AS became smaller and narrower. For the northern AS, the distribution pattern skewed negatively, giving a standard deviation of 0.88 ‰, falling exactly in the range of 7–9 ‰ for $\delta^{15}N_{NO_3}$ (7–9 ‰) at the bottom of the euphotic zone. As a result, the corrected nitrogen isotopic signals in sediments more truthfully represent the $\delta^{15}N_{NO_3}$ value at the bottom depth of the euphotic zone. Meanwhile, the statistically significant difference in $\delta^{15}N_{bulk}$ distribution between the northern and southern AS further confirms the feasibility of our classification by using DO isopleth of $25 \, \mu mol \, kg^{-1}$ at 150 m.

5.3 Bottom-depth effect during different climate stages

In order to better decipher the history of $\delta^{15}N_{NO_3}$ in the bottom the euphotic zone of the water column, we synthesized almost all available $\delta^{15}N_{bulk}$ of sediment cores reported for the AS (see Figs. 1a and 1b for locations). Similar to modern surface sediments, northern and southern groups were

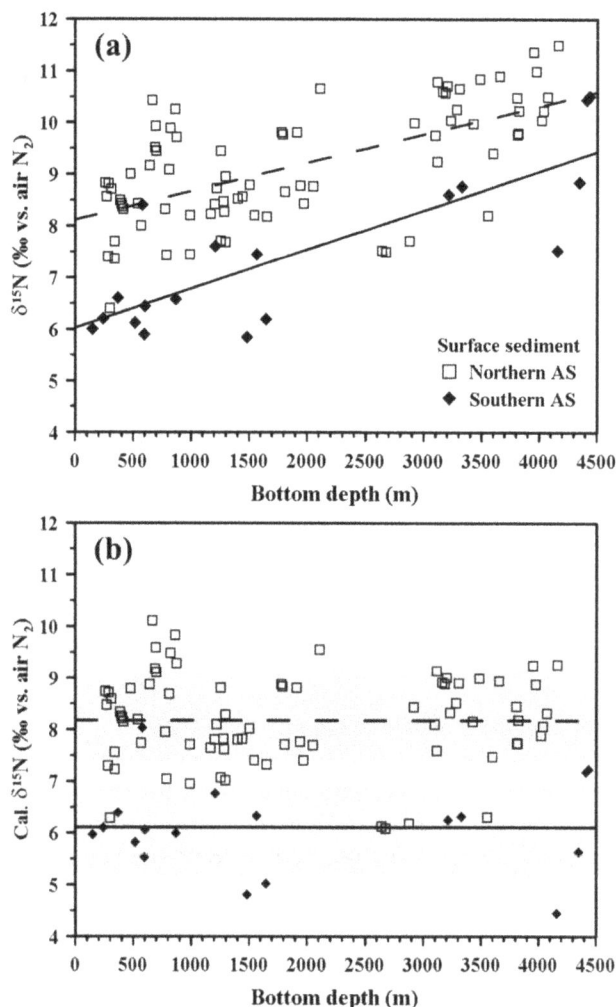

Figure 7. (a) Non-corrected $\delta^{15}N$ values of modern surface sediments against corresponding bottom depth in the northern and southern Arabian Sea (see text for N–S boundary). Regression lines were shown in dashed and solid lines, respectively, for the northern and southern AS. **(b)** Corrected surface sedimentary $\delta^{15}N$ values against water depth.

defined by the contour line of $25 \, \mu mol \, kg^{-1}$ DO. To keep data consistency in the temporal scale, we focused on the last

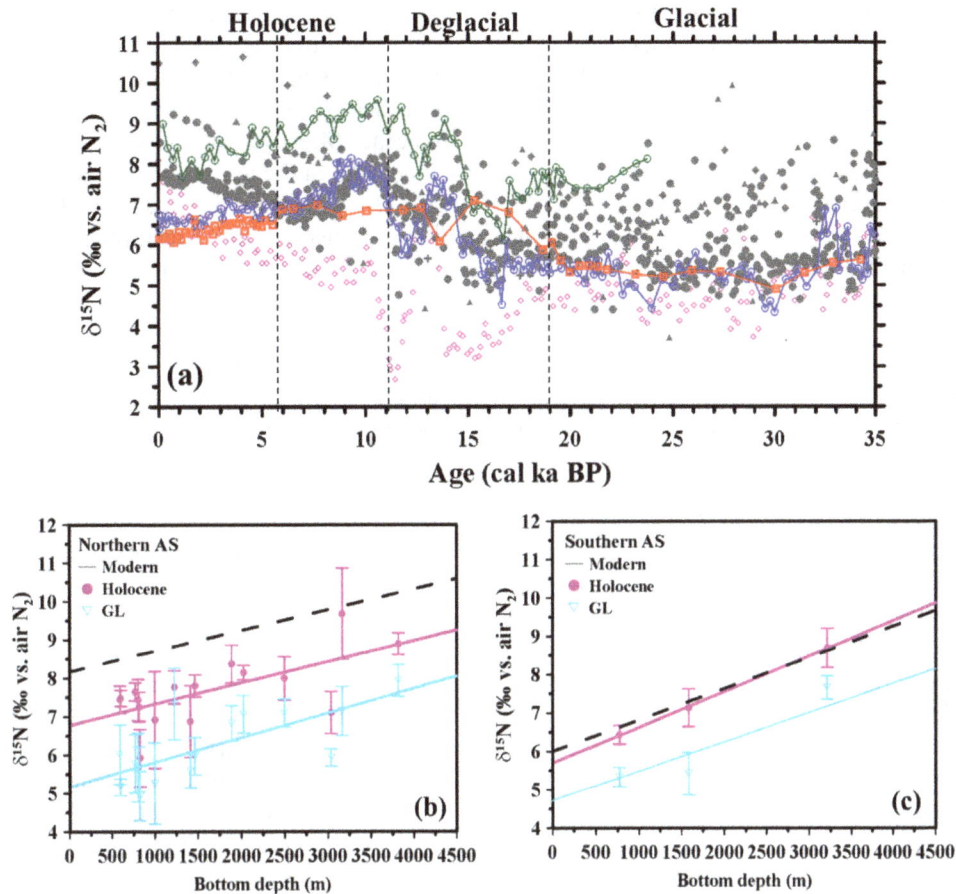

Figure 8. (a) Temporal variations of non-corrected $\delta^{15}N_{bulk}$ values of all reported cores in the AS. Data shown in curves are for cores in the southern Arabian Sea (red for SK177/11, blue for NIOP 905 and green for SO42-74KL), dots in gray are for the northern part and pink dots are for core MD-04-2876. Mean values of $\delta^{15}N$ for fixed periods against corresponding water depths for cores in the (**b**) northern and (**c**) southern Arabian Sea. Pink and indigo blue are for the Holocene and glacial periods, respectively. Error bars represent the standard deviation for mean $\delta^{15}N_{bulk}$. The dashed regression lines for modern surface sediments are shown for reference.

35 ka (Fig. 8a). Unfortunately, data points were less in 0–6 ka and there were only three sediment cores in the southern AS: SK177/11 in this study, and NIOP 905 and SO42-74KL in previous studies.

As shown in Fig. 8a, the original $\delta^{15}N_{bulk}$ from the northern (gray dots) and southern AS (green, blue and red curves) scatter in a wide range from 4.5 to 10.5‰ over the entire 35 ka. The pink dots are for the data from core MD-04-2876, which is peculiar since the relatively low $\delta^{15}N_{bulk}$ values deviated from all other reports in the northern AS. Pichevin et al. (2007) excluded the influences from incomplete nitrate utilization and terrestrial input, thus we still include this core in our statistical analyses. As for the southern cores, the temporal variations of $\delta^{15}N_{bulk}$ in core SK177/11 and NIOP 905 (red and blue) had a very similar trend distributing at the lower bound of the whole data set. The mean $\delta^{15}N_{bulk}$ values for SK177/11 and NIOP 905 during the glacial period were almost identical, and the deviation in the Holocene was as small as 0.7‰. By contrast, the temporal pattern for $\delta^{15}N_{bulk}$

of core SO42-74KL (green) resembles that of NIOP 905, yet with an enrichment in ^{15}N by ~ 2‰ for the entire period. The core SO42-74KL is retrieved from a depth of 3212 m, which is the deepest of the three cores in the southern AS; the positive offset is apparently caused by the bottom-depth effect. Thus, inference should be made with caution when compare sediment cores from different depths.

Below we consider two time spans – 0 ∼ 11 ka (Holocene) and 19 ∼ 35 ka (glacial) – to examine the bottom-depth effect at different climate stages. We ignore the transgression period, which is shorter with more variable in $\delta^{15}N_{bulk}$, to avoid bias caused by dating uncertainties in different studies. Also, we will discuss the peculiar patters for 0–6 ka later. The mean and standard deviation of reported $\delta^{15}N_{bulk}$ values for the specific time span were plotted against the corresponding depth of the core. Accordingly, we obtained the correction factors for glacial and early Holocene, respectively, for the northern and southern AS (Fig. 8b and c). Since only 35 ka was applied in this practice, the long-term alteration (Reichart et al., 1998; Altabet et al., 1999) is ignored. The regression curves for the modern day (dashed lines) were plotted for comparison.

The difference among regressions of three climate stages in the northern AS (Table 2) is not significant (0.55×10^{-3} km^{-1} to 0.70×10^{-3} km^{-1}). However, the regression slopes for the northern AS are significantly lower compared with those obtained from the southern AS for all climate states. This might indicate that the oxygen content in the northern AS is always lower, resulting in a lower degree of alteration of δ^{15}N$_{bulk}$. On the other hand, we may not exclude the effect by sedimentation rate changes over these two stages, which also affect the oxygen exposure time; unfortunately, insufficient sedimentation rate data in the northern AS in previous reports prevent us from implementing further analysis.

As for the southern AS, correction factors are always higher than those in the northern AS. The overall spatial–temporal patterns are consistent with the oxygen distribution in the Arabian Sea (Olson et al., 1993; Morrison et al., 1999; Pichevin et al., 2007) and agree with the view that DO concentration was the dominant factor for organic matter preservation (Aller, 2001; Zonneveld et al., 2010). Meanwhile, the regression slopes remained high from 0.76×10^{-3} to 1.01×10^{-3} km^{-1} over different climate stages in the southern AS, suggesting that environmental situations, and thus those correction factor, change less relatively to that in the northern AS. For SK177/11, the sedimentation rate in Holocene is two-fold higher compared to that in the glacial period. However, the influence caused by the sedimentation rate changes is likely not significant enough to alter the regression slopes for the southern AS, based on the small changes in the slope (0.93×10^{-3} and 1.01×10^{-3} km^{-1}).

5.4 Insights from temporal changes in geographic δ^{15}N$_{bulk}$ distribution

Based on the earlier comparison among δ^{15}N$_{NO_3}$, sinking particles and surface sediments, we recognized that the regression intercept is representative of the nitrogen isotope of nitrate source at a depth of 100 m. Therefore, the regression-derived intercepts given in Table 2 can be used to infer the δ^{15}N$_{NO_3}$ source at different climate stages, while the slopes can be used as correction factors to eliminate the positive shift in δ^{15}N$_{bulk}$ caused by bottom depth; by doing this, we can get the original signal of δ^{15}N$_{bulk}$ prior to alteration. We applied the correction factor to be equal to (bottom depth -100 m) $\times$ slope, ignoring the sea level changes during the different climate stages.

Noticeably, the regression intercepts for both the northern and southern AS are higher in the Holocene compared to those in the glacial period, indicating the intensified isotopic enrichment in δ^{15}N$_{NO_3}$ in the entire AS in Holocene. Such increment is almost the same to be ~ 1.5 ‰, which is similar to the increase in the eastern tropical North Pacific,

but slightly smaller than that in the eastern tropical South Pacific (Galbraith et al., 2012). The 120 m sea level increase, which may induce only a 0.1 ‰ offset, cannot be the reason for such a significant increase of average δ^{15}N$_{bulk}$ during the Holocene. Moreover, deviations between the northern and southern AS at the respective climate stage are almost identical (1.0 ‰ for Holocene and 0.9 ‰ for glacial), indicating a synchronous shift in the relative intensity of denitrification and N$_2$ fixation over the basin to keep such a constant latitudinal gradient of subsurface δ^{15}N$_{NO_3}$.

The intermediate water formation near the polar region controls the oxygen supply to the intermediate water and thus the extent of denitrification on global scale and the stoichiometry of nutrient source to the euphotic zone (Galbraith et al., 2004). Lower glacial-stage sea surface temperature may increase oxygen solubility, while stronger winds in high-latitude regions enhance the rate of thermocline ventilation. The resultant colder and rapidly flushed thermocline thus lessened the spatial extent of denitrification and, consequently, N fixation (Galbraith et al., 2004). Therefore, such a basin of wide synchronous increase in δ^{15}N$_{bulk}$ is likely a global control. The lower intercepts in glacial time (4.3 ‰ for the south and 5.3 ‰ for the north), which are similar to the global mean δ^{15}N$_{NO_3}$ (4.5–5 ‰, Sigman et al., 1997), illustrate a better ventilation of intermediate water during glacial time in the Arabian Sea (Pichevin et al., 2007). In fact, the AAIWs penetrate further northward over 5° N in the present day and even during the late Holocene (You, 1998; Pichevin et al., 2007). Since the δ^{13}C of autochthonous particulate organic carbon is negatively correlated to [CO$_2$ (aq)] in the euphotic zone (Rau et al., 1991), the sharp decrease of δ^{13}C$_{TOC}$ in SK177/11 at the start of deglaciation (Fig. 3b) may infer the timing of a rapid accumulation of dissolved inorganic carbon driven by the shrinking of oxygenated intermediate water (Pichevin et al., 2007) or enhanced monsoon-driven upwelling (Ganeshram et al., 2000); both facilitate the promotion of denitrification. Nevertheless, the mirror image between δ^{15}N and δ^{13}C$_{TOC}$ pro?les revealed their intimate relation, of which the variability was attributable to the change of physical processes.

The intercepts of the northern AS increase continuously from 5.2 to 8.1 from glacial through to modern day, indicating the strengthened intensity of denitrification relative to nitrogen fixation in the northern AS (Altabet, 2007). When we take a close look at the temporal pattern of corrected δ^{15}N$_{bulk}$ for long cores (Fig. 9), we can see an amplified deviation since 6 ka, during which δ^{15}N$_{bulk}$ increases continuously in the northern AS, whereas it decreases in the southern AS. Note that the northern most core, MD-04-2876, also followed the increasing trend in recent 6 ka even though its δ^{15}N$_{bulk}$ values deviated from all other cores. Such opposite trends indicate that the controlling factors on the nitrogen cycle in the northern AS were different from that in the southern AS, which means that localized enhancement in specific process had occurred.

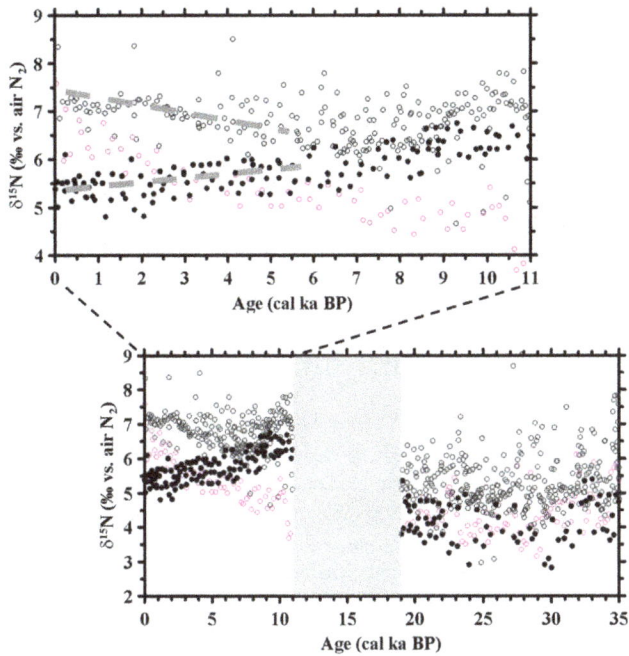

Figure 9. Temporal variations of corrected $\delta^{15}N_{bulk}$ values of all reported cores in the AS. Gray and black dots are for the northern and southern AS, respectively. Pink dots are specifically for core MD-04-2876. The deglacial period is in shadow because non proper equations for bottom-depth effect correction. The upper panel is the blow-up for the Holocene period. The intensified deviation trends since 6 ka were marked by bold dashed lines.

Besides the oxygen supply to the intermediate water, the intensity of water column denitrification varies with primary productivity (Altabet, 2006; Naqvi et al., 2006). Strong summer monsoon and winter monsoon drive upwelling or convective mixing enhances primary productivity, which, in turn, intensifies denitrification (Altabet et al., 2002; Ganeshram et al., 2002). However, it was also reported that primary productivity did not correlate well with water column denitrification underneath during the Holocene in some parts of the northern AS (Banakar et al., 2005, and references therein). Regardless of the declining summer monsoon strength since 5500 ka (Hong et al., 2003), primary productivity in the northern AS seemed to have increased. Similar to the patterns observed for TOC and TN in this study, productivity indicators (TOC and Ba / Al ratios), reported by Rao et al. (2010) in the core SK148/4 located near our SK177/11, also increased gradually since the Holocene. Incomplete nitrate consumption can hardly explain the decreasing pattern for all three cores in the southern AS, where upwelling intensity is much less relative to that in the north. Moreover, lower TOC / TN ratios observed in Holocene in SK177/11, as mentioned earlier, rule out the influence of terrestrial organic input. Therefore, a spatial coupling of denitrification-dependent N_2 fixation is the more plausible cause of the decreasing $\delta^{15}N_{bulk}$ pattern (Deutsch et al., 2007).

We suggested that the intensified supply of excess phosphorous (phosphorus in stoichiometric excess of fixed nitrogen) toward the southern AS to stimulate N_2 fixation, subsequently responsible for the decreasing $\delta^{15}N_{bulk}$ pattern in the southern basin. The intensification in excess phosphorous supply can be driven by enhanced upwelling or intensified subsurface water column denitrification or both. According to the increasing pattern in $\delta^{15}N_{bulk}$ and primary productivity in the northern AS, synergetic processes are suggested. The upwelled water in the northern AS basin brings up low N / P water to the surface for non-diazotrophs to uptake. If we assume complete consumption, the remaining excess phosphorous after complete consumption will be transported toward the south by clockwise surface circulation and advection. Therefore, N_2 fixation in the southern AS acts as feedback to balance denitrification changes in the northern AS. This phenomenon is similar to the illustration for the spatial coupling of nitrogen inputs and losses in the Pacific Ocean, proposed by Deutsch et al. (2007). The question as to why such forcing to expand the N–S deviation had not occurred before 6 ka warrants more studies.

6 Conclusions

The available data showed that values of $\delta^{15}N_{NO_3}$ at the bottom of the euphotic zone ($\sim 150\,m$) were similar to $\delta^{15}N_{SP}$, implying that the source of nutrients for sinking particulate organic matter was largely derived from a depth of around 150 m. Values of sedimentary $\delta^{15}N_{bulk}$ were obviously higher than $\delta^{15}N_{SP}$ in surrounding areas, suggesting that such a shift of sedimentary $\delta^{15}N_{bulk}$ occurred after deposition. It is necessary to remove site-specific bias of $\delta^{15}N_{bulk}$ values caused by bottom depth to retrieve the original signal before alteration. As a result, the corrected nitrogen isotopic signal in sediments could be representative of the value of $\delta^{15}N_{NO_3}$ at the bottom depth of the euphotic zone. The bottom-depth effects in the northern AS vary during different climate stages, but the variation is always lower than such effects in the southern AS in general. The modern surface $\delta^{15}N_{bulk}$ values can be separated statistically into northern and southern AS groups, reflecting a special coupling of denitrification to the north, and N_2 fixation to the south. This phenomenon is supported by the reported modern-day N^* distribution. As for historical records, the offset in $\delta^{15}N_{bulk}$ between the southern and northern AS remained relatively constant (1.0‰ for early Holocene and 0.9‰ for glacial) prior to 6 ka, indicating a synchronous shift in the relative intensity of denitrification and N_2 fixation over the basin to keep such a constant latitudinal gradient of subsurface $\delta^{15}N_{NO_3}$. However, this offset expanded gradually since 6 ka, likely due to more localized intensifications in denitrification and N_2 fixation had occurred, respectively, in the northern and southern Arabian Seas. The spatial coupling of nitrogen inputs and losses in the Arabian Sea was proposed, yet the question as to why the driving force did not expand the N–S deviation before 6 ka warrants more studies.

Acknowledgements. This research was supported by the National Natural Science Foundation of China (NSFC 41176059, 91328207, and 41273083) and Shanhai Fund (2013SH012). K. Selvaraj personally thanks the director of the National Center for Antarctic and Ocean Research, Goa and the Secretary, the Department of Ocean Development, New Delhi, for providing the ship time, and also the crew of *ORV Sagar Kanya* for coring operation. K. Selvaraj also thanks V. Yoganandan for onboard help of subsampling, and the coordinator of Ocean Science and Technology Cell of Mangalore University, for their kind encouragement.

Edited by: N. Ohkouchi

References

Agnihotri, R., Altabet, M. A., and Herbert, T.: Influence of marine denitrification on atmospheric N_2O variability during the Holocene, Geophys. Res. Lett., 33, , L13704, doi:10.1029/2006GL025864, 2006.

Aller, R. C.: Transport and reactions in the bioirrigated zone, The benthic boundary layer: Transport processes and biogeochemistry, edited by: B. Boudreau, and Jørgensen, B. B., Oxford University Press, Oxford, UK, 269–301, 2001.

Altabet, M.: Variations in nitrogen isotopic composition between sinking and suspended particles: Implications for nitrogen cycling and particle transformation in the open ocean, Deep-Sea Res. Pt. A, 35, 535–554, 1988.

Altabet, M. and Francois, R.: Sedimentary nitrogen isotopic ratio as a recorder for surface ocean nitrate utilization, Global Biogeochem. Cy., 8, 103–116, 1994.

Altabet, M., Francois, R., Murray, D. W., and Prell, W. L.: Climate-related variations in denitrification in the Arabian Sea from sediment $^{15}N/^{14}N$ ratios, Nature, 373, 506–509, 1995.

Altabet, M., Murray, D. W., and Prell, W. L.: Climatically linked oscillations in Arabian Sea denitrification over the past 1 my: Implications for the marine N cycle, Paleoceanography, 14, 732–743, 1999.

Altabet, M., Higginson, M. J., and Murray, D. W.: The effect of millennial-scale changes in Arabian Sea denitrification on atmospheric CO_2, Nature, 415, 159–162, 2002.

Altabet, M.: Isotopic tracers of the marine nitrogen cycle: Present and past, in: Marine organic matter: biomarkers, isotopes and DNA, edited by: Volkman, J. K., Springer-Verlag Berlin Heidelberg, 251–293, 2006.

Altabet, M.: Constraints on oceanic N balance/imbalance from sedimentary ^{15}N records, Biogeosciences, 4, 75–86, 2007, http://www.biogeosciences.net/4/75/2007/.

Banakar, V., Oba, T., Chodankar, A., Kuramoto, T., Yamamoto, M., and Minagawa, M.: Monsoon related changes in sea surface productivity and water column denitrification in the Eastern Arabian Sea during the last glacial cycle, Mar. Geol., 219, 99–108, 2005.

Bassinot, F. C., Marzin, C., Braconnot, P., Marti, O., Mathien-Blard, E., Lombard, F., and Bopp, L.: Holocene evolution of summer winds and marine productivity in the tropical Indian Ocean in response to insolation forcing: data-model comparison, Clim. Past, 7, 815–829, doi:10.5194/cp-7-815-2011, 2011.

Brandes, J. A., Devol, A. H., Yoshinari, T., Jayakumar, D., and Naqvi, S.: Isotopic composition of nitrate in the central Arabian Sea and eastern tropical North Pacific: A tracer for mixing and nitrogen cycles, Limnol. Oceanogr., 43, 1680–1689, 1998.

Brummer, G., Kloosterhuis, H., and Helder, W.: Monsoon-driven export fluxes and early diagenesis of particulate nitrogen and its $\delta^{15}N$ across the Somalia margin, Geological Society, London, Special Publications, 195, 353–370, 2002.

Capone, D. G., Zehr, J. P., Paerl, H. W., Bergman, B., and Carpenter, E. J.: Trichodesmium, a globally significant marine cyanobacterium, Science, 276, 1221-1229, 1997.

Capone, D. G., Subramaniam, A., Montoya, J. P., Voss, M., Humborg, C., Johansen, A. M., Siefert, R. L., and Carpenter, E. J.: An extensive bloom of the N_2-fixing cyanobacterium Trichodesmium erythraeum in the central Arabian Sea, Mari. Ecol.-Prog. Ser., 172, 281–292, 1998.

Codispoti, L. and Christensen, J.: Nitrification, denitrification and nitrous oxide cycling in the eastern tropical South Pacific Ocean, Mar. Chem., 16, 277–300, 1985.

Cowie, G. L., Mowbray, S., Lewis, M., Matheson, H., and McKenzie, R.: Carbon and nitrogen elemental and stable isotopic compositions of surficial sediments from the Pakistan margin of the Arabian Sea, Deep-Sea Res. Pt. II, 56, 271–282, 2009.

Deutsch, C., Sarmiento, J. L., Sigman, D. M., Gruber, N., and Dunne, J. P.: Spatial coupling of nitrogen inputs and losses in the ocean, Nature, 445, 163–167, 2007.

Falkowski, P. G. and Godfrey, L. V.: Electrons, life and the evolution of Earth's oxygen cycle, Philosoph. Trans. Roy. Soc. B, 363, 2705–2716, 2008.

Galbraith, E. D., Kienast, M., Pedersen, T. F., and Calvert, S. E.: Glacial-interglacial modulation of the marine nitrogen cycle by high-latitude O_2 supply to the global thermocline, Paleoceanography, 19, PA4007, doi:10.1029/2003PA00100, 2004.

Galbraith, E. D., Kienast, M., Albuquerque, A. L., Altabet, M., Batista, F., Bianchi, D., Calvert, S., Quintana, S. C., Crosta, X., Holz, R. D. P., Dubois, N., Etourneau, J., Francois, R., Hsu, T.-C., Ivanochko, T., Jaccard, S., Kao, S.-J., Kiefer, T., Kienast, S., Lehmann, M. F., Martinez, P., McCarthy, M., Meckler, A. N., Mix, A., Möbius, J., Pedersen, T., Pichevin, L., Quan, T. M., Robinson, R. S., Ryabenko, E., Schmittner, A., Schneider, R., Schneider-Mor, A., Shigemitsu, M., Sinclair, D., Somes, C., Studer, A., Tesdal, J. E., Thunell, R., and Yang, J.-Y.: The acceleration of oceanic denitrification during deglacial warming, Nat. Geosci., 5, 151–156, 2012.

Ganeshram, R. S., Pedersen, T. F., Calvert, S. E., and Murray, J. W.: Large changes in oceanic nutrient inventories from glacial to interglacial periods, Nature, 376, 755–758, 1995. Ganeshram, R. S., Pedersen, T. F., Calvert, S. E., McNeill, G. W., and Fontugne, M. R.: Glacial-interglacial variability in denitrification in the world's oceans: Causes and consequences, Paleoceanography, 15, 361–376, 2000.

Ganeshram, R. S., Pedersen, T. F., Calvert, S., and François, R.: Reduced nitrogen fixation in the glacial ocean inferred from changes in marine nitrogen and phosphorus inventories, Nature, 415, 156–159, 2002.

Gaye, B., Wiesner, M., and Lahajnar, N.: Nitrogen sources in the South China Sea, as discerned from stable nitrogen isotopic ratios in rivers, sinking particles, and sediments, Marine Chemistry, 114, 72–85, 2009.

Gaye-Haake, B., Lahajnar, N., Emeis, K. C., Unger, D., Rixen, T., Suthhof, A., Ramaswamy, V., Schulz, H., Paropkari, A., and Guptha, M.: Stable nitrogen isotopic ratios of sinking particles and sediments from the northern Indian Ocean, Mar. Chem., 96, 243–255, 2005.

Gouretski, V. and Koltermann, K. P.: WOCE global hydrographic climatology, Berichte des BSH, 35, 1–52, 2004.

Gruber, N. and Sarmiento, J. L.: Biogeochemical/physical interactions in elemental cycles, in: The sea: Biological-Physical Interactions in the Oceans, edited by: Robinson, A. R., McCarthy, J. J., and Rothschild, B. J., John Wiley and Sons, New York, 337–399, 2002.

Gruber, N.: The dynamics of the marine nitrogen cycle and its influence on atmospheric CO_2 variations, in: The ocean carbon cycle and climate, edited by: Follows, M., and Oguz, T., Springer Netherlands, 97–148, 2004.

Gruber, N. and Galloway, J. N.: An Earth-system perspective of the global nitrogen cycle, Nature, 451, 293–296, 2008.

Haug, G. H., Pedersen, T. F., Sigman, D. M., Calvert, S. E., Nielsen, B., and Peterson, L. C.: Glacial/interglacial variations in production and nitrogen fixation in the Cariaco Basin during the last 580 kyr, Paleoceanography, 13, 427–432, 1998.

Holmes, M. E., Schneider, R. R., Müller, P. J., Segl, M., and Wefer, G.: Reconstruction of past nutrient utilization in the eastern Angola Basin based on sedimentary $^{15}N/^{14}N$ ratios, Paleoceanography, 12, 604–614, 1997.

Hong, Y., Hong, B., Lin, Q., Zhu, Y., Shibata, Y., Hirota, M., Uchida, M., Leng, X., Jiang, H., and Xu, H.: Correlation between Indian Ocean summer monsoon and North Atlantic climate during the Holocene, Earth Planet. Sci. Lett., 211, 371–380, 2003.

Ivanochko, T. S., Ganeshram, R. S., Brummer, G. J. A., Ganssen, G., Jung, S. J. A., Moreton, S. G., and Kroon, D.: Variations in tropical convection as an amplifier of global climate change at the millennial scale, Earth Planet. Sci. Lett., 235, 302–314, 2005.

Kao, S., Lin, F., and Liu, K.: Organic carbon and nitrogen contents and their isotopic compositions in surficial sediments from the East China Sea shelf and the southern Okinawa Trough, Deep-Sea Res. Pt. II, 50, 1203–1217, 2003.

Kao, S., Dai, M., Wei, K., Blair, N., and Lyons, W.: Enhanced supply of fossil organic carbon to the Okinawa Trough since the last deglaciation, Paleoceanography, 23, PA2207, doi:10.1029/2007PA001440, 2008.

Kienast, M., Higginson, M., Mollenhauer, G., Eglinton, T. I., Chen, M. T., and Calvert, S. E.: On the sedimentological origin of down-core variations of bulk sedimentary nitrogen isotope ratios, Paleoceanography, 20, PA2009, doi:10.1029/2004PA0018081, 2005.

Liu, K.-K. and Kaplan, I. R.: The eastern tropical Pacific as a source of ^{15}N-enriched nitrate in seawater off southern California, Limnol. Oceanogr, 34, 820–830, 1989.

Meyers, P. A.: Organic geochemical proxies of paleoceanographic, paleolimnologic, and paleoclimatic processes, Organ. Geochem., 27, 213–250, 1997.

Möbius, J., Gaye, B., Lahajnar, N., Bahlmann, E., and Emeis, K.-C.: Influence of diagenesis on sedimentary $\delta^{15}N$ in the Arabian Sea over the last 130 kyr, Mar. Geol., 284, 127–138, 2011.

Mollenhauer, G., Kienast, M., Lamy, F., Meggers, H., Schneider, R. R., Hayes, J. M., and Eglinton, T. I.: An evaluation of ^{14}C age relationships between co-occurring foraminifera, alkenones, and total organic carbon in continental margin sediments, Paleoceanography, 20, PA1016, doi:10.1029/2004PA001103, 2005.

Morrison, J., Codispoti, L., Gaurin, S., Jones, B., Manghnani, V., and Zheng, Z.: Seasonal variation of hydrographic and nutrient fields during the US JGOFS Arabian Sea Process Study, Deep-Sea Res. Pt. II, 45, 2053–2101, 1998.

Morrison, J., Codispoti, L., Smith, S. L., Wishner, K., Flagg, C., Gardner, W. D., Gaurin, S., Naqvi, S., Manghnani, V., and Prosperie, L.: The oxygen minimum zone in the Arabian Sea during 1995, Deep-Sea Res. Pt. II, 46, 1903–1931, 1999.

Nair, R., Ittekkot, V., Manganini, S., Ramaswamy, V., Haake, B., Degens, E., Desai, B. t., and Honjo, S.: Increased particle flux to the deep ocean related to monsoons, Nature, 338, 749–751, 1989.

Naqvi, S., Noronha, R. J., and Reddy, C.: Denitrification in the Arabian Sea, Deep-Sea Res. Pt. A, 29, 459–469, 1982.

Naqvi, S.: Denitrification processes in the Arabian Sea, Proc. Ind. Acad. Sci.-Earth Plane. Sci., 103, 279–300, 1994.

Naqvi, S., Naik, H., Pratihary, A., D'Souza, W., Narvekar, P., Jayakumar, D., Devol, A., Yoshinari, T., and Saino, T.: Coastal versus open-ocean denitrification in the Arabian Sea, Biogeosciences, 3, 621–633, 2006, http://www.biogeosciences.net/3/621/2006/.

Olson, D. B., Hitchcock, G. L., Fine, R. A., and Warren, B. A.: Maintenance of the low-oxygen layer in the central Arabian Sea, Deep-Sea Res. Pt. II, 40, 673–685, 1993.

Pandarinath, K., Subrahmanya, K., Yadava, M., and Verma, S.: Late Quaternary Sedimentation Records on the Continental Slope Off Southwest Coast of India-Implications for Provenance, Depositional and Paleomonsoonal Conditions, J. Geol. Soc. India, 69, 1285–1292, 2007.

Parab, S. G. and Matondkar, S.: Primary productivity and nitrogen fixation by *Trichodesmium* spp. in the Arabian Sea, J. Mar. Syst., 105, 82–95, 2012.

Paulmier, A. and Ruiz-Pino, D.: Oxygen minimum zones (OMZs) in the modern ocean, Prog. Oceanogr., 80, 113–128, 2009.

Pichevin, L., Bard, E., Martinez, P., and Billy, I.: Evidence of ventilation changes in the Arabian Sea during the late Quaternary: Implication for denitrification and nitrous oxide emission, Global Biogeochem. Cy., 21, GB4008, doi:10.1029/2006GB002852, 2007.

Rao, V. P., Kessarkar, P. M., Thamban, M., and Patil, S. K.: Paleoclimatic and diagenetic history of the late quaternary sediments in a core from the Southeastern Arabian Sea: Geochemical and magnetic signals, J. Oceanogr., 66, 133–146, 2010.

Rau, G., Froelich, P. N., Takahashi, T., and Des Marais, D.: Does sedimentary organic $\delta^{13}C$ record variations in Quaternary ocean $[CO_2(aq)]$, Paleoceanography, 6, 335–347, 1991.

Reichart, G.-J., Lourens, L., and Zachariasse, W.: Temporal variability in the northern Arabian Sea Oxygen Minimum Zone (OMZ) during the last 225,000 years, Paleoceanography, 13, 607–621, 1998.

Reimer, P. J., Baillie, M. G., Bard, E., Bayliss, A., Beck, J. W., Blackwell, P. G., Ramsey, C. B., Buck, C. E., Burr, G. S., and Edwards, R. L.: IntCal09 and Marine09 radiocarbon age calibration curves, 0–50 000 yeats cal BP, Radiocarbon, 51, 1111–1150, 2009.

Rixen, T., Haake, B., Ittekkot, V., Guptha, M., Nair, R., and Schlüssel, P.: Coupling between SW monsoon-related surface and deep ocean processes as discerned from continuous particle flux measurements and correlated satellite data, J. Geophys. Res., 101, 28569–28528, 1996.

Robinson, R. S., Brunelle, B. G., and Sigman, D. M.: Revisiting nutrient utilization in the glacial Antarctic: Evidence from a new method for diatom-bound N isotopic analysis, Paleoceanography, 19, PA3001, doi:10.1029/2003PA000996, 2004.

Robinson, R. S., Kienast, M., Luiza Albuquerque, A., Altabet, M., Contreras, S., De Pol Holz, R., Dubois, N., Francois, R., Galbraith, E., and Hsu, T. C.: A review of nitrogen isotopic alteration in marine sediments, Paleoceanography, 27, PA4203, doi:10.1029/2012PA002321, 2012.

Schäfer, P. and Ittekkot, V.: Seasonal variability of δ^{15}N in settling particles in the Arabian Sea and its palaeogeochemical significance, Naturwissenschaften, 80, 511–513, 1993.

Schmittner, A., Galbraith, E. D., Hostetler, S. W., Pedersen, T. F., and Zhang, R.: Large fluctuations of dissolved oxygen in the India and Pacific oceans during Dansgaard-Oeschger oscillations caused by variations of North Atlantic Deep Water subduction, Paleoceanography, 22, PA3207, doi:10.1029/2006PA001384, 2007.

Schubert, C. J. and Calvert, S. E.: Nitrogen and carbon isotopic composition of marine and terrestrial organic matter in Arctic Ocean sediments: implications for nutrient utilization and organic matter composition, Deep-Sea Res. Pt. I, 48, 789–810, 2001.

Schulte, S., Rostek, F., Bard, E., Rullkötter, J., and Marchal, O.: Variations of oxygen-minimum and primary productivity recorded in sediments of the Arabian Sea, Earth Planet. Sci. Lett., 173, 205–221, 1999.

Sigman, D., Altabet, M., Michener, R., McCorkle, D., Fry, B., and Holmes, R.: Natural abundance-level measurement of the nitrogen isotopic composition of oceanic nitrate: an adaptation of the ammonia diffusion method, Mar. Chem., 57, 227–242, 1997.

Sigman, D., Altabet, M., McCorkle, D., Francois, R., and Fischer, G.: The δ^{15}N of nitrate in the Southern Ocean: nitrogen cycling and circulation in the ocean interior, J. Geophys. Res.-Oceans (1978–2012), 105, 19599–19614, 2000.

Sigman, D. M., Karsh, K. L., and Casciotti, K. L.: Nitrogen Isotopes in the Ocean, in: Encyclopedia of Ocean Sciences, edited by: John, H. S. (Editor-in-Chief), Academic Press, Oxford, 1884–1894, 2001.

Stuiver, M. and Braziunas, T. F.: Anthropogenic and solar components of hemispheric ^{14}C, Geophys. Res. Lett., 25, 329–332, 1998.

Suthhof, A., Ittekkot, V., and Gaye-Haake, B.: Millennial-scale oscillation of denitrification intensity in the Arabian Sea during the late Quaternary and its potential influence on atmospheric N_2O and global climate, Global Biogeochem. Cy., 15, 637–649, 2001.

Thunell, R. C., Sigman, D. M., Muller-Karger, F., Astor, Y., and Varela, R.: Nitrogen isotope dynamics of the Cariaco Basin, Venezuela, Global Biogeochem. Cy., 18, GB3001, doi:10.1029/2003GB002185, 2004.

You, Y.: Intermediate water circulation and ventilation of the Indian Ocean derived from water-mass contributions, J. Mar.Res., 56, 1029–1067, 1998.

Zonneveld, K. A. F., Versteegh, G. J. M., Kasten, S., Eglinton, T. I., Emeis, K.-C., Huguet, C., Koch, B. P., de Lange, G. J., de Leeuw, J. W., Middelburg, J. J., Mollenhauer, G., Prahl, F. G., Rethemeyer, J., and Wakeham, S. G.: Selective preservation of organic matter in marine environments; processes and impact on the sedimentary record, Biogeosciences, 7, 483–511, doi:10.5194/bg-7-483-2010, 2010.

2

The effects of river inflow and retention time on the spatial heterogeneity of chlorophyll and water–air CO_2 fluxes in a tropical hydropower reservoir

F. S. Pacheco[1], M. C. S. Soares[2], A. T. Assireu[3], M. P. Curtarelli[4], F. Roland[2], G. Abril[5], J. L. Stech[4], P. C. Alvalá[1], and J. P. Ometto[1]

[1]Earth System Science Center, National Institute for Space Research, São José dos Campos, 12227–010, São Paulo, Brazil
[2]Laboratory of Aquatic Ecology, Federal University of Juiz de Fora, Juiz de Fora, 36036–900, Minas Gerais, Brazil
[3]Institute of Natural Resources, Federal University of Itajubá, Itajubá, 37500–903, Minas Gerais, Brazil
[4]Remote Sense Division, National Institute for Space Research, São José dos Campos, 12227–010, São Paulo, Brazil
[5]Laboratoire Environnements et Paléoenvironnements Océaniques et Continentaux (EPOC), CNRS, Université Bordeaux 1, Avenue des Facultés, 33405 Talence, France

Correspondence to: F. S. Pacheco (felipe.pacheco@inpe.br)

Abstract. Abundant research has been devoted to understanding the complexity of the biogeochemical and physical processes that are responsible for greenhouse gas (GHG) emissions from hydropower reservoirs. These systems may have spatially complex and heterogeneous GHG emissions due to flooded biomass, river inflows, primary production and dam operation. In this study, we investigated the relationships between the water–air CO_2 fluxes and the phytoplanktonic biomass in the Funil Reservoir, which is an old, stratified tropical reservoir that exhibits intense phytoplankton blooms and a low partial pressure of CO_2 (pCO_2). Our results indicated that the seasonal and spatial variability of chlorophyll concentrations (Chl) and pCO_2 in the Funil Reservoir are related more to changes in the river inflow over the year than to environmental factors such as air temperature and solar radiation. Field data and hydrodynamic simulations revealed that river inflow contributes to increased heterogeneity during the dry season due to variations in the reservoir retention time and river temperature. Contradictory conclusions could be drawn if only temporal data collected near the dam were considered without spatial data to represent CO_2 fluxes throughout the reservoir. During periods of high retention, the average CO_2 fluxes were 10.3 mmol m^{-2} d^{-1} based on temporal data near the dam versus -7.2 mmol m^{-2} d^{-1} with spatial data from along the reservoir surface. In this case, the use of solely temporal data to calculate CO_2 fluxes results in the reservoir acting as a CO_2 source rather than a sink. This finding suggests that the lack of spatial data in reservoir C budget calculations can affect regional and global estimates. Our results support the idea that the Funil Reservoir is a dynamic system where the hydrodynamics represented by changes in the river inflow and retention time are potentially a more important force driving both the Chl and pCO_2 spatial variability than the in-system ecological factors.

1 Introduction

Over the last two decades, hydropower reservoirs have been identified as potentially important sources of greenhouse gas (GHG) emissions (St Louis et al., 2000; Rosa et al., 2004; Demarty et al., 2011). In tropical regions, high temperatures and forest flooding have intensified GHG emissions (Abril et al., 2005; Fearnside and Pueyo, 2012). However, emissions are larger in tropical Amazonian (Abril et al., 2013) than in tropical non-Amazonian reservoirs (Ometto et al., 2011) and are larger in younger than in older reservoirs (Barros et al., 2011). Large hydroelectric reservoirs, particularly those created by impounded rivers, are morphometrically complex

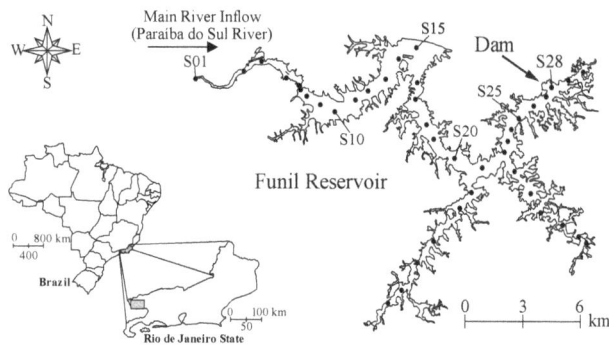

Figure 1. Map of the Funil Reservoir showing the geographic location and sampling stations.

and spatially heterogeneous (Roland et al., 2010; Teodoru et al., 2011; Zhao et al., 2013). Different regions of the reservoir may have different CO_2 dynamics because of flooded biomass, river input of organic matter, primary production and dam operations. Furthermore, both heterotrophic and autotrophic activities influence the CO_2 concentrations along reservoirs located in subtropical (Di Siervi et al., 1995), tropical (Roland et al., 2010; Kemenes et al., 2011) and temperate areas (Richardot et al., 2000; Lauster et al., 2006; Finlay et al., 2009; Halbedel and Koschorreck, 2013).

As sedimentation and light availability increase along a reservoir, the biomass of the primary producers may increase. Phytoplankton is distributed in patches along the reservoir due to differences in the nutrient distribution, light availability and stratification (Serra et al., 2007). Additionally, hydrodynamics factors, such as retention time and river inflow, may influence the phytoplankton communities and their growth (Vidal et al., 2012; Soares et al., 2008). Intense phytoplankton primary production has been identified as the main regulator of carbon (C) budgets in temperate eutrophic lakes (Finlay et al., 2010; Pacheco et al., 2014); however, the impact of these communities on tropical hydropower reservoirs is still unclear.

River inflows may affect the biogeochemical patterns in river valley reservoirs (Kennedy, 1999). Density differences of incoming stream and lake water, the stream and lake hydraulics, the strength of stratification and mixing patterns are all features that control how river water will flow when it reaches the reservoir (Fischer and Smith, 1983; Fischer et al., 1979). As a result of density differences between river and lake water, a river enters a lake and can flow large distances as a gravity-driven density current (Ford, 1990; Martin and McCutcheon, 1998). The interactions of large nutrient loads injected by rivers and the dynamics of river inflow can determine the spatial heterogeneity of phytoplankton distribution (Vidal et al., 2012). Consequently, the river inflow may affect primary production along a river/dam axis in hydropower reservoirs that are strongly influenced by rivers with high nutrient levels.

In this study, we investigated the relationships between phytoplanktonic biomass and water–air CO_2 fluxes in an old, stratified tropical reservoir (Funil, state of Rio de Janeiro, Brazil) where intense phytoplankton blooms and low pCO_2 are observed in the water. We combined fieldwork and modeling to analyze the impact of meteorological and hydrological factors on the spatial and temporal dynamics of phytoplankton and the intensity of CO_2 fluxes. We demonstrate the effect of river inflow on the heterogeneity of pCO_2 and Chl in the Funil Reservoir. We also compare temporal data of the pCO_2 collected near the dam with a high density of spatial data. Our hypothesis is that the seasonal and spatial variability of pCO_2 and Chl in the Funil Reservoir is related more to the river inflow and retention time than to external environmental factors such as air temperature and solar radiation. We highlight that very different conclusions can be drawn about carbon cycling in reservoirs if the spatial heterogeneity is not adequately considered.

2 Materials and methods

2.1 Study site

The Funil Reservoir is an old impoundment constructed at the end of the 1960s that is located on the Paraíba do Sul River, in the southern part of the state of Rio de Janeiro, Brazil ($22°30'$ S, $44°45'$ W; Fig. 1). The site is 440 m above sea level, with wet-warm summers and dry-cold winters. The Funil Reservoir mainly functions for energy production, but the reservoir is also used for irrigation and recreation. It has a surface area of $40\,km^2$, a mean and maximum depth of 22 and 74 m, respectively, and a total volume of $890 \times 10^6\,m^3$. The maximum and minimum reservoir water level occurs at the end of the rainy season (April) and dry season (October), respectively. From October 2011 to September 2012, the difference between the minimum and maximum water level was 15.6 m and the average retention time was 32 days.

The Funil Reservoir has a catchment area of $12\,800\,km^2$ that is one of the most highly industrialized regions in Brazil. There are approximately 2 million people living inside the catchment area and 39 cities that depend on the Paraíba do Sul River for their water supply. These cities represent 2 % of Brazil's gross domestic product (GDP) (IBGE, 2010). In this area, 46 % of the sewage is untreated (AGEVAP, 2011), and the Paraíba do Sul River receives a large portion of the sewage from one of the most populated regions in Brazil (20–50 inhabitants per square kilometer; IBGE, 2010). Consequently, the river exerts a large influence on the reservoir's water quality, and the reservoir has experienced tragic eutrophication in recent decades, resulting in frequent and intense cyanobacterial blooms (Klapper, 1998; Branco et al., 2002; Rocha et al., 2002). The river inflow is affected by the water supply and the operation of upstream dams. In general, the Funil Reservoir is a turbid, eutrophic system, with high

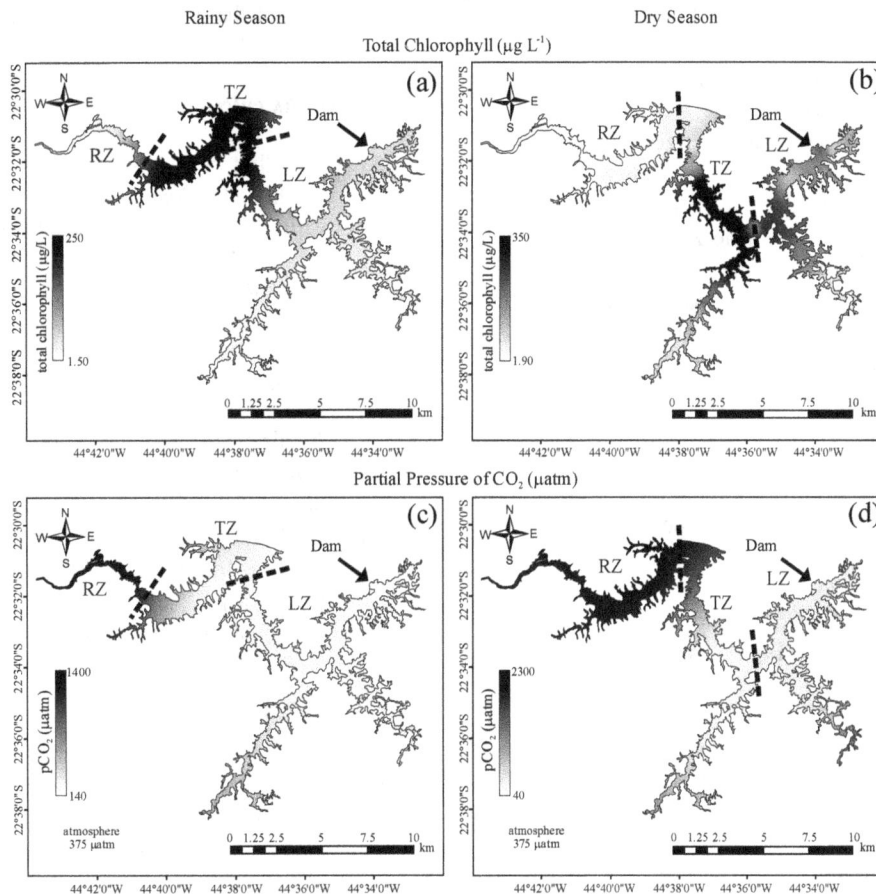

Figure 2. Map of pCO_2 and Chl as expressed by a color gradient obtained from interpolation of measured data using ordinary kriging statistics. The root-mean-square error (RMSE) of the kriging prediction, calculated by comparing observed and calculated values, was 90 μatm and 15 μg L^{-1} for pCO_2 and Chl, respectively. Lighter gray represents low Chl (**a, b**) and low pCO_2 (**c, d**). RZ: riverine zone; TZ: transition zone; LZ: lacustrine zone.

levels of phytoplankton (cyanobacteria) biomass (Soares et al., 2012; Rangel et al., 2012).

2.2 Field sampling

Spatial data: we considered 42 stations in the Funil Reservoir (28 were located along the main body of the reservoir, Fig. 1) for the spatial analyses. Water samples to determine Chl and pCO_2 were obtained between 9:00 and 12:00 local time (LT: UTC/GMT -3 h) on 1 March 2012 (at the end of the rainy season, at high water levels) and on 20 September 2012 (at the end of the dry season, at low water levels). Samples were taken from the surface (0.3 m) on the same day to limit the effect of diurnal variations on the results. We measured Chl using a compact version of PHYTO-PAM (Heinz Walz GmbH, PHYTO-ED, Effelrich, Germany). The pCO_2 data were determined using a water–air equilibration method. In a marble-type equilibrator (Abril et al., 2014, 2006), the water was pumped directly from the lake and flowed from the top to the bottom (0.8 L min^{-1}), whereas a constant volume of air (0.4 L min^{-1}) flowed from the bottom to the top. The

large gas exchange surface area promoted by contact with the marbles accelerated the pCO_2 water–air equilibrium. The air pump conducted the air from the top of the equilibrator through a drying tube containing a desiccant (Drierite), then to an infrared gas analyzer (IRGA; LI-840, LICOR, Lincoln, NE, USA) and back to the bottom of the equilibrator (closed air circuit; Abril et al., 2006). For each station, the lake water and air were pumped through the system for 2 min before the pCO_2 from the IRGA stabilized to a constant value.

Color maps were created to represent the spatial distribution of Chl and pCO_2 (Fig. 2). We used a variogram analysis to describe the spatial correlation among the samples and to spatially interpolate with kriging (Bailey and Gatrell, 1995). The empirical variograms were fitted to different mathematical models using the Akaike's information criterion (AIC; Akaike, 1974) to evaluate the best fit. The best variogram model was used for interpolation with ordinary kriging. We used the Spring software (National Institute for Space Research, São José dos Campos, SP, Brazil; Câmara et al.,

1996) version 5.1.8 to conduct a spatial analysis and to produce in situ pCO_2 and Chl maps.

In this study, we used Chl as a parameter to separate the reservoir into three zones. The riverine zone is characterized by low Chl ($< 5\,\mu g\,L^{-1}$). The transition zone begins where the Chl starts to increase ($> 5\,\mu g\,L^{-1}$) and ends when the Chl decreases to levels close to the Chl in the lacustrine zone ($< 60\,\mu g\,L^{-1}$). Finally, the lacustrine zone is characterized by intermediate Chl levels (> 5 and $< 60\,\mu g\,L^{-1}$); however, peaks of Chl were observed in some parts of the lacustrine zone. We estimated the size of each zone (riverine, transition, lacustrine) of the reservoir during the dry and rainy seasons using the results from the spatial interpolation of the Chl data. After the interpolation, we used a pixel classification method to determine the boundaries of each zone (class). We checked the boundary locations with observed data. Finally, we determined the area by multiplying the number of pixels of each class by the area of each pixel. The boundary of each zone is represented by dashed lines on the maps (Fig. 2).

Time series data: the temporal data were collected at a single station (S28) near the dam (Fig. 1). Wind speed and direction, solar radiation, pH, dissolved oxygen (DO), air temperature and temperature profiles (2, 5, 20 and 40 m depth) were collected hourly and transmitted by satellite in quasi-real time by the Integrated System for Environmental Monitoring (SIMA). SIMA is a set of hardware and software developed for data acquisition and real-time monitoring of hydrological systems (Alcantara et al., 2013; Stevenson et al., 1993). SIMA consists of an independent system formed by an anchored buoy containing data storage systems, sensors (air temperature, wind direction and intensity, pressure, incoming and reflected radiation, and a thermistor chain), a solar panel, a battery and a transmission antenna. A sonde probe (YSI model 6600, Yellow Spring, OH, USA) was attached to the SIMA buoy to collect hourly surface data on temperature, conductivity, pH and oxygen. The sonde was calibrated every 15 days based on the YSI Environmental Operations Manual (http://www.ysi.com/ysi/support). We used data collected between 25 October 2011 and 25 October 2012 for our analyses.

Samples for alkalinity (ALK), total phosphorous (TP) and nitrogen (TN) analyses were taken monthly. ALK was determined by the titration method (APHA, 2005). For TP, the samples were oxidized by persulfate and were then analyzed as soluble reactive phosphorus. TN was determined as the sum of the organic fraction measured with the Kjedahl method and the dissolved inorganic nutrients. A laboratory analysis of TP and NP was performed based on standard spectrophotometric techniques (Wetzel and Likens, 2010).

We calculated the pCO_2 from the surface water over 1 year near the dam using the measured pH and alkalinity. The calculations included the dependence on temperature for the dissociation constants of carbonic acid (Millero et al., 2002) and the solubility of CO_2. We used the hourly data of pH and

temperature and the monthly data of alkalinity collected at station S28 (Fig. 1).

2.3 CO_2 flux calculation

The air–water flux of CO_2 ($mmol\,m^{-2}\,d^{-1}$) was calculated according to Eq. (1). Positive values of CO_2 fluxes denote the net gas flux from the lake to the atmosphere,

$$F(CO_2) = k\alpha\Delta pCO_2, \tag{1}$$

where k is the gas transfer velocity of CO_2 (in $m\,h^{-1}$), α is the solubility coefficient of CO_2 (in $mmol\,m^{-3}\,\mu atm^{-1}$) as a function of water temperature (Weiss, 1974), and ΔpCO_2 is the air–water gradient of pCO_2 (in μatm). The atmospheric pCO_2 measured during the rainy and dry season was $375\,\mu atm$, and this atmospheric value was used for all of the flux calculations. The gas transfer velocity k was calculated from the gas transfer velocity normalized to a Schmidt number of 600 (k_{600}) that corresponds to the CO_2 at $20\,°C$ (Eq. 2) (Jahne et al., 1987), as follows:

$$k = k_{600}\left(\frac{Sc}{600}\right)^{-0.5}, \tag{2}$$

where Sc is the Schmidt number of a given gas at a given temperature (Wanninkhof, 1992). k_{600} is the normalized gas transfer velocity calculated from the wind speed (MacIntyre et al., 2010) using different equations under cooling and heating conditions (Eqs. 3, 4). We also evaluated a wind-speed formulation by Cole and Caraco (1998) to investigate the importance of different formulation of k_{600} (Eq. 5). A more detailed description for these equations is in Staehr et al. (2012). The k_{600} was calculated in $cm\,h^{-1}$ and were converted to $m\,d^{-1}$.

$$k_{600} = 2.04\,U_{10} + 2.0 \tag{3}$$

(under cooling; MacIntyre et al., 2010),

$$k_{600} = 1.74\,U_{10} - 0.15 \tag{4}$$

(under heating; MacIntyre et al., 2010),

$$k_{600} = 2.07 + 0.21\,U_{10}^{1.7} \tag{5}$$

(Cole and Caraco, 1998),

where U_{10} is the wind speed at 10 m height. The wind speed was obtained from SIMA data at 3 m height and was calculated at 10 m height (Smith, 1985).

In the riverine zone, we considered the k_{600} as a function of wind and water currents. The contribution of the water current to the gas transfer velocity was estimated using the water current (w, $cm\,s^{-1}$) and depth (h, m) and the equations in Borges et al. (2004) (Eq. 6):

$$k_{600} = 1.719w^{0.5}h^{-0.5}. \tag{6}$$

2.4 Temperature profile

Temperature profiles were collected using a thermistor chain deployed at station S09 during the rainy season and station S14 during the dry season to determine the thermal structure in the transition zone. Eleven thermistors (Hobo, U22 Water Temp Pro v2, Bourne, MA, USA) were placed every 0.5 m up to 4 m and every 1 m from 5 to 7 m. We also deployed a thermistor chain at the riverine zone at station S05 with thermistors placed every 2 m. The thermistors were programmed to record the temperature every 10 min. During the rainy season, the thermistor chain was deployed on 29 February 2012 at 18:30 LT and was recovered after 40 h. In the dry season, the thermistor chain was deployed on 20 September 2012 at 11:30 LT and was recovered after 25 h.

In our analysis, the temperature was considered to be the factor controlling water density. The use of temperature is justified by the low conductivity and turbidity in the river. The turbidity values of 29 and 11 NTU that were measured during the rainy and dry seasons, respectively, would have affected density < 5 % relative to temperature (Gippel, 1989).

2.5 Numerical model description and setup

Numerical simulations of lake hydrodynamics were conducted with the Estuary and Lake Computer Model (EL-COM; Hodges et al., 2000). This model solves the 3-D hydrostatic, Boussinesq, Reynolds-averaged Navier–Stokes and scalar transport equations, separating mixing of scalars and momentum from advection. The hydrodynamic algorithms that are implemented in the ELCOM use a Euler–Lagrange approach for the advection of momentum adapted from the work of Casulli and Cheng (1992), whereas the advection of scalars (i.e., tracers, conductivity and temperature) is based on the ULTIMATE QUICKEST method proposed by Leonard (1991). The thermodynamics model considers the penetrative (i.e., shortwave radiation) and nonpenetrative components (i.e., longwave radiation, sensible and latent heat fluxes) (Hodges et al., 2000). The vertical mixing model uses the transport equations of turbulent kinetic energy (TKE) to compute the energy available from wind stirring and shear production for the mixing process (Spigel and Imberger, 1980). A complete description of the formulae and numerical methods used in the ELCOM was given by Hodges et al. (2000).

Hydrodynamic simulations of the Funil Reservoir were conducted under realistic forcing conditions (e.g., inflow, outflow, atmospheric temperature and radiation). These simulations were aimed at testing hypotheses about river inflows in transition zones during the rainy and dry seasons in the Funil Reservoir. The simulations started 4 days before our study period. This was necessary to let the model equilibrate beyond the initial physical conditions. The digital representation of the reservoir bathymetry (numerical domain) was defined based on the bathymetric data collected from 27 to 29 February 2012. The numerical domain was discretized in a uniform horizontal grid containing 100 m x 100 m cells. The vertical grid resolution was set to a uniform 1 m of thickness, resulting in 72 vertical layers. The water albedo was set to 0.03 (Slater, 1980), and the bottom drag coefficient was set to 0.001 (Wüest and Lorke, 2003). The attenuation coefficient for PAR was set to $0.6\,\mathrm{m^{-1}}$ based on Secchi disk measurements. Based on a previous study conducted in another tropical reservoir (Pacheco et al., 2011), a value of $5.25\,\mathrm{m^2\,s^{-1}}$ was chosen for the horizontal diffusivity of temperature and for the horizontal momentum.

Because of the presence of persistent unstable atmospheric conditions over tropical reservoirs (Verburg and Antenucci, 2010), an atmospheric stability sub-model was activated during the simulation; this procedure was adequate because the meteorological sensors were placed within the atmospheric boundary layer (ABL) over the surface of the lake and the data were collected at sub-daily intervals (Imberger and Patterson, 1990). In this manner, at each model time step, the heat and momentum transfer coefficients were adjusted based on the stability of the ABL. The stability of the ABL is evaluated through the stability parameter, which was derived from the Monin–Obukhov length scale. ELCOM uses similarity functions presented in Imberger and Patterson (1990) in both stable (negative values stability parameter) and unstable conditions (positive values). The Coriolis sub-model was also activated during the simulation and the Coriolis force was then considered in the Navier–Stokes equation. This force causes the deflection of moving objects (in this case, the water currents) when they are viewed in a rotating reference frame (e.g., the Earth).

We defined two sets of boundary cells to force the inflow (Paraíba do Sul River) and outflow (the water intake at the dam). The meteorological driving forces over the free surface of the reservoir were considered uniform. The model was forced using hourly meteorological data acquired by SIMA, the daily inflow and outflow provided by Eletrobrás-Furnas, and the river temperatures extracted from thermistor chain data. To complement river inflow temperature data collected in situ, we used the Moderate Resolution Imaging Spectroradiometer (MODIS; Justice et al., 2002) level 3 land surface temperature (LST) product (named M*D11A1; see Wan, 2008, for more details) to estimate the temporal variation of temperature at the reservoir's inflow. The M*D11A1 is a standard remote-sensing-based product, generated using a split-window algorithm and seven spectral MODIS bands located in the regions of the shortwave infrared and thermal infrared bands. This algorithm is based on the differential absorption of adjacent bands in the infrared region (Wan and Dozier, 1996). The M*D11A1 product has been validated at stage 2 by a series of field campaigns conducted between 2000 and 2007 and over more locations and time periods during radiance-based validation studies. Accuracy is better than 1 °C (0.5 °C in most cases). This product is generated up to four times each day (i.e., 10:30, 13:30, 23:30 and 02:30) and

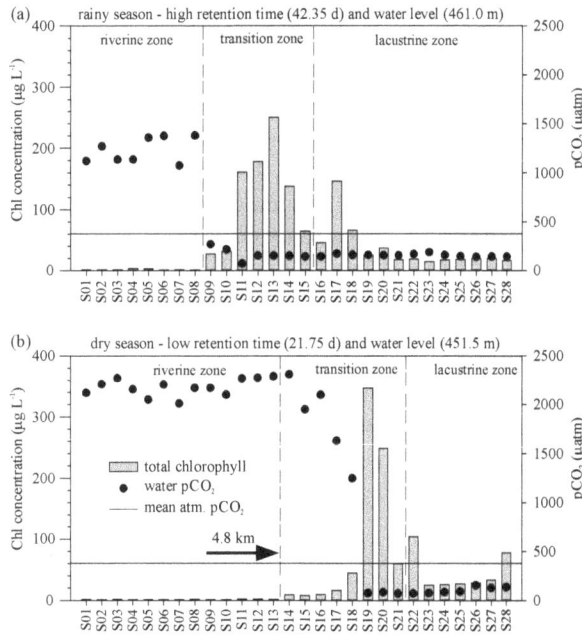

Figure 3. Lotic–lentic gradient of pCO_2 and Chl at the 28 sampling stations in the main reservoir body during the rainy season (**a**) and dry season (**b**). The water level was 461.0 and 451.5 m during the rainy season and dry season, respectively. Three zones can be clearly defined (riverine, transition and lacustrine zones). The arrow shows that the transition zone starts 4.8 km down-reservoir during a period of low water levels.

is delivered in a georeferenced grid with 1 km of spatial resolution in a sinusoidal projection by the National Aeronautics and Space Administration Land Processes Distributed Active Archive Center (NASA LP DAAC).

The cloud cover fraction over the Funil Reservoir was estimated using a MODIS level 2 cloud mask product (named M*D35L2; see Ackerman et al., 1998, for more details). The algorithm used to generate the M*D35L2 product employs a series of visible and infrared threshold and consistency tests to specify confidence that an unobstructed view of the Earth's surface is observed. This product is generated up to four times each day (i.e., 10:30, 13:30, 23:30 and 02:30) and is delivered in a georeferenced grid with 1 km of spatial resolution in a sinusoidal projection.

The MODIS products were acquired online (http://reverb.echo.nasa.gov/reverb/) and were preprocessed using the MODIS Reprojection Tool (available at https://lpdaac.usgs.gov). The data were first resampled to a 100 m spatial resolution (compatible with the bathymetric grid). Next, they were re-projected into the universal transverse mercator (UTM) coordinate system (zone 23 south) with the World Geodetic System (WGS-84) datum as reference; then, they were converted into a raster image. Finally, MATLAB® routines were used to calculate the river inflow temperature and cloud cover fraction time series. The river inflow temperature (°C) time series was computed

using the preprocessed M*D11A1 data by extracting the temperature values from the pixel located within the Paraíba do Sul River channel near the Funil Reservoir entrance. The cloud cover fraction (dimensionless) time series was obtained using the preprocessed M*D35L2 data by computing the ratio between the cloudy pixels and the total pixels covering the reservoir surface. Two periods were simulated: one to represent the rainy season (25 February 2012 to 4 March 2012) and one to represent the dry season (15 to 23 September 2012).

3 Results

3.1 Spatial variability

Based on the spatial data of Chl and pCO_2, a typical zonation pattern that is usually found in reservoirs was observed in the main sections of the Funil Reservoir (riverine, transition and lacustrine zones) (Fig. 2). Although the boundaries are influenced by many factors and are not easily determined, these regions have distinct physical, chemical and biological features. The riverine zone (RZ) has a high input of nutrients coming from terrestrial systems and human activities, but primary production is limited by high turbidity and turbulence. As the sedimentation and light availability increase along the reservoir, the biomass of the primary producers increases in the transition zone (TZ). The lacustrine zone (LZ) is characterized by nutrient limitations and reduced phytoplankton biomass (Thornton, 1990).

The Funil Reservoir was spatially heterogeneous with seasonal differences in the Chl and pCO_2 (Fig. 2). There was only high spatial variation in the main body of the reservoir, whereas the southern part was undersaturated in CO_2 during the rainy and dry seasons (Fig. 2a, b). The spatial averages of pCO_2 during the rainy and dry season were 259 ± 221 and $881 \pm 900\,\mu\text{atm}$, respectively. The pCO_2 varied from 140 to $1376\,\mu\text{atm}$ during the rainy season and from 43 to $2290\,\mu\text{atm}$ during the dry season. Higher values of pCO_2 in the riverine zone of the reservoir and a drastic decrease in the transition zone were observed in both sample periods (Fig. 3a, b). In the lacustrine zone, undersaturation of CO_2 was prevalent at all sample sites in the rainy and dry seasons. Considering all of the sample sites, there were significant differences between the rainy and dry seasons ($t = 1.99$, $p < 0.05$). The Chl values were similar in the transition and lacustrine zone in the rainy season ($t = 2.01$, $p > 0.05$) and were higher in the transition zone during the dry season ($t = 2.01$, $p < 0.05$; Fig. 3a, b; Table 1). Furthermore, the average concentration in the transition zone was 2.5 times higher than the reservoir average (129.2 and $52.0\,\mu\text{g}\,\text{L}^{-1}$, respectively). Unlike the pCO_2, the Chl data showed no significant difference between the rainy and dry season considering all of the spatial data ($t = 1.99$, $p > 0.05$).

Table 1. Average CO_2 fluxes $(mmol\ m^{-2}\ d^{-1})$ calculated using spatial and temporal data. Positive fluxes denote net gas fluxes from the lake into the atmosphere. In the last column, the different letters represent significant differences (t test, $p < 0.05$). Lowercase letters represent differences between the fluxes in the reservoir zones, and uppercase letters represent the differences between the fluxes during the seasons.

		CO_2 fluxes mmol m^{-2} d^{-1}				
		k_{600} (MacIntyre et al., 2010)		k_{600} (Cole and Caraco, 1998)		
		Average	SD	Average	SD	Significant differences
	Area (km^2)	Spatial data				
Rainy summer						
Entire reservoir	36.0	−10.1	26.8	−7.2	21.9	
Riverine zone	5.7	44.5	6.5	37.6	5.5	a
Transition zone	9.3	−24.8	15.3	−19.1	11.7	b, e
Lacustrine zone	20.9	−18.3	9.1	−14.1	7.0	b
Dry winter						
Entire reservoir	34.3	24.6	61.5	22.1	50.8	
Riverine zone	13.7	93.0	13.3	78.7	11.2	c
Transition zone	7.6	−4.7	51.5	−2.0	42.1	d
Lacustrine zone	13.1	−29.7	18.1	−22.9	13.9	e
		At the dam				
All data throughout the year		−0.1	39.8	−0.9	33.1	
Rainy spring		−28.6	24.6	−27.1	18.5	A
Rainy summer		8.1	41.8	7.6	35.6	B
Dry autumn		23.7	39.2	19.6	29.9	C
Dry winter		−0.4	33.0	−0.6	25.5	D

The calculated CO_2 fluxes from the spatial data varied from −46.5 to 52.2 mmol m^{-2} d^{-1} and −61.9 to 103.16 mmol m^{-2} d^{-1} during the rainy and dry season, respectively. In both the rainy and dry seasons, the maximum emissions were observed in the riverine zone and the minimum was observed in the transition zone. The spatial average was −10.1 and 24.6 mmol m^{-2} d^{-1} during the rainy and dry season, respectively (Table 1).

3.2 Temporal variability

The pCO_2 calculated from the multi-parameter sonde data (temperature and pH) and the alkalinity showed a large seasonal variability over the year at the station near the dam (Table 2). The pCO_2 varied from 35 to 4058 μatm, with an average of $624 \pm 829\,\mu$atm and median of 165 μatm. The pCO_2 supersaturation was prevalent between April and June, whereas pCO_2 undersaturation was prevalent during all other periods (Fig. 4a). The lowest median of pCO_2 was observed between October and December (43 μatm). Considering all of the temporal data throughout the year, 59.8 % of the data measured below atmospheric equilibrium and 1.1 % were within 5 % of the atmospheric equilibrium.

In the Funil Reservoir, the seasonal pCO_2 variation over the year at the station near the dam agreed with the variation in the retention time (Fig. 4). The yearly average of the reservoir retention time was 32.6 days over the research year. The lower retention time occurred between October and December, when the water level was low and the reservoir was ready to stock water coming from the watershed and rain during the rainy season (October to March).

Because we sampled temperature on a sub-daily scale over the year, we used the equations proposed by MacIntyre et al. (2010) to calculate k_{600}, which also incorporates the turbulence from heat loss. The turbulence from heat loss, especially overnight, often exceeds that from wind mixing in tropical lakes that tend to have low wind. However, the estimates using Cole and Caraco (1998) formulations to calculate k_{600} did not significantly change our results (Table 1). Due to the large sample size of the temporal data (hourly data), significant differences were observed between the estimates, primarily in the dry autumn, when the surface temperature decreased after the warm-summer ($t = 1.96$, $p < 0.05$). The CO_2 flux over the year at the station near the dam varied from −104.7 to 175.88 mmol m^{-2} d^{-1}. The average flux was -0.1 ± 39.8 mmol m^{-2} d^{-1} and the median was −7.4 mmol m^{-2} d^{-1}. We observed a substantial uptake of CO_2 between October and December (rainy spring) (Table 1). From January to July, the lake lost substantial CO_2 via degassing (Table 1). The uptake of CO_2 from the atmosphere

Table 2. Average and standard deviations of environmental and chemical variables from station S28 (near the dam) and river. *Cumulative precipitation over 3 months.

Months	Oct–Dec		Jan–Mar		Apr–Jun		Jul–Sep	
Season	Rainy autumn		Rainy summer		Dry spring		Dry winter	
	Average	SD	Average	SD	Average	SD	Average	SD
Air temperature ($^{\circ}$C)	22.5	4.0	24.0	3.3	20.7	3.1	19.6	4.0
Alkalinity (mg L^{-1} as CaCO$_3$)	11.0	0.2	15.5	4.6	11.3	3.7	12.5	3.0
Chlorophyll (mg L^{-1})	12.9	12.8	23.8	20.6	3.0	0.2	23.2	35.0
Total phosphorus (μg L^{-1})	42.3	8.5	41.7	12.2	18.4	8.6	33.7	28.0
Total nitrogen (μg L^{-1})	1264.6	357.1	1143.2	305.3	1505.6	454.3	1203.3	299.7
Maximum depth (m)	65.1	1.8	69.3	1.4	71.6	2.5	69.1	4.4
Mean reservoir depth (m)	19.3	0.4	20.3	0.4	20.9	0.7	20.3	1.1
pCO$_2$ (μatm)	68.9	118.6	848.9	1027.5	1111.8	907.5	521.9	618.5
Precipitation (mm)*	547.0		420.2		230.2		71.6	
Retention time (days)	27.9	7.7	33.0	9.0	36.4	6.4	33.2	7.4
Max daily solar radiation (W m^{-2})	937.7	276.1	958.1	246.8	716.9	227.2	758.0	189.7
Surface water temperature ($^{\circ}$C)	24.7	1.1	27.1	1.0	24.1	1.7	22.0	1.0
Wind speed (m s^{-1})	–	–	1.6	1.2	1.4	1.3	1.6	1.5
River total phosphorus (mg L^{-1})	80.6	–	77.1	–	42.4	–	88.3	–
River total nitrogen (mg L^{-1})	1535.5	–	2072.5	–	1524.2	–	1972.6	–
River total carbon (mg L^{-1})	12.9	2.0	13.3	1.8	13.7	2.5	12.1	2.9
Downstream total carbon (mg L^{-1})	12.4	2.3	11.8	0.3	13.7	2.6	11.9	1.6
Inflow (m^3 s^{-1})	224.2	58.9	236.4	74.1	234.1	36.7	168.9	28.7
Outflow (m^3 s^{-1})	223.6	57.2	236.4	74.1	226.0	30.9	219.1	10.7

* Cumulative precipitation over 3 months.

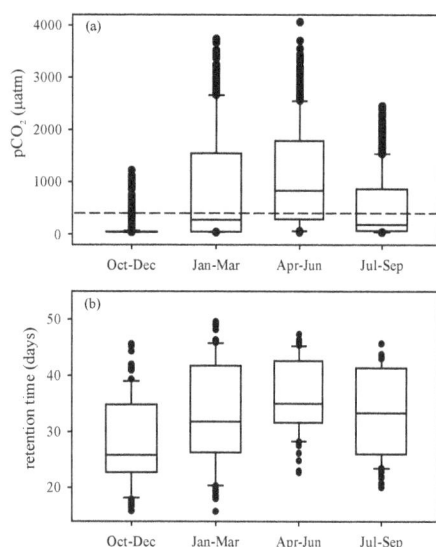

Figure 4. Box plot of the pCO$_2$ at station S28 near the dam (**a**) and the mean reservoir retention time (**b**) over the studied year. The dashed line represents the average pCO$_2$ in the atmosphere (375 μatm). The data are subdivided into four seasons: rainy spring (October–December), rainy summer (January–March), dry autumn (April–June) and dry winter (July–September).

was also prevalent between July and September (dry winter). A summary of all other data collected over the study period is shown in Table 2.

3.3 Thermal structure of the transition and riverine zone

We observed significant differences between the thermal structures during the rainy and dry season (Fig. 5). During the rainy season, thermal stratification only occurred in the transition zone during the daytime, at approximately 16:30 LT, when a maximum of 33.1 $^{\circ}$C and a minimum of 27.8 $^{\circ}$C was observed at the surface and bottom, respectively (Fig. 5a). In contrast, the temperature was vertically homogeneous at nighttime. The daily temperature oscillation during the rainy season at the surface reached up to 5 $^{\circ}$C. During the dry season, the water temperature was lower compared to the rainy season in the transition zone. Stratification occurred at approximately 14:00 LT during the dry season, when we observed a maximum of 25.7 $^{\circ}$C and a minimum of 23.1 $^{\circ}$C at the bottom. The daily temperature oscillation reached up to 3 $^{\circ}$C at the surface and stratus layers with different temperatures were observed every 2.5 m (Fig. 5b). The river temperature varied from 27.7 to 28.7 $^{\circ}$C and 23.6 to 24.1 $^{\circ}$C in the rainy and dry season, respectively (Table 3). The average temperature difference between the river and reservoir

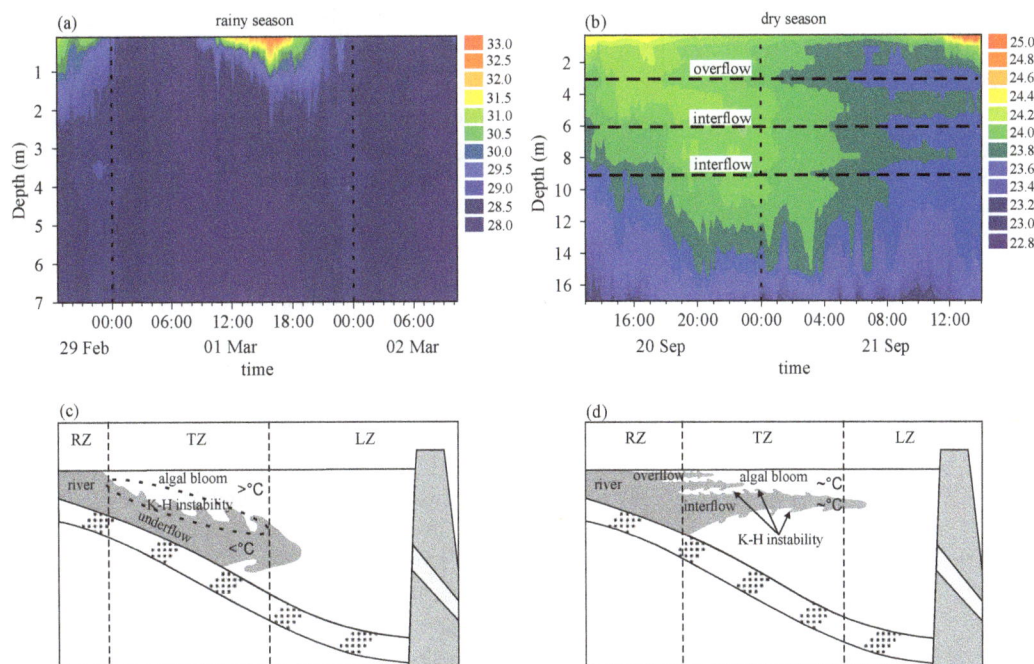

Figure 5. The temperature profile collected at station S09 during the rainy season (**a**) and at station S14 during the dry season (**b**). The dashed line represents the depths where the river flows as overflow or interflows. During the rainy season, the river plunges and flows under the reservoir (underflow) due to differences in density (**c**). The waves and billows develop along the interface due to the shear velocity (Kelvin–Helmholtz instability) and facilitate vertical mixing (see text). During the dry season, the river flows as overflow or interflow (**d**) because the difference in the density between the river and reservoir is low. In this situation, the river can influence the reservoir surface water more 5 km toward the dam. RZ: riverine zone; TZ: transition zone; LZ: lacustrine zone.

surface waters was 2.1 and 0.3 °C during the rainy and dry season, respectively.

3.4 Simulations

We first compared the simulated and real temperatures at stations S09 and S14 from the rainy and dry season, respectively. The RMSE, calculated by comparing the data every 20 min, was 1.4 °C for the rainy season and 1.1 °C for the dry season. These results, obtained for both seasons, were comparable with previous modeling exercises found in the literature (Jin et al., 2000; Vidal et al., 2012). We also analyzed the ability of the model to reproduce inflow, using data from drifters released in the riverine and transition zones of the reservoir from 1 March and 20 September (data not shown). Although the vertical thermal structures observed during the dry season (Fig. 5b) were not well represented, the model reproduced the behavior of the inflow as underflow during the rainy season (Fig. 6a) and as interflow and overflow during the dry season (Fig. 6b) as anticipated by the schematic representation (Fig. 5c,d). The river flowed mainly at 6 m depth near the bottom of the Funil Reservoir after the river plunge point during the rainy season. During the dry season, the river flowed mainly at 3 m depth at night and 4 m during the daytime.

The daily oscillation of the neutral buoyancy observed occurs because of the variation of the reservoir surface and river temperatures (Vidal et al., 2012, Curtarelli et al., 2013). The level of neutral buoyancy, where the densities of the flowing current and the ambient fluid are equal, represents the depth at which the river water spreads laterally in the reservoir. During the rainy season, the river flowed as underflow (Fig. 6a); however, when the river reached its maximum temperature, at approximately 21:00 LT (Table 3), the temperature difference between the river and surface water decreased, the level of river neutral buoyancy moved upward and the maximum flow was observed between 4 and 6 m (Fig. 6a). During the dry season, the river overflowed, but it plunged down to 4 to 6 m depth when high surface temperatures during the day coincided with a period of the lowest river temperatures (Table 3) and neutral buoyancy moved downward (Fig. 6b). The change in river flow patterns between September 20 and 21 occurred due to a decrease in the river temperature during a rainfall that occurred at approximately 16:00 LT on 20 September 2012 (Fig. 6b).

Table 3. Average of the hourly temperature profile collected by the thermistor chain located at station S05 (river) on 29 February 2012 (rainy season) and 20 September 2012 (dry season).

Rainy season					
Hour (LT)	River temp. (°C)		Hour (LT)	River temp. (°C)	
	Average	SD		Average	SD
00:00	28.39	0.04	12:00	27.71	0.03
01:00	28.28	0.04	13:00	27.72	0.04
02:00	28.17	0.05	14:00	27.79	0.11
03:00	28.07	0.03	15:00	27.97	0.06
04:00	28.00	0.02	16:00	28.03	0.02
05:00	27.91	0.04	17:00	28.16	0.09
06:00	27.85	0.04	18:00	28.34	0.09
07:00	27.77	0.05	19:00	28.49	0.06
08:00	27.73	0.00	20:00	28.63	0.04
09:00	27.72	0.01	21:00	28.70	0.01
10:00	27.71	0.02	22:00	28.67	0.03
11:00	27.69	0.01	23:00	28.55	0.05
Max	28.70 (21:00)				
Min	27.69 (11:00)				
Dry season					
Hour (LT)	River temp. (°C)		Hour (LT)	River temp. (°C)	
	Average	SD		Average	SD
00:00	23.90	0.02	12:00	23.80	0.08
01:00	23.88	0.02	13:00	23.82	0.02
02:00	23.80	0.06	14:00	23.87	0.04
03:00	23.74	0.04	15:00	23.89	0.04
04:00	23.71	0.04	16:00	24.00	0.04
05:00	23.66	0.01	17:00	23.97	0.05
06:00	23.64	0.01	18:00	23.99	0.08
07:00	23.60	0.04	19:00	24.08	0.02
08:00	23.57	0.03	20:00	24.03	0.02
09:00	23.59	0.01	21:00	24.00	0.02
10:00	23.62	0.02	22:00	23.96	0.02
11:00	23.65	0.02	23:00	23.95	0.02
Max	24.08 (19:00)				
Min	23.57 (08:00)				

4　Discussion

4.1　pCO_2 driven by phytoplankton

Primary production associated with high Chl levels was the main regulator of the CO_2 concentration at the surface of the Funil Reservoir (Fig. 7). Spatially, the pCO_2 levels were negatively correlated with the Chl ($r^2 = 0.71$). In old hydropower reservoirs, where the C source from flooded soil after impounding has become negligible, primary production may become a significant element of the C budget. Intense primary production fueled by high levels of nutrients reduces the CO_2 concentrations to levels below the atmospheric equilibrium in transition and lacustrine zones of the Funil Reservoir (Fig. 3). The high pCO_2 in the riverine zone may be explained by the terrestrial ecosystem respiration entering the river as dissolved soil CO_2, the oxidation of allochthonous and emergent autochthonous organic carbon, the acidification of buffered waters, the precipitation of carbonate min-

erals, and the direct pumping of root respiration CO_2 from riparian vegetation (Butman and Raymond, 2011).

Low pCO_2 levels observed at the station near the dam over the year was associated with the following: (1) high primary production due to higher temperatures and solar radiation that promoted water column stability and stratification and (2) constant high nutrient availability. Because nutrient availability in the Funil Reservoir was high during the entire year (Table 2), the phytoplankton growth was not limited by nutrients in the lacustrine zone. However, seasonal variation in factors that control stability and stratification, such as temperature, wind and mixing zone depth, may inhibit algal growth near the dam, especially between April and June.

Due to phytoplankton productivity, we observed a net uptake of CO_2 over the year at the station near the dam, especially between October and December (Table 1). However, the fate of carbon fixed by phytoplankton in the Funil Reservoir is still unclear. The higher flux of methane (CH_4) from sediment to water that was observed in the Funil Reservoir compared with other tropical reservoir (Ometto et al., 2013) suggests that a substantial fraction of the carbon fixed by the phytoplankton reached the sediment and was further mineralized as CH_4. However, in the lacustrine zone, the higher depth and high temperatures may promote the decomposition of dead phytoplankton generating CO_2 or CH_4 in the water column before it reaches the sediment.

It is important to note that the CO_2 production in the sediments can leave an imprint on the pCO_2 of the surface water, especially in the dry season, when the reservoir is not stratified. During periods of water stratification, the carbon coming from organic carbon mineralization in the sediment may be trapped in the hypolimnion and may not contribute to the CO_2 flux from the water to the atmosphere (Cardoso et al., 2013). Furthermore, it is important to highlight that the contribution of carbon mineralization in the sediment to the pCO_2 at the surface can also be regulated by other factors such as CO_2 saturation in the water and the depth of the reservoir (Guérin et al., 2006). Moreover, when the river plunges and flows at the bottom of the reservoir, the water flow can disturb the sediment and enhance the carbon flux from the sediment to the hypolimnion, which can affect the contribution of organic carbon mineralized on the sediment to the amount of carbon emitted by the reservoir.

By considering that the outflow exported the same amount of carbon as that from the watershed (Table 2), we suggest that a high sedimentation rate offsets the uptake of CO_2 from the atmosphere to close the carbon budget. Although there are no data to support this statement, we hypothesize that both (i) the burial of organic carbon composed of phytoplankton and (ii) methanogenesis are important carbon pathways for the carbon fixed by the phytoplankton in the Funil Reservoir, as reported in natural eutrophic lakes (Downing et al., 2008).

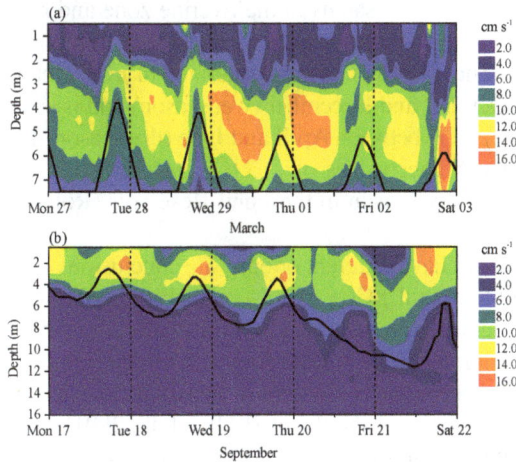

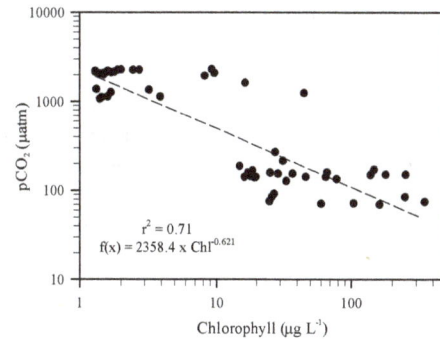

Figure 7. Relationship between the spatial $p\mathrm{CO}_2$ and Chl data in the Funil Reservoir. The regression is represented by a dashed line ($r^2 = 0.71$, $p < 0.001$).

Figure 6. Simulated velocity profile using realistic forcing. Higher velocities represent the depth where the river flows though the transition zone. The river flows as underflow during the rainy season when a denser (colder) river plunges beneath the surface and flows downward along the bottom as a gravity-driven density current (**a**). The river flows as overflow during the dry season, when the temperature from the river and reservoir are similar (**b**). As overflow, the river characteristics can be found many kilometers toward the dam at the water surface. The black line represents the depth of neutral buoyancy as estimated from temperature records, presuming that the lake and river water do not mix. The anomaly observed in the river flow and depth of neutral buoyancy between 20 and 21 September 2012 occurred due to a decrease in the river temperature during a rainfall that occurred at approximately 16:00 LT on 20 September.

4.2   Physical features and spatial distribution

The Funil Reservoir's retention time is strongly driven by the operation of the dam. The volume of water that flows through the turbine depends on the energy demands and inflow from the Paraíba do Sul River. Periods of low retention time and water levels do not necessarily correspond to periods of low precipitation. In fact, the highest retention time and water level is often observed in the middle of the dry season, when the reservoir is full to ensure enough water to produce energy during the entire dry season. This suggests that these processes are not only driven by natural factors but that they may also be regulated by dam operation in the Funil Reservoir.

The position of the transition zone of the reservoir moves as a result of the season (Fig. 3). At the end of the rainy season, the retention time and water level was high and the influence of the river in the surface water of the reservoir was restricted to a small area (Fig. 2a, c). However, when the water level and retention time was low, the transition zone moved toward the dam and the river inflow influenced the surface Chl and $p\mathrm{CO}_2$ over more than 40 % of the total reservoir surface area (Fig. 2b, d). As previously reported, when the reten-

tion time is short, a reservoir can become a fluvial-dominated system (Straškraba, 1990).

The size of the river-influenced area over the reservoir surface water also depends on the water density. Differences in the river and reservoir temperatures, the total dissolved solids, and the suspended solids can cause a density gradient in the water column. Depending on the water density differences between the inflow and the reservoir, the river can flow into the downstream area as overflow, underflow or interflow (Martin and McCutcheon, 1998). During the rainy season in the Funil Reservoir, due to the substantial difference between the river and reservoir surface temperatures ($\sim 4\,^{\circ}\mathrm{C}$), the river water progressively sinks down (underflow) and contributes to the thermal stability of the water column (Fig. 5a, Assireu et al., 2011). The denser river water flows under the lighter reservoir water and waves and billows develop along the interface due to the shear velocity. This behavior is indicative of Kelvin–Helmholtz instability, in which waves made up of fluid from the current (river) promote mixing with the reservoir water (Thorpe and Jiang, 1998; Corcos and Sherman, 2005) (Fig. 5c). This mixing, and the high nutrient concentration coming from Paraíba do Sul River (Table 2), may explain the high Chl levels observed in the transition zone (Fig. 3).

Many cold fronts pass through the Brazilian midwest and southeast during the dry seasons. (Lorenzzetti et al., 2005, Alcântara et al., 2010). As a result, the decrease in the reservoir surface temperature (Table 2) and consequent decrease in the density difference between the river and reservoir surfaces leads to river inflow that is characterized by inter-overflow (Fig. 5b, d). In an inter-overflow, the riverine characteristics of high turbulence, $p\mathrm{CO}_2$ and low Chl are observed in the reservoir surface 5 km toward the dam (Fig. 3a, b). Although there are high nutrient concentrations in the transition zone (Table 1) between S19 and the river, the surface water is dominated by river flow with low Chl concentrations (Fig. 3). Favorable conditions for phytoplankton blooming will only exist down-reservoir in the transition

Table 4. Comparison between CO_2 fluxes ($mmol\,m^{-2}\,d^{-1}$) calculated during periods of low retention time and high retention time. Positive fluxes denote net gas fluxes from the lake to the atmosphere. Statistical analyses showed significant differences between temporal and spatial data and between low and high retention times (t test, $p < 0.05$). * We considered data for low retention and high retention time as values less than 25 days and more than 38 days, respectively. The average CO_2 fluxes during periods of intermediate retention time were close to 0 ($0.5\,mmol\,m^{-2}\,d^{-1}$).

| | CO_2 fluxes $mmol\,m^{-2}\,d^{-1}$ | | | |
| | Low retention time | | High retention time | |
	Average	SD	Average	SD
Temporal data	−18.6	30.3	14.5	33.6
Spatial data	24.6	61.5	−10.1	26.8

zone, where the inflow mixes with the reservoir and loses velocity (Vidal et al., 2012).

The simulation of the rainy season (Fig. 6) indicated a minimal influence from the river inflow on the surface water, which was suggested by the thermal stability in the transition zone (Fig. 5a). The simulation of the dry season represented the overflow, particularly at night (Fig. 6b). However, the simulation did not represent intrusions of the river water on the different depths (every 2.5 m) as suggested by the temperature profile in the transition zone (Fig. 5b). The variation in the river inflow over the day (Fig. 6) occurred in response to the lag change in the temperature of the river and the reservoir. During the rainy season, this oscillation enhanced the intake of nutrients into the euphotic zone when the reservoir surface temperature decreased and the river temperature reached its maximum at the end of the day (Table 3). During the day, when the river temperature dropped, the large peak of Chl in the transition zone (Fig. 3a) could be a result of developing diurnal stratification (Fig. 5). During the dry season, peak Chl occurs 5 km further downstream (Fig. 3b) because the inflow never plunges due to lower temperature differences between the river and reservoir surfaces.

4.3 Spatial and temporal heterogeneity

As a result of the phytoplankton growth associated with these physical features, there are large spatial and temporal variations in the CO_2 fluxes in the Funil Reservoir. Several studies of hydropower reservoirs have suggested that significant CO_2 is emitted from these systems into the atmosphere at a global scale (St Louis et al., 2000; Roehm and Tremblay, 2006; Barros et al., 2011; Fearnside and Pueyo, 2012). However, recent studies have shown that the growing nutrient enrichment caused by human activities (eutrophication) can reverse this pattern in some hydropower reservoirs (Roland et al., 2010) and natural lakes (Pacheco et al., 2014). Our study indicates that the Funil Reservoir is spatially heterogeneous,

with high CO_2 emissions in the riverine zone and high CO_2 uptake in the transition and lacustrine zones. Temporally, the reservoir near the dam is undersaturated in pCO_2 mainly between October and December (wet season) and is supersaturated in pCO_2 between April and June (dry season, Table 1). Similarly, higher values of pCO_2 were previously reported during the dry season in the Funil Reservoir (Roland et al., 2010).

We could have different or opposite conclusions if the spatial and temporal pCO_2 data were analyzed separately. Previous studies have suggested that, in natural small lakes, a single sample site would be adequate to determine whether a lake is above or below equilibrium with the atmosphere and the intensity of the fluxes (Kelly et al., 2001). However, large spatial heterogeneity of pCO_2 and CO_2 emissions into the atmosphere have been observed in boreal (Teodoru et al., 2011) and tropical (Roland et al., 2010) reservoirs. Our temporal data at the dam station exhibited lower pCO_2 in October, November and December, when the retention time was extremely low (Table 4); however, this observation did not represent the entire reservoir. The spatial data collected at low water levels showed low pCO_2 in the dam as well; however, almost half of the reservoir is supersaturated due to river influences (Fig. 2d). The average pCO_2 during low retention time was 881 μatm over the whole reservoir area, contrasting with only 69 μatm near the dam. Furthermore, if we considered only one station near the dam to estimate the CO_2 flux between the lake surface and atmosphere, the conclusion would be contradictory. For example, during periods of low retention time, the calculated CO_2 flux indicated that the CO_2 flux would be $-17.6\,mmol\,m^{-2}\,d^{-1}$ (CO_2 sink) for one-spot temporal data and $22.1\,mmol\,m^{-2}\,d^{-1}$ (CO_2 source) for the whole reservoir (Table 4).

The same contradictory conclusions can be expressed when studies with low sample site numbers are considered for spatial heterogeneity. Previous studies looking at the heterogeneity of the Funil Reservoir indicated no phytoplankton biomass peak in the transition zone (Soares et al., 2012). In our study, the Chl data collected every 1000 m as a proxy were able to show a clear transition zone within the reservoir. Furthermore, data analysis in Soares et al. (2012), considering four sampling stations, exhibited high spatial heterogeneity in periods of high retention times (high water level). In contrast, we found high spatial heterogeneity in low retention times, corresponding to periods of high influence from the river on the surface water. Thus, the different conclusions in Soares et al. (2012) may be explained by the variation in the spatial distribution of the transition zone location once the retention time and inflow are used as key parameters to define its location (Fig. 2c, d).

5   Conclusions

In summary, the seasonal and spatial variability of the Chl and CO_2 fluxes in the Funil Reservoir are mainly related to river inflow and retention times. However, the relationship between the pCO_2 and Chl suggests that primary production regulates the surface CO_2 fluxes in the transition and lacustrine zone. The average spatial data showed CO_2 emissions into the atmosphere during periods of low retention time (even with higher Chl) due to the river's influence on the surface water and CO_2 uptake during periods of high retention time when the river plunges and flows under the reservoir. However, the retention time threshold that seals the transition between a source and sink of CO_2 could not be determined. A comparison between the spatial (42 stations) and temporal data (1 station) indicated that different conclusions may be drawn if spatial heterogeneity is not adequately considered. Furthermore, transition zone location changes over the year must be considered when a low number of stations are used to represent spatial heterogeneity. The lack of spatial information about CO_2 fluxes could lead to erroneous conclusions about the importance of hydropower reservoirs in the freshwater carbon cycle. The Funil Reservoir is a dynamic system where the hydrodynamics linked to the river inflow and retention time control both pCO_2 and Chl spatial variability and seem to regulate most ecological processes.

Acknowledgements. This work was supported by the project "Carbon Budgets of Hydroelectric Reservoirs of Furnas Centrais Elétricas S. A.". Thanks to the Center for Water Research (CWR) and its director, Jörg Imberger, for making ELCOM available for this study. We also thank the São Paulo State Science Foundation for financial support (FAPESP process no. 2010/06869-0 and 2014/06556-3). G. Abril is a visiting special researcher from the Brazilian CNPq program Ciência Sem Fronteiras (process #401726/2012-6)

Edited by: T. J. Battin

References

Abril, G., Guerin, F., Richard, S., Delmas, R., Galy-Lacaux, C., Gosse, P., Tremblay, A., Varfalvy, L., Dos Santos, M. A., and Matvienko, B.: Carbon dioxide and methane emissions and the carbon budget of a 10-year old tropical reservoir (Petit Saut, French Guiana), Global Biogeochem. Cy., 19, GB4007, doi:10.1029/2005gb002457, 2005.

Abril, G., Richard, S., and Guerin, F.: In situ measurements of dissolved gases (CO_2 and CH_4) in a wide range of concentrations in a tropical reservoir using an equilibrator, Sci. Total Environ., 354, 246–251, doi:10.1016/j.scitotenv.2004.12.051, 2006.

Abril, G., Parize, M., Perez, M. A. P., and Filizola, N.: Wood decomposition in Amazonian hydropower reservoirs: An additional source of greenhouse gases, J. S. Am. Earth Sci., 44, 104–107, doi:10.1016/j.jsames.2012.11.007, 2013.

Abril, G., Martinez, J.-M., Artigas, L. F., Moreira-Turcq, P., Benedetti, M. F., Vidal, L., Meziane, T., Kim, J.-H., Bernardes, M. C., Savoye, N., Deborde, J., Souza, E. L., Alberic, P., Landim de Souza, M. F., and Roland, F.: Amazon River carbon dioxide outgassing fuelled by wetlands, Nature, 505, 395–398, doi:10.1038/nature12797, 2014.

Ackerman, S. A., Strabala, K. I., Menzel, W. P., Frey, R. A., Moeller, C. C., and Gumley, L. E.: Discriminating clear sky from clouds with MODIS, J. Geophys. Res.-Atmos., 103, 32141–32157, doi:10.1029/1998JD200032, 1998.

AGEVAP: Relatório Técnico - Bacia do Rio Paraíba Do Sul - Subsídios às Ações de Melhoria da Gestão 2011, Associação Pró-Gestão das Águas da Bacia Hidrográfica do Rio Paraíba do Sul, 255, Resende, 2011.

Akaike, H.: New look at statistical-model identification, IEEE T. Automat. Contr., 19, 716–723, doi:10.1109/tac.1974.1100705, 1974.

Alcantara, E., Curtarelli, M., Ogashawara, I., Stech, J., and Souza, A.: Hydrographic observations at SIMA station Itumbiara in 2013, in: Long-term environmental time series of continuously collected data in hydroelectric reservoirs in Brazil, edited by: Alcantara, E., Curtarelli, M., Ogashawara, I., Stech, J., and Souza, A., PANGAEA, Bremerhaven, 1–3, 2013.

Alcântara, E. H., Bonnet, M. P., Assireu, A. T., Stech, J. L., Novo, E. M. L. M., and Lorenzzetti, J. A.: On the water thermal response to the passage of cold fronts: initial results for Itumbiara reservoir (Brazil), Hydrol. Earth Syst. Sci. Discuss., 7, 9437–9465, doi:10.5194/hessd-7-9437-2010, 2010.

APHA: Standard Methods for the Examination of Water and Wastewater, 21 ed., Washington, DC, 1368 pp., 2005.

Assireu, A. T., Alcântara, E., Novo, E. M. L. M., Roland, F., Pacheco, F. S., Stech, J. L., and Lorenzzetti, J. A.: Hydrophysical processes at the plunge point: an analysis using satellite and in situ data, Hydrol. Earth Syst. Sci., 15, 3689–3700, doi:10.5194/hess-15-3689-2011, 2011.

Bailey, T. C. and Gatrell, A. C.: Interactive spatial data analysis, in, Essex: Longman Scientific & Technical, 432 pp., 1995.

Barros, N., Cole, J. J., Tranvik, L. J., Prairie, Y. T., Bastviken, D., Huszar, V. L. M., del Giorgio, P., and Roland, F.: Carbon emission from hydroelectric reservoirs linked to reservoir age and latitude, Nat. Geosci., 4, 593–596, doi:10.1038/Ngeo1211, 2011.

Borges, A. V., Vanderborght, J.-P., Schiettecatte, L. S., Gazeau, F., Ferrón-Smith, S., Delille, B., and Frankignoulle, M.: Variability of the gas transfer velocity of CO_2 in a macrotidal estuary (the Scheldt), Estuaries, 27, 593–603, doi:10.1007/BF02907647, 2004.

Branco, C. W. C., Rocha, M. I. A., Pinto, G. F. S., Gômara, G. A., and Filippo, R.: Limnological features of Funil Reservoir (R.J., Brazil) and indicator properties of rotifers and cladocerans of the zooplankton community, Lakes Reserv. Res. Manag., 7, 87–92, doi:10.1046/j.1440-169X.2002.00177.x, 2002.

Butman, D. and Raymond, P. A.: Significant efflux of carbon dioxide from streams and rivers in the United States, Nat. Geosci., 4, 839–842, doi:10.1038/ngeo1294, 2011.

Câmara, G., Souza, R. C. M., Freitas, U. M., and Garrido, J.: Spring: Integrating remote sensing and gis by object-oriented data modelling, Comput. Graph., 20, 395–403, doi:10.1016/0097-8493(96)00008-8, 1996.

Cardoso, S. J., Vidal, L. O., Mendonça, R. F., Tranvik, L. J., Sobek, S., and Roland, F.: Spatial variation of sediment mineralization supports differential CO_2 emissions from a tropical hydroelectric reservoir, Front. Microbiol., 4, 101, doi:10.3389/fmicb.2013.00101, 2013.

Casulli, V. and Cheng, R. T.: Semiimplicit Finite-Difference Methods for 3-Dimensional Shallow-Water Flow, Int. J. Numer. Meth. Fl., 15, 629–648, doi:10.1002/fld.1650150602, 1992.

Cole, J. J. and Caraco, N. F.: Atmospheric exchange of carbon dioxide in a low-wind oligotrophic lake measured by the addition of SF6, Limnol. Oceanogr., 43, 647–656, 1998.

Corcos, G. M. and Sherman, F. S.: The mixing layer: deterministic models of a turbulent flow, J. Fluid Mech., 139, 29–65, 2005.

Curtarelli, M. P., Alcântara, E., Renno, C. D., Assireu, A. T., Stech, J. L., and Bonnet, M. P.: Modelling the surface circulation and thermal structure of a tropical reservoir using three-dimensional hydrodynamic lake model and remote-sensing data, Water Environ. J., 28, 516–525, 2013.

Demarty, M., Bastien, J., and Tremblay, A.: Annual follow-up of gross diffusive carbon dioxide and methane emissions from a boreal reservoir and two nearby lakes in Québec, Canada, Biogeosciences, 8, 41–53, doi:10.5194/bg-8-41-2011, 2011.

Di Siervi, M. A., Mariazzi, A. A., and Donadelli, J. L.: Bacterioplankton and phytoplankton production in a large Patagonian reservoir (Republica Argentina), Hydrobiologia, 297, 123–129, 1995.

Downing, J. A., Cole, J. J., Middelburg, J. J., Striegl, R. G., Duarte, C. M., Kortelainen, P., Prairie, Y. T., and Laube, K. A.: Sediment organic carbon burial in agriculturally eutrophic impoundments over the last century, Global Biogeochem. Cy., 22, Gb1018, doi:10.1029/2006gb002854, 2008.

Fearnside, P. M. and Pueyo, S.: Greenhouse-gas emissions from tropical dams, Nature Clim. Change, 2, 382–384, 2012.

Finlay, K., Leavitt, P. R., Wissel, B., and Prairie, Y. T.: Regulation of spatial and temporal variability of carbon flux in six hard-water lakes of the northern Great Plains, Limnol. Oceanogr., 54, 2553–2564, doi:10.4319/lo.2009.54.6_part_2.2553, 2009.

Finlay, K., Leavitt, P. R., Patoine, A., and Wissel, B.: Magnitudes and controls of organic and inorganic carbon flux through a chain of hard-water lakes on the northern Great Plains, Limnol. Oceanogr., 55, 1551–1564, doi:10.4319/lo.2010.55.4.1551, 2010.

Fischer, H. B. and Smith, R. D.: Observations of transport to surface waters from a plunging inflow to Lake Mead, Limnol. Oceanogr., 28, 258–272, 1983.

Fischer, H. B., List, E. J., Koh, R. C. Y., Imberger, J., and Brooks, N. H.: Mixing in inland and coastal waters, Academic Press, New York, 483 pp., 1979.

Ford, D. E.: Reservoir Transport Processes, in: Reservoir Limnology: Ecological Perspectives, edited by: Thornton, K. W., Kimmel, B. L., and Payne, F. E., Wiley-Interscience, New York, 15–41, 1990.

Gippel, C. J.: The use of turbidimeters in suspended sediment research, Hydrobiologia, 176, 465–480, doi:10.1007/bf00026582, 1989.

Guérin, F., Abril, G., Richard, S., Burban, B., Reynouard, C., S eyler, P., and Delmas, R.: Methane and carbon dioxide emissions from tropical reservoirs: Significance of downstream rivers, Geophys. Res. Lett., 33, L21407, doi:10.1029/2006GL027929, 2006.

Halbedel, S. and Koschorreck, M.: Regulation of CO_2 emissions from temperate streams and reservoirs, Biogeosciences, 10, 7539–7551, doi:10.5194/bg-10-7539-2013, 2013.

Hodges, B. R., Imberger, J., Saggio, A., and Winters, K. B.: Modeling basin-scale internal waves in a stratified lake, Limnol. Oceanogr., 45, 1603–1620, 2000.

IBGE: Instituto Brasileiro de Geografia e Estatística. Censo Demográfico 2010, Rio de Janeiro, 2010.

Imberger, J. and Patterson, J. C.: Physical Limnology, Adv. Appl. Mech., 27, 303–475, 1990.

Jahne, B., Munnich, K. O., Bosinger, R., Dutzi, A., Huber, W., and Libner, P.: On the parameters influencing airwater gas-exchange, J. Geophys. Res.-Oceans, 92, 1937–1949, doi:10.1029/JC092iC02p01937, 1987.

Jin, K., Hamrick, J., and Tisdale, T.: Application of Three-Dimensional Hydrodynamic Model for Lake Okeechobee, J. Hydraul. Eng., 126, 758–771, doi:10.1061/(ASCE)0733-9429(2000)126:10(758), 2000.

Justice, C. O., Vermote, E., Townshend, J. R. G., DeFries, R., Roy, D. P., Hall, D. K., Salomonson, V. V., Privette, J. L., Riggs, G., Strahler, A., Lucht, W., Myneni, R. B., Knyazikhin, Y., Running, S. W., Nemani, R. R., Zhengming, W., Huete, A. R., Van Leeuwen, W., Wolfe, R. E., Giglio, L., Muller, J. P., Lewis, P., and Barnsley, M. J.: The Moderate Resolution Imaging Spectroradiometer (MODIS): land remote sensing for global change research, IEEE T. Geosci. Remote Sens., 36, 1228–1249, doi:10.1109/36.701075, 1998.

Kelly, C. A., Fee, E., Ramlal, P. S., Rudd, J. W. M., Hesslein, R. H., Anema, C., and Schindler, E. U.: Natural variability of carbon dioxide and net epilimnetic production in the surface waters of boreal lakes of different sizes, Limnol. Oceanogr., 46, 1054–1064, 2001.

Kemenes, A., Forsberg, B. R., and Melack, J. M.: CO_2 emissions from a tropical hydroelectric reservoir (Balbina, Brazil), J. Geophys. Res.-Biogeo., 116, G03004, doi:10.1029/2010jg001465, 2011.

Kennedy, R. H.: Reservoir design and operation: limnological implications and management opportunities, in: Theoretical reservoir ecology and its applications, edited by: Tundisi, J. G. and Straškraba, M., Backhuys Publishers, Leiden, 1–28, 1999.

Klapper, H.: Water quality problems in reservoirs of Rio de Janeiro, Minas Gerais and Sao Paulo, Int. Rev. Hydrobiol., 83, 93–101, 1998.

Lauster, G. H., Hanson, P. C., and Kratz, T. K.: Gross primary production and respiration differences among littoral and pelagic habitats in northern Wisconsin lakes, Can. J. Fish. Aquat. Sci., 63, 1130–1141, doi:10.1139/f06-018, 2006.

Leonard, B. P.: The Ultimate Conservative Difference Scheme Applied to Unsteady One-Dimensional Advection, Comput. Method. Appl. M., 88, 17–74, doi:10.1016/0045-7825(91)90232-U, 1991.

Lorenzzetti, J. A., Stech, J. L., Assireu, A. T., Novo, E. M. L. D., and Lima, I. B. T.: SIMA: a near real time buoy acquisition and telemetry system as a support for limnological studies., in: Global warming and hydroelectric reservoirs., edited by: Santos, M. A. and Rosa, L. P., COPPE, Rio de Janeiro, 71–79, 2005.

MacIntyre, S., Jonsson, A., Jansson, M., Aberg, J., Turney, D. E., and Miller, S. D.: Buoyancy flux, turbulence, and the gas transfer coefficient in a stratified lake, Geophys Res Lett, 37, L24604, doi:10.1029/2010GL044164, 2010.

Martin, J. L. and McCutcheon, S. C.: Hydrodynamics and Transport for Water Quality Modeling, CRC Press, Boca Raton, 1998.

Millero, F. J., Pierrot, D., Lee, K., Wanninkhof, R., Feely, R., Sabine, C. L., Key, R. M., and Takahashi, T.: Dissociation constants for carbonic acid determined from field measurements, Deep-Sea Res. Pt.-I, 49, 1705–1723, doi:10.1016/s0967-0637(02)00093-6, 2002.

Ometto, J. P., Cimbleris, A. C. P., dos Santos, M. A., Rosa, L. P., Abe, D., Tundisi, J. G., Stech, J. L., Barros, N., and Roland, F.: Carbon emission as a function of energy generation in hydroelectric reservoirs in Brazilian dry tropical biome, Energ. Policy, 58, 109–116, doi:10.1016/j.enpol.2013.02.041, 2013.

Ometto, J. P. H. B., Pacheco, F. S., Cimbleris, A. C. P., Stech, J. L., Lorenzzetti, J., Assireu, A. T., Santos, M. A., Matvienko, B., Rosa, L. P., Sadigisgalli, C., Donato, A., Tundisi, J. G., Barros, N. O., Mendonca, R., and Roland, F.: Carbon Dynamic and Emissions in Brazilian Hydropower Reservoirs, in: Energy Resources: Development, Distribution and Exploitation, edited by: Alcantara, E., Nova Science Publishers, Hauppauge, 155–188, 2011.

Pacheco, F. S., Assireu, A. T., and Roland, F.: Drifters tracked by satellite applied to freshwater ecosystems: study case in Manso Reservoir, in: New technologies for the monitoring and study of large hydroelectric reservoirs and lakes, edited by: Alcantara, E. H., Stech, J. L., and Novo, E. M. L. M., Parêntese, Rio de Janeiro, 193–218, 2011.

Pacheco, F. S., Roland, F., and Downing, J. A.: Eutrophication reverses whole-lake carbon budgets, Inland Waters, 4, 41–48, doi:10.5268/iw-4.1.614, 2014.

Rangel, L. M., Silva, L. H. S., Rosa, P., Roland, F., and Huszar, V. L. M.: Phytoplankton biomass is mainly controlled by hydrology and phosphorus concentrations in tropical hydroelectric reservoirs, Hydrobiologia, 693, 13–28, doi:10.1007/s10750-012-1083-3, 2012.

Richardot, M., Debroas, D., Jugnia, L. B., Tadonleke, R., Berthon, L., and Devaux, J.: Changes in bacterial processing and composition of dissolved organic matter in a newly-flooded reservoir (a three-year study), Arch. Hydrobiol., 148, 231–248, 2000.

Rocha, M. I. A., Branco, C. W. C., Sampaio, G. F., Gômara, G. A., and de Filippo, R.: Spatial and temporal variation of limnological features, Microcystis aeruginosa and zooplankton in a eutrophic reservoir (Funil Reservoir, Rio de Janeiro), Acta Limnol. Bras., 14, 73–86, 2002.

Roehm, C., and Tremblay, A.: Role of turbines in the carbon dioxide emissions from two boreal reservoirs, Quebec, Canada, J. Geophys. Res.-Atmos., 111, D24101, doi:10.1029/2006jd007292, 2006.

Roland, F., Vidal, L. O., Pacheco, F. S., Barros, N. O., Assireu, A., Ometto, J. P. H. B., Cimbleris, A. C. P., and Cole, J. J.: Variability of carbon dioxide flux from tropical (Cerrado) hydroelectric reservoirs, Aquat. Sci., 72, 283–293, doi:10.1007/s00027-010-0140-0, 2010.

Rosa, L. P., dos Santos, M. A., Matvienko, B., dos Santos, E. O., and Sikar, E.: Greenhouse gas emissions from hydroelectric reservoirs in tropical regions, Climatic Change, 66, 9–21, doi:10.1023/B:Clim.0000043158.52222.Ee, 2004.

Serra, T., Vidal, J., Casamitjana, X., Soler, M., and Colomer, J.: The role of surface vertical mixing in phytoplankton distribution in a stratified reservoir, Limnol. Oceanogr., 52, 620–634, 2007.

Slater, P. G.: Remote sensing, optics and optical systems, Addison-Wesley Pub. Co., Reading, 575 pp., 1980.

Smith, S. V.: Physical, chemical and biological characteristics of CO_2 gas flux across the air water interface, Plant Cell Environ., 8, 387–398, doi:10.1111/j.1365-3040.1985.tb01674.x, 1985.

Soares, M. C. S., Marinho, M. M., Huszar, V. L. M., Branco, C. W. C., and Azevedo, S. M. F. O.: The effects of water retention time and watershed features on the limnology of two tropical reservoirs in Brazil, Lakes Reserv. Res. Manag., 13, 257–269, doi:10.1111/j.1440-1770.2008.00379.x, 2008.

Soares, M. C. S., Marinho, M. M., Azevedo, S. M. O. F., Branco, C. W. C., and Huszar, V. L. M.: Eutrophication and retention time affecting spatial heterogeneity in a tropical reservoir, Limnologica – Ecology and Management of Inland Waters, 42, 197–203, doi:10.1016/j.limno.2011.11.002, 2012.

Spigel, R. H. and Imberger, J.: The Classification of Mixed-Layer Dynamics in Lakes of Small to Medium Size, J. Phys. Oceanogr., 10, 1104–1121, doi:10.1175/1520-0485(1980)010<1104:Tcomld>2.0.Co;2, 1980.

Staehr, P. A., Christensen, J. P. A., Batt, R. D., and Read, J. S.: Ecosystem metabolism in a stratified lake, Limnol. Oceanogr., 57, 1317–1330, doi:10.4319/lo.2012.57.5.1317, 2012.

Stevenson, M. R., Lorenzzetti, J. A., Stech, J. L., and Arlino, P. R. A.: SIMA – An Integrated Environmental Monitoring System, VII Simpósio Brasileiro de Sensoriamento Remoto, Curitiba, 10–14 May, 1993.

St Louis, V. L., Kelly, C. A., Duchemin, E., Rudd, J. W. M., and Rosenberg, D. M.: Reservoir surfaces as sources of greenhouse gases to the atmosphere: A global estimate, Bioscience, 50, 766–775, 2000.

Straškraba, M.: Retention time as a key variable of reservoir limnology, in: Theoretical reservoir ecology and its applications, edited by: Tundisi, T. G., and Straškraba, M., Backhuys Publishers, Leiden, 43–70, 1990.

Teodoru, C. R., Prairie, Y. T., and del Giorgio, P. A.: Spatial Heterogeneity of Surface CO_2 Fluxes in a Newly Created Eastmain-1 Reservoir in Northern Quebec, Canada, Ecosystems, 14, 28–46, doi:10.1007/s10021-010-9393-7, 2011.

Thornton, K. W.: Sedimentary processes, in: Reservoir Limnology: Ecological Perspectives, edited by: Thornton, K. W., Kimmel, B. L., and Payne, F. E., John Wiley & Sons, New York, 43–70, 1990.

Thorpe, S. A. and Jiang, R.: Estimating internal waves and diapycnal mixing from conventional mooring data in a lake, Limnol. Oceanogr., 43, 936–945, 1998.

Verburg, P. and Antenucci, J. P.: Persistent unstable atmospheric boundary layer enhances sensible and latent heat loss in a tropical great lake: Lake Tanganyika, J. Geophys. Res.-Atmos., 115, D11109, doi:10.1029/2009jd012839, 2010.

Vidal, J., Marce, R., Serra, T., Colomer, J., Rueda, F., and Casamitjana, X.: Localized algal blooms induced by river inflows in a canyon type reservoir, Aquat. Sci., 74, 315–327, doi:10.1007/s00027-011-0223-6, 2012.

Wan, Z.: New refinements and validation of the MODIS Land-Surface Temperature/Emissivity products, Remote Sens. Environ., 112, 59–74, doi:10.1016/j.rse.2006.06.026, 2008.

Wan, Z. and Dozier, J.: A generalized split-window algorithm for retrieving land-surface temperature from space, IEEE Trans. Geosci. Remote Sens., 34, 892–905, doi:10.1109/36.508406, 1996.

Wanninkhof, R.: Relationship between wind-speed and gas-exchange over the ocean, J. Geophys. Res.-Oceans, 97, 7373–7382, doi:10.1029/92jc00188, 1992.

Weiss, R. F.: Carbon dioxide in water and seawater: the solubility of a non-ideal gas, Mar. Chemi., 2, 203–215, 1974.

Wetzel, R. G. and Likens, G. E.: Limnological Analyses, Springer, New York, 429 pp., 2000.

Wüest, A. and Lorke, A.: Small-scale hydrodynamics in lakes, Annu. Rev. Fluid. Mech., 35, 373–412, doi:10.1146/annurev.fluid.35.101101.161220, 2003.

Zhao, Y., Wu, B. F., and Zeng, Y.: Spatial and temporal patterns of greenhouse gas emissions from Three Gorges Reservoir of China, Biogeosciences, 10, 1219–1230, doi:10.5194/bg-10-1219-2013, 2013.

3

Upper ocean mixing controls the seasonality of planktonic foraminifer fluxes and associated strength of the carbonate pump in the oligotrophic North Atlantic

K. H. Salmon[1], P. Anand[1], P. F. Sexton[1], and M. Conte[2]

[1]Environment, Earth and Ecosystems, The Open University, UK
[2]Bermuda Institute of Ocean Sciences, St George's GE01, Bermuda

Correspondence to: K. H. Salmon (kate.salmon@open.ac.uk)

Abstract. Oligotrophic regions represent up to 75 % of Earth's open-ocean environments. They are thus areas of major importance in understanding the plankton community dynamics and biogeochemical fluxes. Here we present fluxes of total planktonic foraminifera and 11 planktonic foraminifer species measured at the Oceanic Flux Program (OFP) time series site in the oligotrophic Sargasso Sea, subtropical western North Atlantic Ocean. Foraminifera flux was measured at 1500 m water depth, over two $\sim$ 2.5-year intervals: 1998–2000 and 2007–2010. We find that foraminifera flux was closely correlated with total mass flux, carbonate and organic carbon fluxes. We show that the planktonic foraminifera flux increases approximately 5-fold during the winter–spring, contributing up to $\sim$ 40 % of the total carbonate flux. This was primarily driven by increased fluxes of deeper-dwelling globorotaliid species, which contributed up to 90 % of the foraminiferal-derived carbonate during late winter–early spring. Interannual variability in total foraminifera flux, and in particular fluxes of the deep-dwelling species (*Globorotalia truncatulinoides, Globorotalia hirsuta* and *Globorotalia inflata*), was related to differences in seasonal mixed layer dynamics affecting the strength of the spring phytoplankton bloom and export flux, and by the passage of mesoscale eddies. As these heavily calcified, dense carbonate tests of deeper-dwelling species (3 times denser than surface dwellers) have greater sinking rates, this implies a high seasonality of the biological carbonate pump in oligotrophic oceanic regions. Our data suggest that climate cycles, such as the North Atlantic Oscillation, which modulates nutrient supply into the euphotic zone and the strength of the spring bloom, may also in turn modulate the production and flux of these heavily calcified deep-dwelling foraminifera by increasing their food supply, thereby intensifying the biological carbonate pump.

1 Introduction

Planktonic foraminifera (PF) comprise 23–56 % of the total open marine calcite flux and thus exert an important control on global carbon cycling (Schiebel, 2002). They are used extensively in palaeoceanographic and palaeoclimatic reconstructions via utilisation of their species abundance and assemblage composition (e.g. Lutz, 2011; Sexton and Norris, 2011), geochemical signatures (e.g. Zeebe et al., 2008), shell mass (e.g. Barker and Elderfield, 2002) and in evolutionary and biogeographic studies (e.g. Sexton and Norris, 2008). However, gaps remain in our understanding of the controls on their spatial and temporal distribution in the upper water column. Following the early 1980s when sea surface temperatures (SSTs) were thought to dominantly control PF distributions and abundance (CLIMAP project members, 1994), a number of other environmental parameters have also been shown to exert influence on the distribution and abundance of PF, such as salinity (Kuroyanagi and Kawahata, 2004), productivity, nutrient availability (Schiebel, 2002; Northcote and Neil, 2005; Žarić et al., 2005; Storz et al., 2009; Sexton and Norris, 2011) and water column stability (Hemleben et al., 1989; Lohmann and Schweitzer, 1990; King and Howard, 2003). It is thus imperative to better understand the environmental factors controlling modern-day PF abundance in order to produce accurate interpretations of palaeorecords based on PF assemblages.

The response of PF flux and species composition to environmental and/or oceanographic factors have been studied using plankton tow materials which can give information about living populations' species distribution and depth habitats within the upper ocean (Tolderlund and Be; 1971, Fairbanks et al., 1980; Schiebel, 2002). However, temporal resolution is often limited when using plankton tows. The continuous time series records provided by sediment traps allow a more complete understanding of the seasonal and interannual changes in PF flux and can aid in integrating living assemblages with the sedimentary record.

Earlier studies of planktonic foraminifer flux off Bermuda at the Seasonal Changes in Foraminifera Flux (SCIFF) site (Fig. 1) (Deuser et al., 1981; Hemleben et al., 1985; Deuser, 1987; Deuser and Ross, 1989) were based on a bimonthly sampling interval and provide a general description of foraminifera flux, species composition and seasonality. These studies found that PF $> 125\,\mu$m comprise on average 22 % of the total calcium carbonate flux in the Sargasso Sea (Deuser and Ross, 1989), although this average underestimates the importance of the PF flux contribution during different seasons. Here we utilise a higher resolution bi-weekly sediment trap time series from the Oceanic Flux Program (OFP), ideal for studying the detailed response of PF species flux to physical oceanographic changes because PF species lifespan is approximately 2–3 weeks (Spero, 1998; Erez et al., 1991). These samples also benefit from the availability of upper ocean hydrographic and biogeochemical data collected at the nearby Bermuda Atlantic Time Series (BATS) site, as well as remote sensing data, which allows us to evaluate the environmental factors that control the total foraminifer flux as well as the response of individual species flux. Furthermore, we assess the relative contribution of PF flux to regional carbonate export and explore the implications of our findings for carbonate cycling in the oligotrophic North Atlantic.

2 Oceanographic setting

The Sargasso Sea is located within the North Atlantic gyre, which is characterised by high temperatures and salinities, and weak, variable surface currents (Lomas et al., 2013, and references therein). The OFP and BATS sites are situated in a transition region between the northern eutrophic waters and the relatively oligotrophic subtropical convergence zone in the south (Steinberg et al., 2001, and references therein). Subtropical Mode Water (STMW) forms on the fringes, north of the gyre, owing to convective deep winter mixing and entrainment of nutrients and is characterized by temperatures of 17.8–18.4 °C and salinities of $\sim 36.5 \pm 0.05$

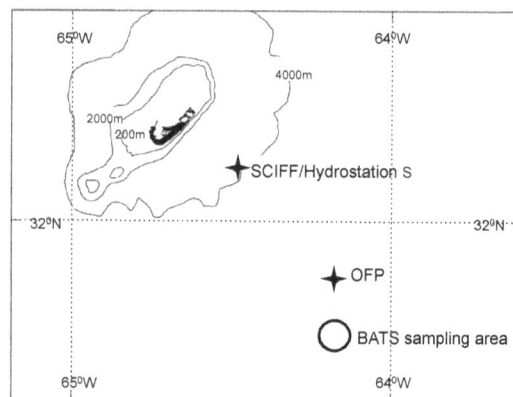

Figure 1. Map to show locations of the Oceanic Flux Program (OFP) mooring (31°50′ N, 64°10′ W) and the Bermuda Atlantic Time Series (BATS) hydrographic station (31°40′ N, 64°10′ W) and Seasonal Changes in Foraminifera Flux (SCIFF) site and Hydrostation S in relation to Bermuda.

(Bates et al., 2002), typically occurring between $\sim$ 250 and 400 m water depth (Bates, 2007).

The hydrography and biogeochemistry of the area have been summarised by Michaels and Knap (1996), Steinberg et al. (2001), Lomas et al. (2013) and references therein. In the absence of large changes in salinity, the 10 °C seasonal change in surface temperatures driven by solar insolation controls the shoaling and erosion of the mixed layer, which reaches a maximum of 250–400 m in late winter, increasing vertical mixing and entraining nutrient-rich waters. The depth of mixing determines the strength of seasonal particulate flux, nutrient concentrations and primary production during the subsequent spring bloom (Michaels and Knap, 1996; Steinberg et al., 2001). With the onset of seasonal stratification in late February–March, a spring bloom develops when phytoplankton biomass and particulate organic carbon standing stocks are maximal. As seasonal stratification intensifies, a nutrient-depleted, shallow surface mixed layer develops which is underlain by a subsurface chlorophyll maximum at approximately 80–100 m depth. Strong stratification in summer and autumn results in low vertical mixing that limits nutrient availability and primary production. Seasonal cooling in late autumn results in erosion and gradual deepening of the mixed layer, with renewed nutrient entrainment into the euphotic zone and an increase in primary production. Mesoscale physical variability in this area is the dominant method of nutrient transport (McGillicuddy et al., 1998). In particular, passage of cyclonic and mode water eddies may lead to nutrient entrainment which generates short-lived phytoplankton blooms and community restructuring (Wiebe and Joyce, 1992; Olaizola et al., 1993; McNeil et al., 1999; Letelier et al., 2000; Seki et al., 2001; Sweeny et al., 2003) which could, in turn, impact higher trophic levels such as planktonic foraminifera. In addition, these blooms often result in short-lived, episodic periods of enhanced export fluxes of labile organic material to depth (Conte et al., 1998, 2003, 2014).

3 Materials and methods

3.1 The OFP sediment trap time series

The OFP mooring is located at 31°50′ N, 64°10′ W, about 55 km southeast of Bermuda at 4200 m water depth (Fig. 1). Three Mark VII Parflux sediment traps (McLane Labs, Falmouth, MA) are deployed at depths of 500, 1500 and 3200 m. The traps (0.5 m² surface area) are programmed to collect a continuous bi-weekly time series of the particle flux. Collected samples were processed according to Conte et al. (2001) and split into < 125, 125–500, 500–1000 and > 1000 μm size fractions. We analysed foraminifera in the 125–500 and 500–1000 μm size fractions of 1500 m trap samples collected during two time periods: 1998–2000 and 2008–2010 (109 samples total). We selected the two equivalent 2.5-year intervals a decade apart to generate a bi-weekly resolved time series which would enable assessment of seasonality as well as interannual variability. Our analyses focused on 11 species that fall within three general groupings: (i) surface-dwelling species living within the upper 50 m water column (*Globigerinoides ruber* var. white/pink *Globigerinella siphonifera*, *Globigerinoides sacculifer*), (ii) intermediate-dwelling species living in the ∼ 50–200 m depth range (*Orbulina universa*, *Globigerinoides conglobatus*, *Neogloboquadrina dutertrei*, *Pulleniatina obliquiloculata*) and (iii) deep-dwelling species (or species that are thought to calcify over a large depth range) living in the ∼ 100–800 m depth range (*Globorotalia inflata*, *Globorotalia crassaformis*, *Globorotalia truncatulinoides*, *Globorotalia hirsuta*). Our assignments of the depth habitats were based on measured species depth distributions and/or inferred distributions based on oxygen isotopic composition (Fairbanks et al., 1980; Anand et al., 2003). The temporal offset between the foraminiferal species fluxes reaching the trap at 1500 m depth versus the timing of these species' growth in overlying waters will vary depending on habitat depths and individual species' sinking rates (Takahashi and Bé, 1984). A surface-dwelling *G. ruber* living at 25 m depth may sink at ∼ 198 m day⁻¹, taking ∼ 7 days to reach the 1500 m trap, whereas a more heavily calcified deeper-dwelling species such as *G. inflata* may sink ∼ 504 m day⁻¹, taking only ∼ 3 days to reach the 1500 m trap. These fast sinking rates are much shorter than the typical lifespans of PF and are thus not anticipated to cause any offset between the hydrographic and sediment trap flux data (Honjo and Manganini, 1993).

On average, ∼ 440 tests were counted in each sample fraction. To generate the flux data, counts of total and individual foraminifera species in the sample aliquots for each size fraction were converted to total counts per sample fraction and then the totals for the two fractions were combined (i.e. total

planktonic foraminifera between 125 and 1000 μm in size). Total counts were then scaled for the processing split (60 %) and converted to flux (tests m⁻² d⁻¹). PF flux data is available through www.pangaea.de.

3.2 BATS and remote sensing data

The BATS site (31°40′ N, 64°10′ W) is located just south of the OFP mooring (Fig. 1). Monthly hydrographic and biogeochemical data collected by the BATS time series were obtained from the BATS website (http://bats.bios.edu). Mixed layer depth (MLD) was available from Lomas et al. (2013) and was calculated from CTD profiles using the variable sigma-*t* criterion equivalent to a 0.2 °C temperature change (Sprintall and Tomczak, 1992). The mesoscale eddy field was assessed using interpolated data on sea surface anomaly available from the CCAR Global Historical Gridded SSH Data Viewer (http://eddy.colorado.edu/ccar/ssh/hist_global_grid_viewer).

4 Total planktonic foraminiferal fluxes

4.1 In relation to other mass fluxes

The seasonal cycle and interannual variability of the PF flux at 1500 m depth is highly correlated with that of the total mass, carbonate and organic carbon fluxes. All fluxes are strongly characterized by an abrupt spring maximum during February–April, which varies significantly on an interannual basis (Fig. 2). For example, the spring PF flux peak ranged from a low of 400 tests m⁻² day⁻¹ in 2008, coinciding with minimal spring mass fluxes, to a high of 900 tests m⁻² day⁻¹ in 2009, coinciding with an extreme peak in spring mass fluxes. All fluxes typically drop to a minimum over the summer months (May–August) and remain low until the following spring bloom. During these minima, the PF flux generally amounts to < 200 tests m⁻² day⁻¹. In some years (e.g. 2009 and, to a lesser extent, 2008), the PF flux displays a smaller, but distinct second peak in the months September–October. This secondary autumn peak can also be seen in the mass flux and carbonate flux in 2009 but is absent in the organic carbon flux. Over the entire record, the correlation between PF flux and mass, carbonate and organic carbon flux is 0.65, 0.64 and 0.55, respectively.

4.2 Relative to upper ocean hydrography

In Fig. 3 we compare interannual variations in bi-weekly resolved total PF flux to ∼ monthly resolved changes in key upper ocean hydrographic parameters, measured at the BATS site. PF flux exhibits an inverse relationship with seasonal variations in SSTs and reaches a maximum when SST is coolest in January–March (Fig. 3a). Of note, is the particularly large and prolonged PF bloom in 2010, which coincided

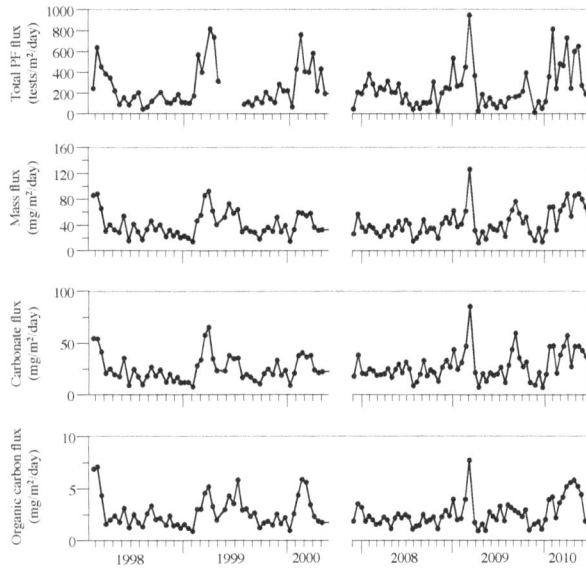

Figure 2. Temporal changes in total planktonic foraminifera flux and mass, carbonate, and organic carbon fluxes at 1500 m depth over the 6-year study period.

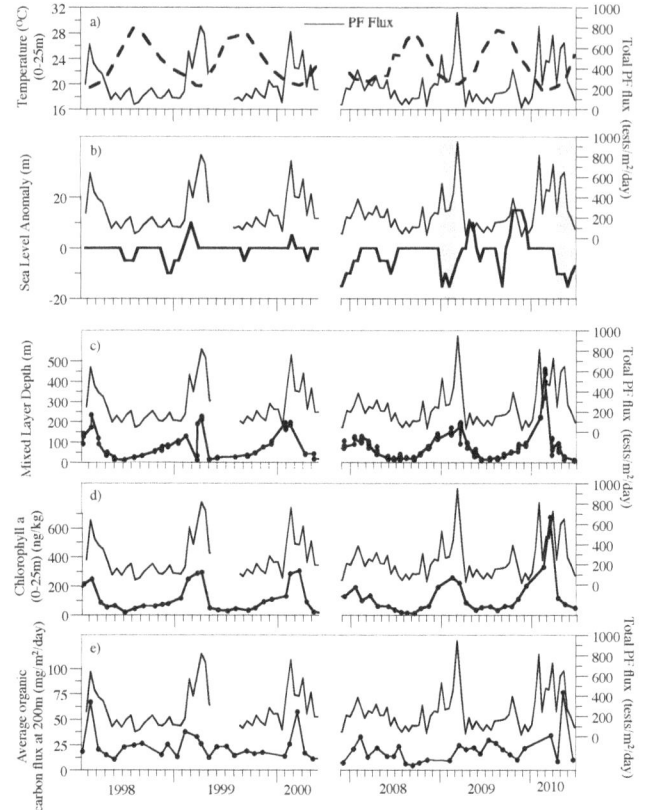

Figure 3. Temporal changes in environmental parameters measured at the BATS site in relation to total planktonic foraminiferal flux in the 1500 m OFP trap (thin, black line): **(a)** sea surface temperature (0–25 m), **(b)** sea level height anomaly; grey bars indicate periods when productive cyclonic eddies influenced the site, **(c)** mixed layer depth, **(d)** chlorophyll a concentration (0–25 m average), and **(e)** average organic carbon flux at 200 m.

with a cyclonic eddy that passed through the area causing the lowest SSTs on record for this site $\sim 18.9\,^{\circ}$C (Fig. 3a–b).

Sea level anomaly (SLA) provides information about eddies passing through the area (Fig. 3b). A negative anomaly is associated with cyclonic eddies and a positive anomaly associated with anticyclonic and mode water eddies. The SLA data show that the particularly high and prolonged PF fluxes, total mass flux and organic carbon flux in spring 2009 and 2010 coincided with the passage of cold, cyclonic eddies (Fig. 2), which enhance nutrient upwelling into the euphotic zone.

The annual and interannual PF flux is in phase with the deepening and shoaling of the mixed layer depth (MLD) (Fig. 3c) and with chlorophyll a concentrations (Fig. 3d). The seasonal PF flux maximum coincides with the chlorophyll a maximum (which is used here as a proxy for the spring phytoplankton bloom) and the organic carbon flux from 200 m, which represents organic carbon export from surface productivity (Fig. 3e), and the deepest MLD during February–March. During April–May, the MLD shoals back towards the surface coinciding with decreasing chlorophyll a concentrations and PF flux. The strong correlation between the seasonality in PF flux and that of primary production and export is demonstrated by the regressions between total PF flux and chlorophyll a concentration (Fig. 4a) and the 1500 m mass flux (Fig. 4b). During the winter–spring period the magnitude of PF flux generally follows the evolution in MLD and is maximal when the MLD is maximal (Fig. 4c). However, when the mixed layer depth shoals to <80 m during the low productivity period in late spring and summer, this correlation is not significant (Fig. 4d).

4.3 Planktonic foraminifera species fluxes

In general, all planktonic foraminifera, and especially deeper-dwelling species, show strong, consistent seasonal variance (Figs. 5–7). Our results demonstrate a clear depth progression towards more pronounced seasonality in the deeper species, compared to a larger intra-seasonal variability in the surface and intermediate dwellers. In addition, the deep-dwelling PF species exhibit repeatable species successions throughout the winter and early spring (Fig. 8, Table 1). Figure 8 shows that *Globorotalia truncatulinoides* dominates the flux of deeper dwellers, and thrives each December, reaching a maximum during January. *G. truncatulinoides* is then followed by *G. hirsuta*, *G. crassaformis*, and *G. inflata*, which all peak between March and April. *G. truncatulinoides* displays large interannual variability (Table 1), ranging from lows of ~ 4000 tests m^{-2} year[1] in 2009–2010 to highs of up to $\sim 14\,000$ tests m^{-2} year[1] in 1999–2000 (Fig. 6). The remaining deeper dwellers (*Globorotalia hirsuta*, *Globorotalia inflata*, *Globorotalia crassaformis*) also

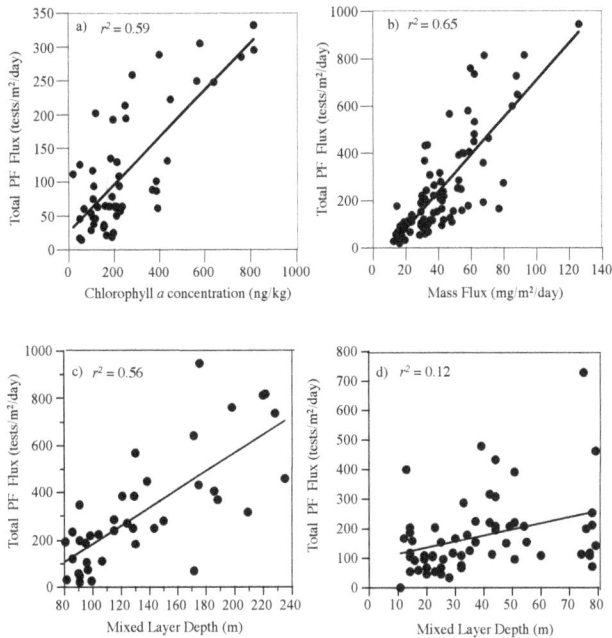

Figure 4. Correlation between total planktonic foraminifera flux in the 1500 m OFP trap (thin, black line) with environmental parameters measured at the BATS site. (**a**) Chlorophyll *a* concentration at 0–25 m. The correlation excludes an anomalous peak in chlorophyll *a* concentration observed in 2010. (**b**) Regression with mass flux (**c**) MLDs > 80 m, excluding the extremely deep MLD observed in 2010. (**d**) MLDs < 80 m.

vary on an interannual basis. Figure 7 and Table 1 show that the largest fluxes of deeper-dwelling species occurred during the winter/spring of 1999–2000 and 2008–2009. Using shell weights from this study averaged with shell weights (125–1000 μm) measured by Deuser (1987) and Deuser and Ross (1989), we estimate that PF flux contributes up to ∼ 40 % of the total carbonate flux during winter–spring but < 10 % during summer (Fig. 9a). Deeper-dwelling species account for 60–90 % of PF carbonate flux (Fig. 9b) and up to 37.5 % of the total carbonate flux (e.g. during the winter–spring of 2000) (Fig. 9c).

5 Discussion

The controls on PF flux in the Sargasso Sea was first introduced by Bé (1960) and later developed by Tolderlund and Bé (1971), who suggested that PF flux is dominantly controlled by the availability of their food phytoplankton. Thus, the environmental factors controlling PF flux should be closely aligned with the factors controlling phytoplankton productivity and export flux.

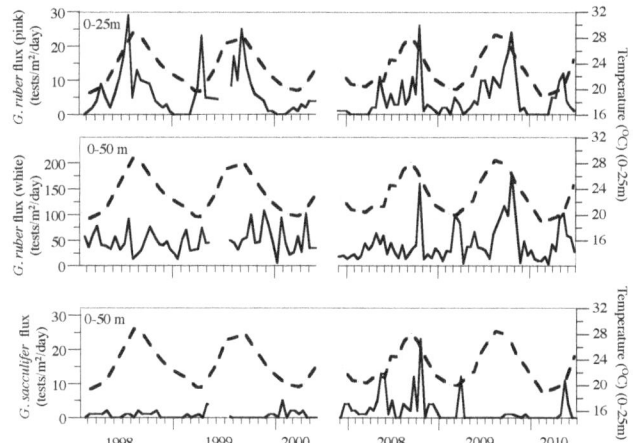

Figure 5. Temporal changes in surface-dwelling planktonic foraminifera fluxes in the 1500 m trap with changes in sea surface temperature (0–25 m) shown in the dashed black line for reference. The approximate depth habitat (Anand et al., 2003) is shown in figures.

5.1 Environmental controls on PF fluxes

5.1.1 Depth of the mixed layer

Previous studies suggest that increased chlorophyll concentrations and larger phytoplankton abundances occur when the MLD deepens (Townsend et al., 1994; Waniek, 2003; Nelson et al., 2004) and the amplitude and timing of MLD deepening determines the size of the following spring bloom (Menzel and Ryther, 1961; Michaels et al., 1994). Here, we also observe a simultaneous seasonal peak in chlorophyll *a* and maximum depth of the MLD, as observed by previous studies at BATS (Steinberg et al., 2001; Cianca et al., 2012), the timing and amplitude of which coincides with the maximum PF flux (Fig. 3c, d). Similarly, seasonal changes in mixed layer depth are closely associated with changes in foraminifer production (Thunell and Reynolds, 1984; Sautter and Thunell, 1989; Pujol and Vergnaud Grazzini, 1995; Schmuker and Schiebel, 2002) and chlorophyll *a* concentrations (King and Howard, 2003, 2005) in other ocean basins. Siegel et al. (2002) proposed that, south of 40° N, the initiation and extent of the spring bloom is dominantly limited by nutrients, and this is supported by the simultaneous increase in phytoplankton concentrations with mixing depth at BATS (Treusch et al., 2012). Vertical mixing in late winter and spring distributes nutrients into the euphotic zone to support the spring phytoplankton bloom, causing the consequent seasonal peak in export fluxes of organic carbon, to fuel symbiont-barren foraminifera production (Fig. 2d). In contrast, no correlation exists between PF flux and MLD during the late spring to autumn when the mixed layer fails to penetrate the minimum depth of the deep chlorophyll maximum layer (∼ 80 m), where many species of planktonic foraminifera reside in association with other zooplankton and

Table 1. Annual fluxes for planktonic foraminifera species at 1500 m depth in 1998–1999, 1999–2000, 2008–2009 and 2009–2010 and the 4-year averages. Fluxes were calculated from the sum of bi-weekly averages between July and June for each year and converted to tests m^{-2} yr^{-1}. Species are listed according to their estimated depth habitats.

| Species | Seasonal flux maximum | Annual flux (tests m^{-2} yr^{-1}) | | | | | |
		1998–1999	1999–2000	2008–2009	2009–2010	Average	3200 m avg (1978–1984)[3]
Surface dwellers:							
G. ruber (pink)	Jul–Sept	2524	1978	1576	2122	2050	1450
G. ruber (white)	Sept–Oct	16 197	19 633	13 917	18 719	17 117	
G. sacculifer	Oct[1], Mar[2]	256	292	1007	348	1903	425
Surface totals		18 977	21 903	16 500	21 189	17 346	
Intermediate dwellers:							
G. siphonifera	[4]	6101	3182	2231	2833	3587	
O. universa	Apr–May[1], Oct–Nov[2]	1429	694	1056	2250	1357	
G. conglobatus	Nov	277	180	0	4	115	300
N. dutertrei	Mar–Apr[1], Nov–Dec[2]	1290	185	471	839	696	876
P. obliquiloculata	Dec–Mar	398	205	708	352	416	762
Intermediate totals		9495	4446	4466	6278	6171	
Deep dwellers:							
G. truncatulinoides	Jan–Feb	5248	13 796	9517	4031	8148	3420
G. hirsuta	Feb–Mar	1784	9888	3859	2770	4575	1520
G. crassaformis	Feb–Mar	26	100	122	139	97	192
G. inflata	Mar–Apr	844	995	1652	1869	1340	1270
Deep totals		7902	24 779	15 150	8809	14 160	5402
Other species	–	51 442	43 704	43 172	70 446	51 191	
Totals	–	87 816	94 831	79 289	106 722	92 165	

[1] Primary peak. [2] Secondary peak. [3] Averages from Deuser and Ross (1989). [4] This species has low seasonality.

algal cells (Fairbanks and Wiebe, 1980) (Fig. 4d). This is also the depth of the nitricline where nitrate concentrations are greater than 0.1 μmol kg^{-1} (Schiebel et al., 2001).

The majority of the increased PF flux in the winter–spring is driven by increased fluxes of deeper-dwelling species, in particular G. truncatulinoides and G. hirsuta (Fig. 9b). These species are symbiont-barren and rely on the flux of phytodetritus and other labile organic carbon as a food source from the spring phytoplankton bloom (Hemleben et al., 1989). The discrepancy in timing of peaks between the deeper-dwelling species (Fig. 8) is likely due to subtle changes in phytoplankton succession related to the species' diets (Deuser and Ross, 1989; Hemleben et al., 1989). Overall, the seasonal PF species succession is broadly similar to previous observations from 1959 to 1963 and 1978 to 1984 (Tolderlund and Bé, 1971; Deuser, 1987; Deuser and Ross, 1989) which suggests that, despite long-term environmental change, species seasonality have remained consistent over the past 50 years.

The correlation observed here between the seasonality in the PF flux, chlorophyll *a* concentration and mass flux at 1500 m (Fig. 4a and b) clearly demonstrates that the seasonality of non-symbiont-bearing foraminifera, such as the globorotaliids, is controlled by phytoplankton production and the export flux of phytodetritus to depth. As

these globorotaliids are up to 3 times denser than surface species (unpublished data), their sinking rates are significantly higher than those of other species. Thus, increased production by these species can accelerate the transfer of carbonate from surface to deep ocean, thereby strengthening the carbonate pump.

In contrast, the surface-dwelling symbiont-bearing foraminifera have lifecycles which strongly benefit from stratified surface waters and shallow mixed layers in order to photosynthesise – allowing them to succeed in low-nutrient conditions (Hemleben et al., 1989). Surface dwellers generally calcify in late summer when sea surface temperatures are at a maximum and dinoflagellates are abundant (Tolderlund and Bé, 1971). We thus conclude that the depth and structure of the mixed layer plays an important role in regulating PF species flux by controlling the abundance and timing of their food availability throughout the seasonal cycle.

5.1.2 MLD deepening and shoaling rates

Current models based on the light-limited higher latitudes (Waniek, 2003; Mao, Y., personal communication, 2013), suggest that if the MLD shoals early and slowly, the

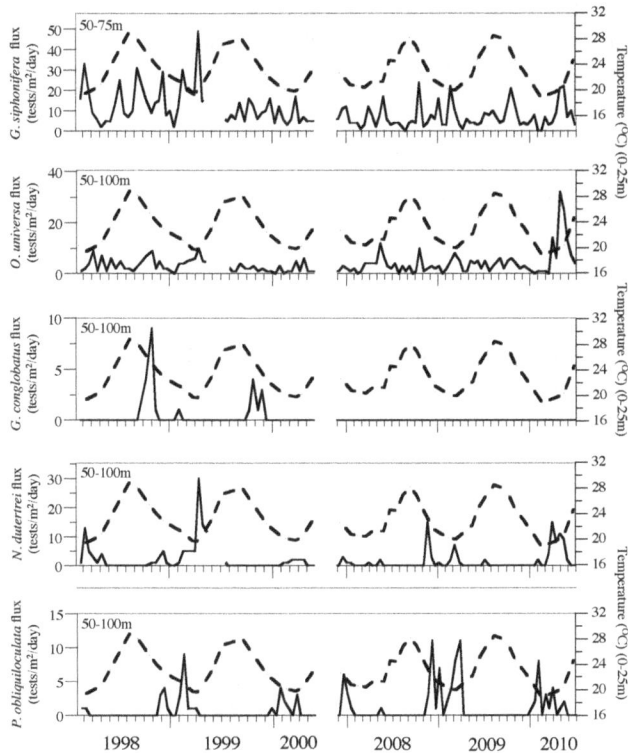

Figure 6. Temporal changes in intermediate-dwelling planktonic foraminifera fluxes in the 1500 m trap with changes in sea surface temperature (0–25 m) for reference. The approximate depth habitat (Anand et al., 2003) is shown in figures.

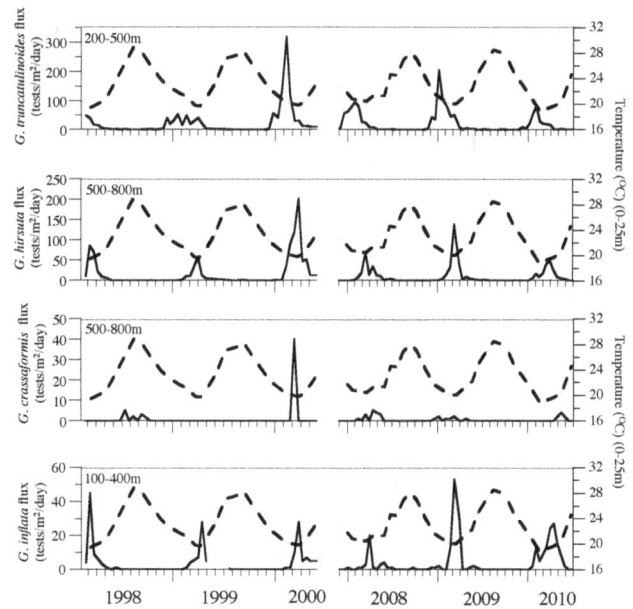

Figure 7. Temporal changes in deeper-dwelling planktonic foraminifera fluxes in the 1500 m trap with changes in sea surface temperature (0–25 m) for reference. The approximate depth habitat (Anand et al., 2003) is shown in figures. Graphs are ordered according to seasonal succession.

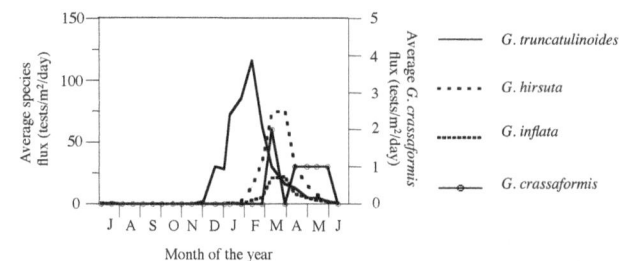

Figure 8. Seasonal succession for deeper-dwelling species averaged over six spring blooms (1998, 1999, 2000, 2008, 2009, 2010) from the 1500 m trap. *G. truncatulinoides*, *G. hirsuta*, and *G. inflata* appear on the left axis and *G. crassaformis* is on the right axis.

consequent bloom will be long and weak compared to if the MLD shoals late and quickly, which causes a short and sharp bloom. At our subtropical study site, the spring bloom is predominantly limited by nutrient input into the euphotic zone, which is determined by the depth of the mixed layer. Increased heat loss and wind stress leading to higher convective mixing during the winter months controls the rate of deepening of the mixed layer, which is strongly correlated to the maximum MLD reached ($r^2 = 0.88$) (Fig. 10a). Years with faster deepening rates have deeper mixed layers and hence larger spring blooms (e.g. winter 2009), whereas slow deepening rates cause shallower mixed layers and smaller spring blooms. There is also some evidence that light limitation could be a secondary control on the peak productivity of the spring bloom at this site (Dutkiewicz et al., 2001; Lomas et al., 2009; Cianca et al., 2012) as the euphotic zone extends to ~ 100 m (Steinberg et al., 2001) and a faster shoaling rate during the spring could concentrate the food available for symbiotic foraminifera in the euphotic zone, resulting in a larger PF flux.

To test whether the rates of mixed layer deepening in early winter and of shoaling in spring affect the PF flux, we computed a mixed layer dynamics index, D_r / S_r, which is the ratio of the rate of deepening to the rate of shoaling, and

compared this to the integrated PF flux (Table 2). The D_r / S_r ratio never exceeds 1, indicating that the shoaling rate always exceeds the deepening rate. For all the years studied, there is a strong inverse relationship between the integrated PF flux over the duration of spring bloom, and the D_r / S_r ratio (Fig. 10b, $r^2 = 0.93$). This relationship is also present in the maximum in chlorophyll *a* concentration and the D_r / S_r ratio (Figure 10c, $r^2 = 0.76$). This correlation indicates that when the MLD shoals more quickly during spring stratification (lower D_r / S_r ratio), the chlorophyll *a* concentrations and PF flux are higher, as supported by a strong correlation ($r^2 = 0.87$) between shoaling rate and integrated PF flux (Fig. 10d).

Table 2. Mixed layer depth and mean rates of mixed layer (ML) deepening and shoaling. The D_r / S_r ratio is a derived value calculated from the rate of ML deepening divided by the rate of ML shoaling (see text). The winter–spring PF flux represents the PF flux integrated over the whole bloom, which varied interannually in length but ranged from December to May. Bold faced years indicate years when a cyclonic eddy was present during the spring bloom period.

Year	MLD max (m)	ML deepening rate (m day^{-1})	ML shoaling rate (m day^{-1})	D_r / S_r ratio (m)	Maximum PF flux (tests m^{-2} day^{-1})	Integrated winter–spring PF flux (tests m^{-2} day^{-1})
1997–1998	235	0.93	1.91	0.49	641	28
1998–1999	222	0.78	7.78	0.10	816	41
1999–2000	197	0.63	Data missing	–	761	30
2007–2008	130	0.55	0.75	0.73	385	17
2008–2009	**198**	**0.95**	**2.21**	**0.43**	**946**	**28**
2009–2010	**464**	**1.76**	**3.82**	**0.46**	**815**	**32**

Years where the shoaling rate is twice as quick as the deepening rate (e.g. winters 1997, 2008, and 2009) have average D_r / S_r ratios, average-length blooms and PF flux ($\sim$ 30 tests m^{-2} day^{-1}, Table 2). Years with comparatively equal rates of shoaling and deepening (e.g. winter 2007) have larger D_r / S_r ratios, longer and slower blooms with shallower MLDs and small PF fluxes. Years when the shoaling rate is much quicker than deepening rate (e.g. winter 1999) have the smallest D_r / S_r ratios and shorter, sharper blooms with greater numbers of intermediate thermocline-dwelling species such as *N. dutertrei, P. obliquiloculata, G. siphonifera*, suggesting that when the rate of shoaling is higher the seasonal thermocline is nearer to the surface for longer, which is beneficial for these symbiont-bearing and symbiont-facultative species. The PF fluxes were large (and prolonged) respectively in winter 2008–2009 and 2009–2010 despite having average D_r /S_r ratios but were probably enhanced by additional factors discussed in the next section.

5.1.3 Eddies

The negative sea level anomalies in spring of 2009 and 2010 indicate that the large (and in 2010 prolonged) PF fluxes in these years were clearly associated with the passage of cyclonic eddies (Fig. 3b). Eddy pumping of nitrate into the euphotic zone has been shown to significantly increase new production (Oschlies and Garçon, 1998; Oschlies, 2002). Cianca et al. (2007) estimate that eddy pumping contributes $\sim$ 50 % of the nutrient input into the euphotic zone in the Sargasso Sea. Studies at the BATS site have demonstrated the influence of cyclonic and mode water eddies in promoting phytoplankton blooms and increased secondary production (Eden et al., 2009; McGillicuddy et al., 2007, 1999; Goldthwait and Steinberg, 2008; Sweeney et al., 2003; Lomas et al., 2013; Cianca et al., 2012) and therefore affecting PF food availability and quality (Schmuker and Schiebel, 2002). Previous studies have found higher fluxes of certain PF species such as *Globigerinita glutinata* associated with cyclonic eddy structures in the Caribbean Sea (Schmuker and Schiebel, 2002) and North Atlantic (Beckman et al.,

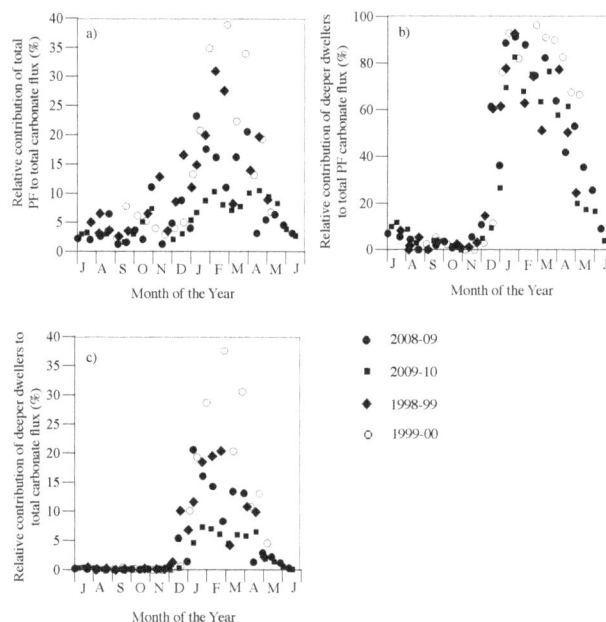

Figure 9. (a) The relative contribution of total PF to total carbonate flux **(b)** The relative contribution of deeper-dwelling planktonic foraminifera (*G. hirsuta, G. truncatulinoides, G. crassaformis, G. inflata*) to the total planktonic foraminiferal carbonate flux **(c)** The relative contribution of total deeper dwellers (*G. hirsuta, G. truncatulinoides, G. crassaformis, G. inflata*) to the total carbonate flux. All graphs show four full years 1998–1999, 1999–2000, 2008–2009 and 2009–2010.

1987), also in conjunction with upwelling frontal regions in the Mexican Pacific (Machain-Castillo et al., 2008) and deep mixed layers during winter in the Mediterranean (Pujol and Vergnaud Grazzini, 1995). Here we observe a similar response during the passage of a cyclonic eddy in spring 2009, particularly for deeper-dwelling species.

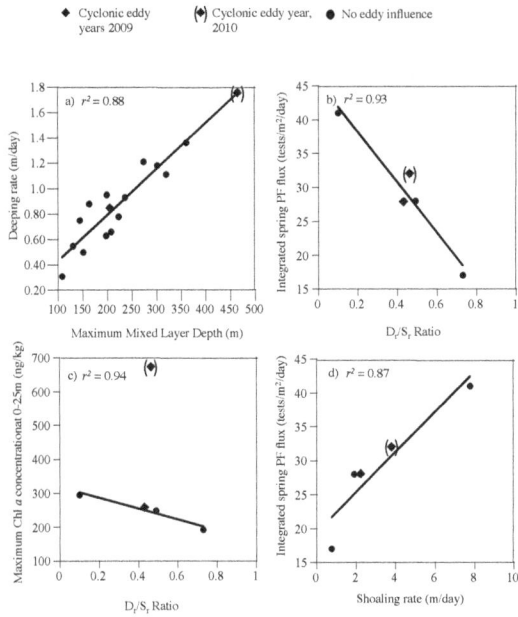

Figure 10. **(a)** Correlation between the maximum mixed layer depth and deepening rate of the mixed layer for years 1995–2011. **(b)** Correlation between the deepening : shoaling rate (D_r / S_r) ratio of the mixed layer depth for all years studies excluding 2000 and integrated PF flux during the spring blooms which ranged from December to May, **(c)** maximum chlorophyll a concentrations in the surface ocean during the spring bloom for all years studied, excluding the anomalous year 2010 in parentheses, **(d)** correlation between the shoaling rate and integrated flux of total PF over the spring bloom period which ranged from December to May. Diamonds indicate years with eddy influence 2009 and diamond with parentheses = 2010. Round points are years without eddy influence.

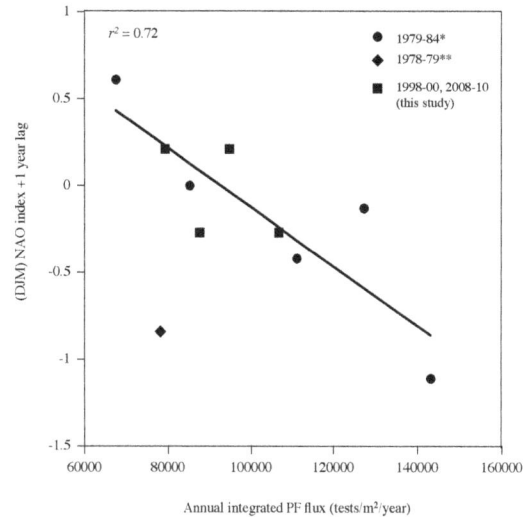

Figure 11. Annual integrated PF flux from this study (1500 m trap, square symbols) and 1979–1984 (*3200 m trap, round symbols, Deuser, 1987; Deuser and Ross, 1989) plotted against wintertime (DJFM) NAO index + 1-year lag. Annual fluxes from both trap depths are comparable. **Annual PF flux from 1978 (diamond symbol) was not included in the regression because it was an anomalously low flux year which could be explained by a shallow MLD and/or possibly the presence of an anticyclonic eddy (no data to test), which may have suppressed the spring bloom and hence PF flux as seen during 1994 at BATS (Lomas et al., 2013). NAO data available at http://www.cpc.ncep.noaa.gov/data/teledoc/nao.shtml.

In fact, the largest PF flux observed over the entire record was associated with this eddy passage, even though the maximum MLD and D_r / S_r were modest (Table 2). Similarly, the mass and organic carbon flux measured during the passage of this eddy (Fig. 2b–d) were the highest fluxes measured over the last 25 years of the OFP time series, indicating that the conditions in this eddy promoted an extremely large export flux to fuel the production of deep-dwelling foraminifera species such as *G. truncatulinoides, G. hirsuta*, and especially *G. inflata* which all experienced higher seasonal fluxes in 2009 (Fig. 7).

This observation is consistent with an exceptionally large increase in the flux of *G. truncatulinoides* (> 600 tests m^{-2} day^{-1}) seen at the OFP traps during the spring of 2007, which was also influenced by the passage of a productive cyclonic eddy (Fang et al., 2010; Conte and Weber, 2014). Both the 2007 and 2009 eddies occurred between January and March during the seasonal flux of the deeper dwellers (Fig. 7), underscoring the importance of the timing of eddy passage in enhancing PF flux. The influence of eddies here is similar to observations from the eastern

Mediterranean where increased numbers of grazing species, such as *G. truncatulinoides* and *G. inflata*, have been found in association with eddy structures and deep mixed layers (Pujol and Vergnaud Grazzini, 1995). These findings suggest that productive cyclonic eddies, when co-occurring with deep MLDs, act to enhance the existing seasonal abundance of deeper-dwelling species through mixing of the water column, which aids their annual reproductive migration in addition to increasing food supply.

Along with the timing of the eddy passage, our observations also suggest that the PF flux response is dependent on whether the eddy is intensifying or weakening. For instance, both cyclonic eddies in 2009 and 2010 intensified over the spring bloom (Fig. 3b) eliciting a large biological response indicated by elevated subsurface Chl a concentrations and increased PF flux. In contrast, the cyclonic eddy in winter 2007–2008 was weakening over the spring bloom and therefore elicited no PF flux response.

Recent studies have found that eddies which are a minimum of 1–2 months in duration are more likely to induce a larger biological response (Mouriño-Carballido and McGillicuddy, 2006, Sweeny et al., 2003). Our observations also suggest that eddies need to be present for at least a month to elicit responses in the flux of PF which have minimum lifecycles of 2 weeks. For instance, in winter 1998–1999 a cyclonic eddy passed over the sediment trap site in only 1 month and elicited no biological response, compared to cyclonic eddies in 2009 and 2010, which both remained over the site for a minimum of 2–3 months and elicited large biological responses (Fig. 3b). These findings suggest that cyclonic eddies which intensify over the spring bloom and last for 1–3 months can elicit a significant biological response and increased PF flux.

6 Implications

Our results show that environmental factors and mesoscale eddy variability play an important role in regulating the planktonic foraminifera fluxes, by regulating the MLD and consequent magnitude of the spring bloom and biological export flux.

An overarching climatological variable affecting this region especially is the North Atlantic Oscillation (NAO), which exerts a strong influence on air temperature, storminess, heat loss, winter mixed layer depth, and, therefore, nutrient injection into the upper ocean during the winter months (Bates, 2012; Bates and Hansell, 2004; Rodwell et al., 1999). Modelling studies have shown that when the NAO is in its low phase, i.e. negative NAO (e.g. winter 2010), there is increased heat loss that intensifies convective mixing and results in enhanced nutrient upwelling into the euphotic zone to support primary production (Oschlies, 2001). The NAO influence on upper ocean productivity and biogeochemical fluxes is demonstrated by the inverse correlation between the wintertime (NDJF) NAO index and the deep particulate nitrogen flux in the OFP traps over a 30-year period (Conte and Weber, 2014) and increased primary productivity in negative wintertime NAO phases (Lomas et al., 2010). If convective mixing and nutrient entrainment into the euphotic zone is stronger during negative NAO years, this could serve to modulate PF flux, and therefore carbonate flux, on decadal timescales. When we compare PF fluxes covering a range of NAO indexes, from this study using the 1500 m trap to the 3200 m trap between 1978 and 1984 (Deuser and Ross, 1989; Deuser, 1987), we find a weak inverse correlation between total PF flux and (DJFM) NAO index in-phase (not significant), but we do find a significant inverse correlation with a (DJFM) NAO with a 1-year lag ($p < 0.005$) (Fig. 11). Cianca et al. (2012) showed that their correlation between winter NAO and total chlorophyll a at BATS improved when applying a + 1-year time lag, but still remained insignificant. They attributed this to variability in the subtropical mode water, which can laterally advect nutrients on interannual timescales (Palter et al., 2005; Patara et al., 2011). We acknowledge that additional longer-term data are needed to test the mechanism behind this correlation, but our results suggest that changes in NAO status and/or mesoscale eddy frequency could significantly modulate planktonic foraminifera

flux and export flux from the surface ocean on interannual timescales.

This study shows that the productivity of the dominant deep-dwelling species *G. truncatulinoides* and *G. hirsuta* is especially responsive to interannual variability in overlying surface water conditions and especially to the transient high-production/flux events that are associated with the passage of productive cyclonic eddies that coincide with their seasonal spring production peak. Our data show that deeper-dwelling species can account for up to ~90 % of the total PF carbonate flux, representing up to ~40 % of the total carbonate flux during winter–spring at the OFP site. Changes in NAO status, which modulates nutrient supply into the euphotic zone and the strength of the spring bloom, also may in turn modulate the production and flux of these heavily calcified deep-dwelling foraminifera by increasing their food supply, thereby intensifying the carbonate pump.

7 Conclusions

Our study demonstrates that the interannual variability in planktonic foraminifera flux can be linked to the MLD and the rate of deepening/shoaling of the mixed layer associated with nutrient injection into the euphotic zone. We find that higher PF fluxes coincide with deeper MLDs, especially when combined with cyclonic eddy-induced nutrient upwelling. In particular, the production of the dominant deep-dwelling species *G. truncatulinoides* and *G. hirsuta* is shown to be particularly responsive to interannual variability in overlying surface water conditions and especially to the transient high-production/flux events that are associated with productive cyclonic eddies. These species dominate the major late winter–early spring pulses of foraminifera and have higher sinking rates than surface-dwelling species because they are up to 3 times denser (unpublished results). We suggest deeper-dwelling species strengthen the carbonate pump by accelerating the transfer of carbonate from surface to deep ocean and contribute up to 40 % of the contemporaneous peak in total carbonate export fluxes. It follows that any increase in fluxes of these deep dwellers arising from climate-induced changes in winter–spring mixed layer dynamics will also increase the average sinking rate of foraminiferal carbonate and intensify the overall carbonate pump. Our findings suggest that the North Atlantic Oscillation, via its influence on mixed layer depth, nutrient upwelling, phytoplankton production and export flux, may also serve to modulate the foraminiferal component of the carbonate pump in the subtropical North Atlantic.

Acknowledgements. We would like to thank two anonymous reviewers for their time and constructive comments that helped improve the manuscript. This research was funded through the U.K. Ocean Acidification Research Program by Natural Environment Research Council grant to P. Anand and P. Sexton (grant NE/I019891/1). We acknowledge the National Science Foundation for its support of the Oceanic Flux Program time series (most recently by grant OCE-1234292) and the Bermuda Atlantic Time Series (most recently by grant OCE-0801991). We thank Mike Lomas for providing MLD data and Yolanda Mao for providing insights and useful discussion on the data. P. Anand is also thankful to Werner Deuser for communication regarding published data.

Edited by: J. Bijma

References

Anand P., Elderfield, H, and Conte, M. H.: Calibration of Mg/Ca thermometry in planktonic foraminifera from a sediment trap time series, Paleoceanography, 18, 1050, doi:10.1029/2002PA000846, 2003.

Barker, S. and Elderfield, H.: Foraminiferal calcification response to Glacial-Interglacial changes in Atmospheric CO_2, Science, 297, 833–836, 2002.

Bates, N. R.: Interannual variability of the oceanic CO_2 sink in the subtropical gyre of the North Atlantic Ocean over the last 2 decades, J. Geophys. Res., 112, C09013, doi:10.1029/2006JC003759, 2007.

Bates, N. R.: Multi-decadal uptake of carbon dioxide into subtropical mode water of the North Atlantic Ocean, Biogeosciences, 9, 2649–2659, doi:10.5194/bg-9-2649-2012, 2012.

Bates, N. R. and Hansell, D. A.: Temporal variability of excess nitrate in the subtropical mode water of the North Atlantic Ocean, Mar. Chem., 84, 225–241, 2004.

Bates N. R., Pequignet, A. C., Johnson, R. J., and Gruber, N. A: short-term sink for atmospheric CO_2 in the subtropical mode water of the North Atlantic Ocean, Nature, 420, 489–493, 2002.

Bé, A. W .H.: Ecology of Recent planktonic foraminifera, Part 2, Bathymetric and seasonal distributions off Bermuda, Micropaleontology, 4, 373–392, 1960.

Beckman, A., Auras, A., and Hemleben, C.: Cyclonic cold-core eddy in the eastern North Atlantic, 111, Zooplankton, Mar. Ecol. Prog. Ser., 39, 165–173, 1987.

Cianca, A., Helmke, P., Mouriño, B., Rueda, M.J., Llinás, O., and Neuer, S.: Decadal analysis of hydrography and in situ nutrient budgets in the western and eastern North Atlantic subtropical gyre, J. Geophys. Res., 112, C07025, doi:10.1029/2006JC003788, 2007.

Cianca, A., Godoy, J. M., Martin, J. M., Perez-Marrero, J., Rueda, M. J., Llinás, O., and Neuer, S.: Interannual variability of chlorophyll and the influence of low-frequency climate modes in the North Atlantic subtropical gyre, Global Biogeochem. Cy., 26, GB2002, doi:10.1029/2010GB004022, 2012.

CLIMAP Project Members: Climap 18k Database: IGBP PAGES/World Data Center-A for Paleoclimatology Data Contribution Series, v. 94–001, 1994,

Conte, M. H. and Weber, J. C.: Particle flux in the deep Sargasso Sea: The 35-year Oceanic Flux Program time series, Oceanography, 27, 142–147, 2014.

Conte, M. H., Weber, J. C., and Ralph, N.: Episodic particle flux in the deep Sargasso Sea: an organic geochemical assessment, Deep-Sea Res.-Pt. I, 45, 1819–1841, 1998.

Conte, M. H., Ralph, N., and Ross, E. H.: Seasonal and interannual variability in deep ocean particle fluxes at the Oceanic Flux Program (OFP)/Bermuda Atlantic Time Series (BATS) site in the western Sargasso Sea near Bermuda, Deep-Sea Res.-Pt. II, 48, 1471–1505, 2001.

Conte, M. H., Dickey, T. D., Weber, J. C., Johnson, R. J., and Knap, A. H.: Transient physical forcing of pulsed export of bioreactive organic material to the deep Sargasso Sea, Deep-Sea Res.-Pt. I, 50, 1157–1187, 2003.

Deuser, W. G.: Seasonal variations in isotopic composition and deep-water fluxes of the tests of perennially abundant planktonic foraminifera of the Sargasso Sea: Results from sediment-trap collections and their paleoceanographic significance, J. Foramin. Res., 17, 14–27, 1987.

Deuser, W. G. and Ross, E. H.: Seasonally abundant planktonic foraminifera of the Sargasso Sea: Succession, deep-water fluxes, isotopic composition, and paleoceanographic implications, J. Foramin. Res., 19, 268–293, 1989.

Deuser, W. G., Ross, E. H., and Anderson, R. F.: Seasonality in the supply of sediment to the deep Sargasso Sea and implications for the rapid transfer of matter to the deep ocean, Deep-Sea Res., 28A, 495–505, 1981.

Dutkiewicz, S., Follows, M., Marshall, J., and Gregg, W. W.: Interannual variability of phytoplankton abundances in the North Atlantic, Deep-Sea Res.-Pt. II, 48, 2323–2344, 2001

Eden, B. R., Steinberg, D. K., Goldthwait, S. A., and McGillicuddy, Jr, D. J. Zooplankton community structure in a cyclonic and mode-water eddy in the Sargasso Sea, Deep-Sea Res.-Pt. I, 56, 1757–1776, 2009.

Erez, J., Almogi-Labin, A., and Avraham, S. On the life history of planktonic foraminifera: lunar reproduction cycle in *Globigerinoides sacculifer* (Brady), Paleoceanography, 6, 295–306, 1991.

Fairbanks, R. G. and Wiebe, P. H.: Foraminifera and Chlorophyll Maximum: Vertical Distribution, Seasonal Succession, and Paleoceanographic Significance, Science, 209, 1524–1526, 1980.

Fairbanks, R. G., Wiebe, P. H., and Bé, A. W. Vertical distribution and isotopic composition of living planktonic foraminifera in the western North Atlantic, Science, 207, 61–63, 1980.

Fang, J., Conte, M. H., and Weber, J. C.: Influence of physical forcing on seasonality of biological components and deep ocean particulate flux in the Sargaddo Sea, Eos, Transactions American Geophysical Union 91, Ocean Sciences Meeting Supplement, Abstract BO24B-02, 2010.

Goldthwait, S. and Steinberg, D. K.: Elevated biomass of mesozooplankton and enhanced fecal pellet flux in cold-core and mode-water eddies in the Sargasso Sea, Deep-Sea Res., 55, 1360–1377, 2008.

Hemleben, C., Spindler, M., Breitinger, I., and Deuser, W. G.: Field and laboratory studies on the ontogeny and ecology of some globorotaliid species from the Sargasso Sea off Bermuda, J. Foramin. Res., 15, 254–272, 1985.

Hemleben, C., Spindler, M., and Anderson, O. R.: Modern Planktonic Foraminifera, Springer, New York, 363 pp., 1989.

Honjo, S. and Manganini, S. J.: Annual biogenic particle fluxes to the interior of the North Atlantic Ocean; studied at 34°N 21°W and 48°N 21°W, Deep-Sea Res.-Pt. I, 40, 587–607, 1993.

King, A. L. and Howard, W. R.: Planktonic foraminiferal flux seasonality in Subantarctic sediment traps: A test for paleoclimate reconstructions, Paleoceanography, 18, 1019, doi:10.1029/2002PA000839, 2003.

King, A. L. and Howard W. R.: δ^{18}O seasonality of planktonic foraminifera from Southern Ocean sediment traps: Latitudinal gradients and implications for paleoclimate reconstructions, Mar. Micropaleontol., 56, 1–24, 2005.

Kuroyanagi, A. and Kawahata, H.: Vertical distribution of living planktonic foraminifera in the seas around Japan, Mar. Micropaleontol., 53, 173–196, doi:10.1016/j.marmicro.2004.06.001, 2004.

Letelier, R. M, Karl, D. M., Abbott, M. R., Flament, P., Freilich, M., Lukas, R., and Strub, T.: Role of late winter mesoscale events in the biogeochemical variability of the upper water column of the North Pacific Subtropical Gyre, J. Goephys. Res., 105, 28 723–28 740, Correction in J. Goephys. Res., 106, 7181–7182, 2000.

Lohmann, G. P. and Schweitzer, P. N.: *Globorotalia truncatulinoides'* Growth and chemistry as probes of the past thermocline: 1. Shell size, Paleoceanography, 5, 55–75, 1990.

Lomas, M. W., Lipschultz, F., Nelson, D. M., Krause, J. W., and Bates N. R.: Biogeochemical responses to late winter storms in the Sargasso Sea I – Pulses of primary and new production, Deep-Sea Res.-Pt. I, 56, 843–860, 2009.

Lomas, M. W., Steinberg, D. K., Dickey, T., Carlson, C. A., Nelson, N. B., Condon, R. H., and Bates, N. R.: Increased ocean carbon export in the Sargasso Sea linked to climate variability is countered by its enhanced mesopelagic attenuation, Biogeosciences, 7, 57–70, doi:10.5194/bg-7-57-2010, 2010.

Lomas, M. W., Bates, N. R., Johnson, R. J., Knap, A. H., Steinberg, D. K., and Carlson, C. A.: Two decades and counting: 24-years of sustained open ocean biogeochemical measurements in the Sargasso Sea, Deep-Sea Res.-Pt. II, 93, 16–32, 2013.

Lutz, B. P.: Shifts in North Atlantic planktic foraminifer biogeography and subtropical gyre circulation during the mid-Piacenzian warm period, Mar. Micropaleontol., 80, 125–149, 2011.

Machain-Castillo, M. L., Monreal-Gómez, M., Arellano-Torres, E., Merino-Ibarra, M., and González-Chávez, G.: Recent planktonic foraminiferal distribution patterns and their relation to hydrographic conditions of the Gulf of Tehuantepec, Mexican Pacific, Mar. Micropaleontol., 66, 103–119, 2008.

McGillicuddy, Jr., D. J., Robinson, A. R., Siegel, D. A., Jannasch, H. W., Johnson, R., Dickey, T. D., McNeil, J., Michaels, A. F., and Knap, A. H.: Influence of mesoscale eddies on new production in the Sargasso Sea, Nature, 394, 263–266, 1998.

McGillicuddy, Jr, D. J., Jonhson, R., Siegel, D. A., Michaels, A. F., Bates, N. R., and Knap, A. H.: Mesoscale variations of biogeochemical properties in the Sargasso Sea, J. Geophys. Res., 104, 13381–13394, 1999.

McGillicuddy D. J., Anderson, L., Bates, N. R., Bibby, T., Buesseler, K. O., Carlson, C. S., Davis, C., Ewart, P. G., Flakowski, S. A., Goldthwait, D., Hansell, Jenkins, W. J., Johnson, R., Kosnyrev, V. K., Ledwell, J., Li, Q., Siegel, D., and Steinberg, D. K.: Eddy/Wind interactions stimulate extraordinary mid-ocean plankton blooms, Science, 316, 1021–1025, 2007.

McNeil, J. D., Jannasch, H. W., Dickey, T., McGillicuddy, D., Brzezinski, M., and Sakamoto, C. M.: New chemical, bio-optical and physical observations of upper ocean response to the passage of a mesoscale eddy off Bermuda, J. Geophys. Res., 104, 15537–15548, 1999.

Menzel, D. W. and Ryther, J. H.: Annual variations in primary production of the Sargasso Sea off Bermuda, Deep-Sea Res., 7, 282–288, 1961.

Michaels, A. F. and Knap, A. H.: Overview of the U.S. JGOFS BATS and Hydrostation S program, Deep-Sea Res.-Pt. II, 43, 157–198, 1996.

Michaels A. F., Knap, A. H., Dow, R. L., Gundersen, K., Johnson, R. J., Sorensen, J., Close, A., Knauer, G. A., Lohrenz, S. E., Asper, V. A., Tuel, M., and Bidigare, R.: Seasonal patterns of ocean biogeochemistry at the U.S. JGOFS Bermuda Atlantic Time-series Study site, Deep-Sea Res.-Pt. I, 41, 1013–1038, 1994.

Mouriño-Carballido, B. and McGillicuddy, D. J.: Mesoscale variability in the metabolic balance of the Sargasso Sea, Limnol. Oceanogr., 51, 2675–2689, 2006.

Nelson, N. B., Siegel, D. A., and Yoder, J. A.: The spring bloom in the northwestern Sargasso Sea: spatial extent and relationship with winter mixing, Deep-Sea Res.-Pt. II, 51, 987–1000, 2004.

Northcote, L. C. and Neil, H. L.: Seasonal variations in foraminiferal flux in the Southern Ocean, Campbell Plateau, New Zealand, Mar. Micropaleontol. 56, 122–137, 2005.

Olaizola, M., Ziemann, D. A., Bienfang, P. K., Walsh, W. A., and Conquest, L. D.: Eddy-induced oscillations of the pycnocline affect the floristic composition and depth distribution of phytoplankton in the subtropical Pacific, Mar. Biol., 116, 533–542, 1993.

Oschlies, A.: NAO-induced long-term changes in nutrient supply to the surface waters of the North Atlantic, Geophys. Res. Lett., 28, 1751–1754, 2001.

Oschlies, A.: Can eddies make ocean deserts bloom?, Global Biogeochem. Cy., 16, 1106, doi:10.1029/2001GB001830, 2002.

Oschlies, A. and Garçon, V.: Eddy-induced enhancement of primary production in a model of the North Atlantic Ocean, Nature, 394, 266–269, 1998.

Palter, J. B., Lozier, M. S., and Barber, R. T.: The effect of advection on the nutrient reservoir in the North Atlantic subtropical gyre, Nature, 437, 687–692, 2005.

Patara, L., Visbeck, M., Masina, S., Krahmann, G., and Vichi, M.: Marine biogeochemical responses to the North Atlantic Oscillation in a coupled climate model, J. Geophys. Res., 116, C07023, doi:10.1029/2010JC006785, 2011.

Pujol, C. and Vergnaud Grazzini, C.: Distribution patterns of live planktic foraminifers as related to regional hydrography and productive systems of the Mediterranean Sea, Mar. Micropaleontol., 25, 187–217, 1995.

Rodwell, M. J., Rowell, D. P., and Folland, C. K.: Oceanic forcing of the wintertime North Atlantic Oscillation and European Climate, Nature, 398, 320–323, 1999.

Sautter, L. and Thunell, R. C.: Seasonal succession of planktonic foraminifera: Results from a four-year time series sediment trap experiment in the northeast Pacific, J. Foramin. Res., 19, 253–267, 1989.

Schiebel, R.: Planktic foraminiferal sedimentation and the marine calcite budget, Global Biogeochem. Cy., 16, 1065, doi:10.1029/2001GB001459, 2002.

Schmuker B. and Schiebel, R.: Planktic foraminifers and hydrography of the eastern and northern Caribbean Sea, Mar. Micropaleontol., 46, 387–403, 2002.

Seki, M. P., Polovina, J. J., Brainard, R. E., Bidigare, R. R., Leonard, C. L., and Foley, D. G.: Biological enhancement at cyclonic eddies tracked with GOES thermal imagery in Hawaiian waters, Geophys. Res. Lett., 28, 1583–1586, 2001.

Sexton, P. F. and Norris, R. D.: Dispersal and biogeography of marine plankton: Long-distance dispersal of the foraminifer Truncorotalia truncatulinoides, Geology, 36, 899–902, 2008.

Sexton, P. F. and Norris, R. D.: High latitude regulation of low latitude thermocline ventilation and planktic foraminifer populations across glacial-interglacial cycles, Earth Planet. Sci. Lett., 311, 69–81, 2011.

Siegel, D. A., Doney, S. C., and Yoder, J. A.: The North Atlantic Spring, Phytoplankton Bloom and Sverdrup's Critical Depth Hypothesis, Science, 296, 730–733, 2002.

Storz, D., Schulz, H., Waniek, J. J., Schulz-Bull, D. E., and Kučera, M.: Seasonal and interannual variability of the planktic foraminiferal flux in the vicinity of the Azores Current, Deep-Sea Res.-Pt. I, 56, 107–124, 2009.

Spero, H. J.: Life history and stable isotope geochemistry of planktonic foraminifera, edited by: Norris, R. D. and Corfield, R. M., in: Isotope Paleobiology and Paleoecology, Paleontological Society Papers, Special Publication, 7–36, 1998.

Sprintall, J. and Tomczak, M.: Evidence of the barrier layer in the surface layer of the tropics, J. Geophys. Res.-Oceans, 97, 7305–7316, 1992.

Steinberg, D. K., Carlson, C. A., Bates, N. R., Rodney, J. J., Michaels, A. F., and Knap, A. H.: Overview of the US JGOFS Bermuda Atlantic Time-series Study (BATS): a decade-scale look at ocean biology and biogeochemistry, Deep-Sea Res.-Pt. II, 48, 1405–1447, 2001.

Sweeney E. N., McGillicuddy, Jr, D. J., and Buesseler, K. O.: Biogeochemical impacts due to mesoscale eddy activity in the Sargasso Sea as measured at the Bermuda Atlantic Time-series Study (BATS), Deep-Sea Res.-Pt. II, 50, 3017–3039, 2003.

Takahashi, K. and Bé, A. W. H.: Planktonic foraminifera: factors controlling sinking speed, Deep-Sea Res., 31, 1477–1500, 1984.

Thunell, R. C. and Reynolds, L. A.: Sedimentation of planktonic foraminifera: seasonal changes in species flux in the Panama Basin, Micropaleontology, 30, 243–262, 1984.

Tolderlund, D. S. and Bé, A. W. H.: Seasonal distribution of planktonic foraminifera in the western North Atlantic, Micropaleontology 17, 297–329, 1971.

Townsend, D. W., Cammen, L. M., Holligan, P. M., Campbell, D. E., and Pettigrew, N. R.: Causes and consequences of variability in the timing of spring phytoplankton blooms, Deep-Sea Res.-Pt. I, 41, 747–765, 1994.

Treusch, A. H., Demir-Hilton, E., Vergin, K. L., Worden, A. Z., Carlson, C. A., Donatz, M. G., Burton, R. M., and Giovannoni, S. J.: Phytoplankton distribution patterns in the northwestern Sargasso Sea revealed by small subunit rRNA genes from plastids, ISME Journal, 6, 481–492, 2012.

Waniek, J. J.: The role of physical forcing in initiation of spring blooms in the northeast Atlantic, J. Mar. Sys., 39, 57–82, 2003.

Wiebe, P. H. and Joyce, T.: Introduction to interdisciplinary studies of Kuroshio and Gulf Stream rings, Deep-Sea Res., 39, Supplement 1, 5–6, 1992.

Žarić, S., Donner, B., Fischer, G., Mulitza, S., and Wefer, G.: Sensitivity of planktic foraminifera to sea surface temperature and export production as derived from sediment trap data, Mar. Micropaleontol., 55, 75–105, 2005.

Zeebe, R. E., Bijma, J., Hoenisch, B., Sanyal, A., Spero, H. J., and Wolf-Gladrow, D. A.: Vital Effects and Beyond: A Modeling Perspective on Developing Paleoceanographic Proxy Relationships in Foraminifera, The Geological Society, London, Special Publications, in: Biogeochemical Controls on Palaeoceanographic Proxies, edited by: James, R., Austin, W. E. N., Clarke, L., and Rickaby, R. E. M., 303, 45–58, 2008.

4

Technical Note: Large overestimation of $p\text{CO}_2$ calculated from pH and alkalinity in acidic, organic-rich freshwaters

G. Abril[1,2], S. Bouillon[3], F. Darchambeau[4], C. R. Teodoru[3], T. R. Marwick[3], F. Tamooh[3], F. Ochieng Omengo[3], N. Geeraert[3], L. Deirmendjian[1], P. Polsenaere[1], and A. V. Borges[4]

[1]Laboratoire EPOC, Environnements et Paléoenvironnements Océaniques et Continentaux, CNRS, Université de Bordeaux, France
[2]Programa de Geoquímica, Universidade Federal Fluminense, Niterói, Rio de Janeiro, Brazil
[3]Katholieke Universiteit Leuven, Department of Earth & Environmental Sciences, Leuven, Belgium
[4]Unité d'Océanographie Chimique, Université de Liège, Belgium

Correspondence to: G. Abril (g.abril@epoc.u-bordeaux1.fr)

Abstract. Inland waters have been recognized as a significant source of carbon dioxide (CO_2) to the atmosphere at the global scale. Fluxes of CO_2 between aquatic systems and the atmosphere are calculated from the gas transfer velocity and the water–air gradient of the partial pressure of CO_2 ($p\text{CO}_2$). Currently, direct measurements of water $p\text{CO}_2$ remain scarce in freshwaters, and most published $p\text{CO}_2$ data are calculated from temperature, pH and total alkalinity (TA). Here, we compare calculated (pH and TA) and measured (equilibrator and headspace) water $p\text{CO}_2$ in a large array of temperate and tropical freshwaters. The 761 data points cover a wide range of values for TA (0 to $14\,200\,\mu\text{mol L}^{-1}$), pH (3.94 to 9.17), measured $p\text{CO}_2$ (36 to $23\,000$ ppmv), and dissolved organic carbon (DOC) (29 to $3970\,\mu\text{mol L}^{-1}$). Calculated $p\text{CO}_2$ were $> 10\%$ higher than measured $p\text{CO}_2$ in 60 % of the samples (with a median overestimation of calculated $p\text{CO}_2$ compared to measured $p\text{CO}_2$ of 2560 ppmv) and were $> 100\%$ higher in the 25 % most organic-rich and acidic samples (with a median overestimation of 9080 ppmv). We suggest these large overestimations of calculated $p\text{CO}_2$ with respect to measured $p\text{CO}_2$ are due to the combination of two cumulative effects: (1) a more significant contribution of organic acids anions to TA in waters with low carbonate alkalinity and high DOC concentrations; (2) a lower buffering capacity of the carbonate system at low pH, which increases the sensitivity of calculated $p\text{CO}_2$ to TA in acidic and organic-rich waters. No empirical relationship could be derived from our data set in order to correct calculated $p\text{CO}_2$ for this bias.

Owing to the widespread distribution of acidic, organic-rich freshwaters, we conclude that regional and global estimates of CO_2 outgassing from freshwaters based on pH and TA data only are most likely overestimated, although the magnitude of the overestimation needs further quantitative analysis. Direct measurements of $p\text{CO}_2$ are recommended in inland waters in general, and in particular in acidic, poorly buffered freshwaters.

1 Introduction

Inland waters (streams, rivers, lakes, reservoirs, wetlands) receive carbon from terrestrial landscapes, usually have a net heterotrophic metabolism, and emit significant amounts of CO_2 to the atmosphere (Kempe 1984; Cole et al., 1994; Raymond et al., 2013). This terrestrial–aquatic–atmosphere link in the global carbon cycle is controlled by complex biogeographical drivers that generate strong spatial and temporal variations in the chemical composition of freshwaters and the intensity of CO_2 outgassing at the water–air interface (e.g. Tamooh et al., 2013; Dinsmore et al., 2013; Abril et al., 2014; Borges et al., 2014). Hence, large data sets are necessary in order to describe the environmental factors controlling these CO_2 emissions and to quantify global CO_2 fluxes from inland waters (Sobek et al., 2005; Barros et al., 2011; Raymond et al., 2013). Dissolved inorganic carbon (DIC) concentration and speciation in freshwaters greatly depend on the lithological nature of watersheds (Meybeck 1987). For instance, rivers draining watersheds rich in carbonate rocks have a high DIC concentration, generally well above $1000 \, \mu\text{mol L}^{-1}$. Bicarbonate ions contribute to most of the total alkalinity (TA) in these waters, which have high conductivities and high pH. In these hard waters, dissolved CO_2 represents a minor fraction (5–15 %) of the DIC compared to bicarbonates. In rivers draining organic-rich soils and noncarbonate rocks, DIC concentrations are lower (typically a few hundred $\mu\text{mol L}^{-1}$) but dissolved organic carbon (DOC) concentrations are higher, and commonly exceed the DIC concentrations. Organic acid anions significantly contribute to TA of these soft waters (Driscoll et al., 1989; Hemond 1990), which have low conductivities and low pH. Dissolved CO_2 represents a large, generally dominant, fraction of DIC in these acidic, organic-rich waters.

Fluxes of CO_2 between aquatic systems and the atmosphere can be computed from the water–air gradient of the concentration of CO_2 and the gas transfer velocity (Liss and Slater, 1974) at local (e.g. Raymond et al., 1997), regional (e.g. Teodoru et al., 2009), and global scales (e.g. Cole et al., 1994; Raymond et al., 2013). The partial pressure of CO_2 (pCO_2) is relatively constant in the atmosphere compared to surface freshwaters pCO_2 that can vary by more than 4 orders of magnitude spatially and temporally (Sobek et al., 2005; Abril et al., 2014). Consequently, water pCO_2 controls the intensity of the air–water flux, together with the gas transfer velocity. At present, both measured and calculated water pCO_2 data are used to compute CO_2 fluxes from freshwater systems, although calculated pCO_2 is overwhelmingly more abundant than directly measured pCO_2 (e.g. Cole et al., 1994; Raymond et al., 2013). pCO_2 can be calculated from the dissociation constants of carbonic acid (which are a function of temperature) and any of the following couples of measured variables: pH/TA, pH/DIC, DIC/TA (Park, 1969). In a majority of cases, calculated pCO_2 is based on the measurements of pH/TA and water temperature. These three parameters are routinely measured by many environmental agencies, and constitute a very large database available for the scientific community. Calculation of pCO_2 from pH and TA was initiated in world rivers in the 1970s (Kempe, 1984) and relies on the dissociation constants of carbonic acid, and the solubility of CO_2, all of which are temperature-dependent (Harned and Scholes, 1941; Harned and Davis, 1943; Millero, 1979; Stumm and Morgan, 1996). Measured pCO_2 is based on water–air phase equilibration either on discrete samples (headspace technique, e.g. Weiss, 1981) or continuously (equilibrator technique, e.g. Frankignoulle et al., 2001) using various systems and devices, followed by direct, generally infrared (IR), detection of CO_2 in the equilibrated gas. Commercial IR gas analysers are becoming cheaper and more accurate, stable and compact, and provide a large range of linear response well adapted to variability of pCO_2 found in freshwaters.

A limited number of studies have compared directly measured pCO_2 to computed pCO_2. Earlier examples provided a comparison between pCO_2 measured by headspace equilibration coupled to gas chromatography (GC) and pCO_2 calculated from pH and DIC (Kratz et al., 1997; Raymond et al., 1997). Reports by these authors in Wisconsin lakes and the Hudson River show that the pCO_2 values were linearly correlated but showed a variability of ± 500 ppmv around the 1 : 1 line, over a range of measured pCO_2 from 300 to 4000 ppmv. Later, Frankignoulle and Borges (2001) reported the first comparison of pCO_2 calculated from pH and TA and pCO_2 measured by equilibration coupled to an IR analyser in an estuary in Belgium. In this high TA $(2500-4800 \, \mu\text{mol L}^{-1})$ and high pH (> 7.4) system, they found a good agreement between the two approaches, calculated pCO_2 being either overestimated or underestimated, but always by less than 7 %. In 2003, concomitant measurements of pH, TA and pCO_2 were performed in acidic, humic-rich ("black" type) waters of the Sinnamary River in French Guiana (Abril et al., 2005, 2006). Calculation of pCO_2 from pH (~ 5) and TA ($\sim 200 \, \mu\text{mol L}^{-1}$) gave unrealistically high values compared to those measured directly with a headspace technique (typically 30000 ppmv vs. 5000 ppmv). Direct measurements of CO_2 and CH_4 outgassing fluxes with floating chambers and the computation of the respective gas transfer velocities of these two gases (Guérin et al., 2007) confirmed that pCO_2 values calculated from pH and TA were overestimated compared to direct measurements in the Sinnamary River. More recently, Hunt et al. (2011) and Wang et al. (2013) provided evidence that organic acid anions in DOC may significantly contribute to TA in some rivers and generate an overestimation of calculated pCO_2. Butman and Raymond (2011) reported higher calculated than measured pCO_2 in some US streams and rivers, but no information was available on the potential role of organic acids in this overestimation. These authors concluded that the low number of samples in their study reflected the need for more research on this topic.

With the growing interest on pCO_2 determination in freshwaters globally, and given the apparent simplicity and low cost of pH and TA measurements, the number of publications that report calculated pCO_2 in freshwaters has increased dramatically in the past decade. Some of these publications report extremely high and potentially biased pCO_2 values in low-alkalinity and high-DOC systems. It has thus become necessary to pay attention to this issue and investigate the occurrence of such potential bias and its magnitude in the different types of freshwaters. Here, we present a large data set of concomitant measurements of temperature, pH, TA, pCO_2, and DOC in freshwaters. This is the first comprehensive data set to investigate the magnitude of the bias between calculated and measured pCO_2, as it covers the entire range of variation of most parameters of the carbonate system in freshwaters. The objective of this paper is to alert the scientific community to the occurrence of a bias in pCO_2

Table 1. Summary of the presented data set. Average, minimum, and maximum values of temperature, DOC, pH (measured on the NBS scale), total alkalinity (TA), and measured partial pressure of CO_2 (pCO_2) in the different freshwater ecosystems.

Country	Watersheds	Temperature (°C)			DOC (μmol L^{-1})			pH (NBS scale)			TA (μmol L^{-1})			Measured pCO_2 (ppmv)			N
		Av.	Min.	Max.	Av.	Min.	Max.	Av.	Min.	Max.	Av.	Min.	Max.	Av.	Min.	Max.	
Brazil	Amazon	30.3	27.4	34.3	352	118	633	6.60	4.53	7.60	385	30	1092	4204	36	18400	155
Kenya	Athi-Galana-Sabaki	25.9	19.8	36.0	307	29	1133	7.69	6.49	8.57	2290	407	5042	2811	608	10405	44
DRC	Congo	26.3	22.6	28.2	1002	149	3968	6.01	3.94	7.22	212	0	576	6093	1582	15571	97
DRC/Rwanda	Lake Kivu	24.0	23.0	24.7	162	142	201	9.05	8.99	9.17	13037	12802	13338	660	537	772	53
France	Leyre	12.5	7.9	19.2	588	142	3625	6.20	4.40	7.41	280	38	1082	4429	901	23047	92
France	Loire	15.5	8.8	19.3	195	167	233	8.70	8.07	9.14	1768	1579	1886	284	65	717	18
Belgium	Meuse	18.1	13.3	25.9	229	102	404	7.89	6.95	8.59	2769	360	7141	2292	176	10033	50
Madagascar	Rianila and Betsiboka	25.4	20.2	29.5	138	33	361	6.84	5.83	7.62	233	76	961	1701	508	3847	36
Kenya	Shimba Hills	25.1	21.9	31.8	214	36	548	7.37	6.22	8.93	1989	227	14244	2751	546	9497	9
French Guiana	Sinnamary	27.1	24.1	28.7	419	213	596	5.50	5.08	6.30	143	66	290	7770	1358	15622	49
Kenya	Tana	26.6	25.0	27.9	321	193	651	7.65	7.32	8.02	1619	1338	2009	2700	845	6014	51
Zambia/Mozambique	Zambezi	26.9	18.8	31.8	252	103	492	7.59	5.06	9.08	1245	52	3134	2695	151	14004	107
Entire data set		24.6	7.9	36.0	408	29	3968	7.00	3.94	9.17	1731	0	14244	3707	36	23047	761

calculation from pH and TA in acidic, poorly buffered and organic-rich freshwaters, to briefly discuss its origin in terms of water chemistry, and to provide the range of pH, TA, and DOC values where pCO_2 calculation should be abandoned and the range where it still gives relatively accurate results.

2 Material and methods

2.1 Sample collection

Our data set consists of 761 concomitant measurements of temperature, pH, TA, water pCO_2, and DOC in 12 contrasting tropical and temperate systems in Europe, Amazonia, and Africa (Fig. 1; Table 1). These samples were obtained in the Central Amazon River and floodplains system in Brazil, the Athi-Galana-Sabaki River in Kenya, the Tana River (Kenya), small rivers draining the Shimba Hills in southeastern Kenya, the Congo River and tributaries in the Democratic Republic of the Congo (DRC), Lake Kivu in Rwanda and DRC, the Leyre River and tributaries in France, the Loire River in France, the Meuse River in Belgium, the Rianila and Betsiboka rivers in Madagascar, the Sinnamary River downstream of the Petit Saut Reservoir in French Guiana, and the Zambezi River in Zambia and Mozambique (Fig. 1). Details on some of the sampling sites can be found in Abril et al. (2005, 2014), Borges et al. (2012, 2014), Marwick et al. (2014a, b), Polsenaere et al. (2013), Tamooh et al. (2013), Teodoru et al. (2014). These watersheds span a range of climates and are occupied by different types of land cover, which include tropical rainforest (Amazon, Congo, Rianila), dry savannah (Tana, Athi-Galana-Sabaki, Betsiboka, Zambezi), temperate pine forest growing on podzols (Leyre), mixed temperate forest, grassland, and cropland (Meuse), and cropland (Loire). Lithology is also extremely contrasted as it includes for instance carbonate-rocks-dominated watershed as for the Meuse, sandstone-dominated silicates (Leyre), and precambrian crystalline magmatic and metamorphic rocks

with a small proportion of carbonate and evaporite rocks for the Congo river.

2.2 Field and laboratory measurements

Although pH measurements might seem almost trivial, highly accurate and precise pH data are in fact not easy to obtain, especially in low-ionic strength waters, where electrode readings are generally less stable. Even though pH measurements in the laboratory might be more accurate, it is crucial to measure pH in situ or immediately after sampling, as pH determination several hours or days after sampling will be affected by CO_2 degassing and/or microbial respiration (Frankignoulle and Borges, 2001). In this work, water temperature and pH were measured in the field with different probes depending on the origin of the data set. However, all the pH data were obtained with glass electrodes and rely on daily calibration with two-point United States National Bureau of Standards (NBS) standards (4 and 7). Measurements were performed directly in the surface water, or in collected water immediately after sampling.

Several techniques were used to measure water pCO_2. Water–gas equilibration was performed with a marbles-type equilibrator (Frankignoulle et al., 2001) for the Amazon, Loire, Leyre, Sinnamary, and Congo rivers (December 2013) as well for Lake Kivu, or with a Liqui-Cel MiniModule membrane contactor equilibrator (see Teodoru et al., 2009, 2014) for the Zambezi and some sites within the Congo basin (December 2012): water was pumped either continuously from a ship, or on an ad hoc basis from the bank of the rivers after waiting ~ 15 min for complete equilibration; air was continuously pumped from the equilibrator to the gas analyser (see e.g. Abril et al., 2014 for a more detailed description of the system). A syringe-headspace technique (Kratz et al., 1997; Teodoru et al., 2009) was used in the field in all African rivers and in the Meuse River: 30 mL volume of atmospheric air was equilibrated with 30 mL volume of river water by vigorously shaking during 5–10 min in four

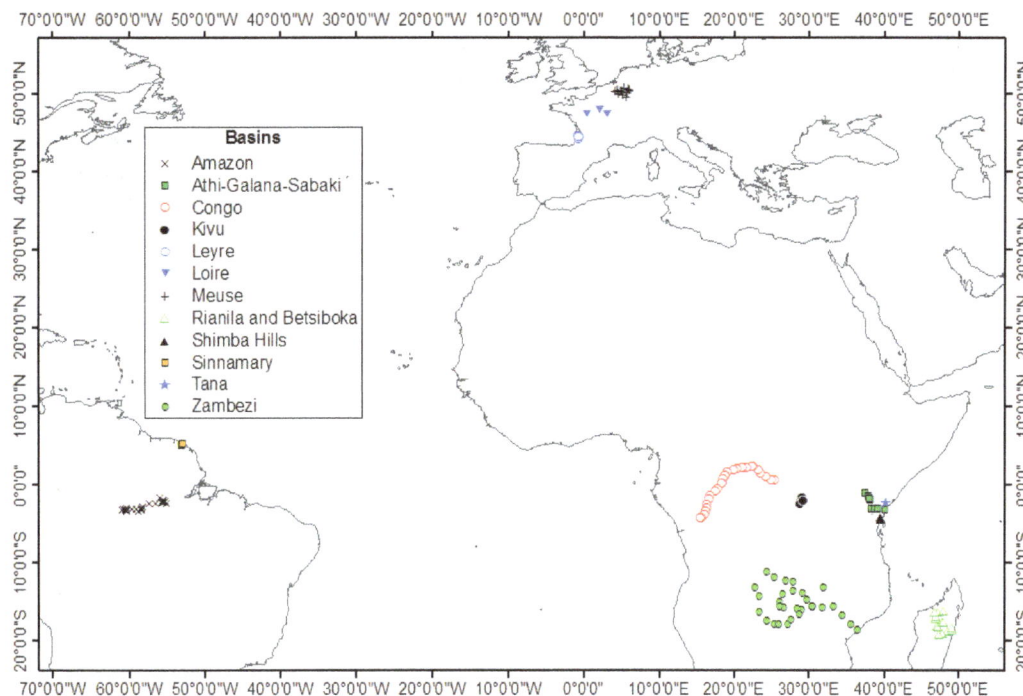

Figure 1. Location of the sampling sites in Africa, Amazonia, and Europe.

replicate gas-tight syringes. The four replicates 30 mL of equilibrated gas and a sample of atmospheric air were injected in an IR gas analyser (Li-Cor® models 820 or 840, or PP systems® model EGM-4); the first gas injection served as a purge for the air circuit and cell and the three other injections were used as triplicate pCO_2 determination (average repeatability of ± 1 %). The pCO_2 in the river water was deduced from that measured in the headspace accounting for the initial pCO_2 in the air used for equilibration, water temperature in the river and in the water at equilibrium in the syringe, and based on Henry's law. Comparison between syringe-headspace and marbles or membrane equilibrator was made during two cruises on the Congo River and three cruises in the Zambezi basin and gave very consistent results, deviation from the 1 : 1 line being always less than 15 % (see Fig. 2). This highlights the consistency of the present data set of direct pCO_2 measurements although different techniques were used. A serum bottle-headspace technique (Hope et al., 1995) was also used on the Sinnamary River; surface water was sampled in 120 mL serum bottles that were poisoned with $HgCl_2$ and sealed excluding air bubbles. Back in the laboratory, a 40 mL headspace was created with pure N_2 (Abril et al., 2005). The CO_2 concentration of equilibrated gas in the headspace was analysed by injecting small volumes (0.5 mL) of gas in a gas chromatograph calibrated with certified gas mixtures.

Immediately after water–gas phase equilibration, CO_2 was detected and quantified in most samples with non-dispersive IR gas analysers (Frankignoulle et al., 2001; Abril et al.,

2014). The gas analysers were calibrated before each field cruise, with air circulating through soda lime or pure N_2 for zero and with a certified gas standard for the span. Depending on the cruises and expected pCO_2 ranges, we used gas standard concentration of 1000–2000 ppmv, or a set of calibration gases at 400, 800, 4000 and 8000 ppmv. Stability of the instrument was checked after the cruise, and deviation of the signal was always less than 5 %. These instruments offer a large range of linear response, depending on manufacturer and model: 0–20 000 ppmv or 0–60 000 ppmv. The linearity of an Li-COR® Li-820 gas analyser was verified by connecting it to a closed circuit of gas equipped with a rubber septum to allow injection of pure CO_2 with a syringe. Linearity was checked by injecting increasing volumes of CO_2 in order to cover the whole range of measurement and was excellent between zero and ~ 20000 ppmv. In addition to the IR analysers generally used in this work, in the Sinnamary River, pCO_2 was also measured with an INNOVA® 1312 optical filter IR photoacoustic gas analyser (range 0–25 000 ppmv) connected to an equilibrator and with a Hewlett Packard® 5890 gas chromatograph equipped with a thermal conductivity detector (TCD); both analysers were calibrated with a gas mixture of 5000 ppmv of CO_2. Both methods gave results consistent at ± 15 % in the 0–13 000 ppmv range (Abril et al., 2006). Sinnamary data reported here are from headspace and GC determination.

TA was analysed by automated electro-titration on 50 mL filtered samples with 0.1N HCl as titrant. Equivalence point was determined with a Gran method from pH between 4

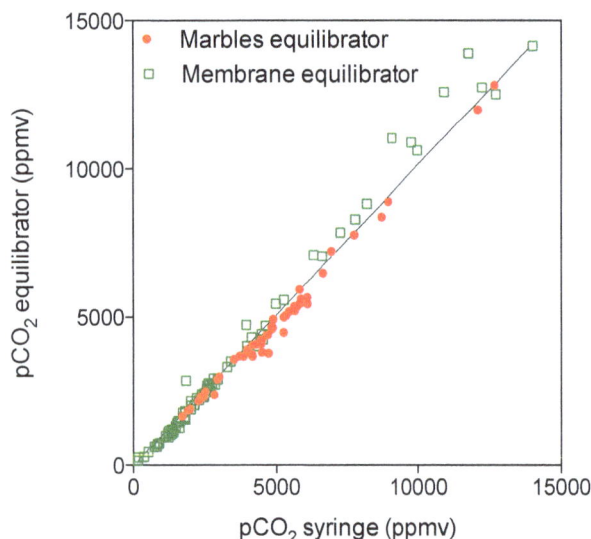

Figure 2. Comparison of results of different water–air equilibration designs for direct pCO_2 measurements; pCO_2 measured with a marbles equilibrator (Congo) and with a membrane equilibrator (Congo and Zambezi) are plotted against pCO_2 measured with a syringe headspace technique. Detection was made with an IR gas analyser.

and 3 (Gran, 1952). Precision based on replicate analyses was better than $\pm 5\,\mu\text{mol L}^{-1}$. TA measurements should be done on filtered samples; otherwise some overestimation would occur in turbid samples, which may contain significant amounts of acid-neutralizing particles (e.g. calcium carbonate). In contrast to TA measurements based on titration to an endpoint of 5.6 (e.g. Wallin et al., 2014), the Gran titration method allows the determination of TA values in samples with in situ pH down to ~ 4.5, i.e. very close to the dissociation constant of HCO_3^- / H_2CO_3. In most acidic samples with low TA, reproducibility was improved by slightly increasing the pH by up to 0.2 units by vigorously stirring during ~ 15 min in order to degas as much CO_2 as possible before starting the titration. DOC was measured on samples filtered through pre-combusted (490 °C) glass fibre filter with a porosity of $0.7\,\mu\text{m}$ and stored acidified with ultrapure H_3PO_4 in borosilicate vials capped with polytetrafluoroethylene stoppers. Analysis was performed with a Shimadzu TOC5000 analyser based on high-temperature catalytic oxidation, after removal of dissolved CO_2 for samples from Amazon, Loire, Leyre, and Sinnamary rivers. DOC concentrations were measured with a customized wet oxidation TOC analyser (Thermo HiperTOC, or IO Analytical Aurora 1030W) coupled to a Delta+XL or Delta V IRMS.

2.3 pCO_2 calculation from pH and TA

We calculated pCO_2 from TA, pH, and temperature measurements using carbonic acid dissociation constants of Millero (1979) (based on those of Harned and Scholes, 1941

and Harned and Davis, 1943) and the CO_2 solubility from Weiss (1974) as implemented in the CO2SYS program. Hunt et al. (2011) reported discrepancy lower than 2 % for pCO_2 computed this way with those obtained with the PHREEQC program (Parkhurst and Appelo, 1999). Differences in software or dissociation constants cannot account for the large bias in calculated pCO_2 compared to measured pCO_2 we report in this paper.

3 Results

3.1 Data ranges and patterns in the entire data set

Measured pCO_2 varied between 36 ppmv in a floodplain of the Amazon River and 23 000 ppmv in a first-order stream of the Leyre River (Table 1). Minimum values of pH and TA occurred in the Congo River (pH = 3.94 and TA = 0) and maximum values in Lake Kivu (pH = 9.16 and TA = $14200\,\mu\text{mol L}^{-1}$). Highest DOC concentrations ($> 3000\,\mu\text{mol L}^{-1}$) were observed in small streams in the Congo basin and in first-order streams draining podzolized soils in the Leyre basin. Lowest DOC concentrations ($< 40\,\mu\text{mol L}^{-1}$) occurred in some tributaries of the Athi-Galana-Sabaki, in the Rianila and Betsiboka rivers, and in the Shimba Hills streams. When considering the whole data set, measured pCO_2 and DOC were negatively correlated with pH, whereas TA was positively correlated with pH (Fig. 3, $p < 0.0001$ for the three variables). This illustrates the large contrast in acid–base properties between acidic, organic-rich, and poorly buffered samples on one hand, and basic, carbonate-buffered samples on the other.

3.2 Comparison between measured and calculated pCO_2

Calculated pCO_2 was more than 10 % lower than measured pCO_2 in 16 % of the samples; the two methods were consistent at ± 10 % in 24 % of the samples; calculated pCO_2 was more than 10 % higher than measured pCO_2 in 60 % of the samples and more than 100 % higher in 26 % of the samples. Absolute values, as expressed in ppmv, were largely shifted towards overestimation, calculated vs. measured pCO_2 data being well above the 1 : 1 line, and calculated minus measured pCO_2 values ranging between -6180 and $+882\,022$ ppmv (Fig. 4). The largest overestimation of calculated pCO_2 occurred in the most acidic samples, whereas underestimations of calculated pCO_2 occurred in neutral or slightly basic samples (Fig. 4b). Ranking the data according to the pH, TA and DOC reveals that overestimation of calculated pCO_2 compared to measured pCO_2 increased in acidic, poorly buffered waters in parallel with an increase in the DOC concentration (Table 2). Discrepancies between calculated and measured pCO_2 were very different from one system to another, depending on the chemical status of the waters. On average at each sampled site, the

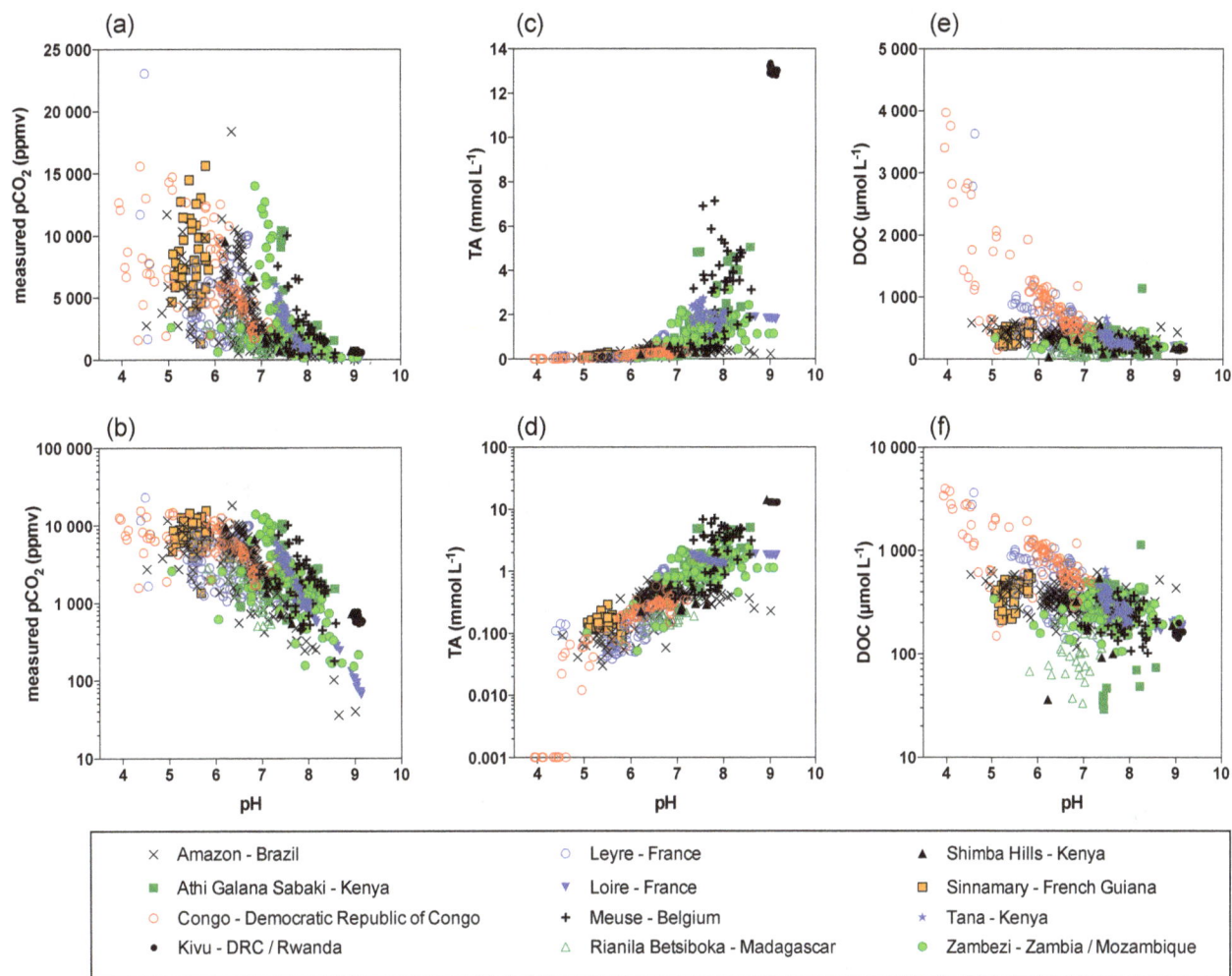

Figure 3. Plot of carbon variables vs. pH in the studied freshwater systems. Top panels are shown with a linear scale and bottom panels with a logarithmic scale; (**a, b**): measured pCO_2; (**c, d**) total alkalinity; (**e, f**) DOC. Zero TA values are plotted as 0.001 in order to be visible on the log pCO_2 scale. Rianila and Betsiboka are plotted together although they belong to different watersheds in Madagascar.

relative overestimation of calculated pCO_2 decreased with pH and TA and increased with DOC (Fig. 5). Overestimation of calculated pCO_2 was on average $< 10\%$ in the Kivu Lake, and the Meuse, Loire, Shimba Hills and Tana rivers, which all have neutral or basic pH, TA $> 1000\,\mu mol\,L^{-1}$ and low to moderate DOC concentrations ($< 400\,\mu mol\,L^{-1}$) (Fig. 5). In contrast, calculated pCO_2 was overestimated by $> 200\%$ on average in the Congo, Leyre, Sinnamary and Amazon rivers, which have acidic pH, TA $< 500\,\mu mol\,L^{-1}$ and highest DOC concentration, reaching $1000\,\mu mol\,L^{-1}$ on average in the Congo. The cases of Athi-Galana-Sabaki, Rianila, Betsiboka, and Zambezi rivers were intermediate in pH, TA and DOC, and with average overestimations of calculated pCO_2 of 50–90 % (Fig. 5).

4   Discussion

4.1   Origin of overestimation of calculated pCO_2

Our data set (Fig. 3; Table 1) probably covers the full range of conditions of carbon speciation that can be encountered in continental surface waters. A pCO_2 overestimation negatively correlated with pH ($p = 0.001$) and TA ($p = 0.005$) and positively correlated with DOC ($p < 0.001$) (Fig. 5) is consistent with the observations of Cai et al. (1998) in the freshwater end-members of some estuaries in Georgia, USA, and of Hunt et al. (2011) in rivers in New England (USA) and New Brunswick (Canada). These authors performed NaOH back-titration in order to measure non-carbonate alkalinity (NCA). They found that NCA accounted for a large fraction (in some cases the greater part) of TA; in addition, the contribution of inorganic species other than carbonate was assumed negligible and most of the NCA was attributed to organic

Table 2. Median and average values of DOC, pH (measured on the NBS scale), total alkalinity (TA), and calculated minus measured pCO_2 in the data set.

	N	% of samples	cal − meas pCO_2 (ppmv)		cal − meas pCO_2 (% of meas pCO_2)		pH		TA (μmol L^{-1})		DOC (μmol L^{-1})	
			Med.	Av.	Med.	Av.	Med.	Av.	Med.	Av.	Med.	Av.
All samples	761	100 %	+611	+10692	+23 %	+194 %	6.94	7.00	467	1731	315	408
Ranked by calculated − measured pCO_2 as % of measured pCO_2												
< −10 %	122	16 %	−540	−890	−34 %	−36 %	7.89	7.85	1269	1766	259	275
±10 %	174	23 %	+15	+50	+2 %	+1 %	7.67	7.78	1576	3735	228	273
> +10 %	465	61 %	+2430	+17710	+72 %	+327 %	6.52	6.49	308	972	360	497
> +50 %	280	37 %	+5490	+28660	+162 %	+526 %	6.18	6.14	192	460	375	567
> +100 %	199	26 %	+9080	+39120	+270 %	+710 %	5.89	5.96	166	364	389	602
Ranked by pH												
pH > 7	368	48 %	+1	+82	+1 %	+15 %	7.82	7.92	1572	3284	231	255
pH < 7	393	52 %	+3280	+20630	+71 %	+362 %	6.30	6.13	232	277	413	558
pH 6–7	256	34 %	+1580	+2710	+40 %	+96 %	6.58	6.55	334	370	350	427
pH < 6	136	18 %	+18410	+54486	+308 %	+864 %	5.50	5.35	93	101	487	828
pH < 5	25	3 %	+115580	+209910	+1645 %	+3180 %	4.53	4.53	41	45	1427	1,843
Ranked by TA												
TA > 2000 μmol L^{-1}	110	14 %	+20	+340	+2 %	+12 %	8.58	8.47	7023	8326	163	202
TA 1000–2000 μmol L^{-1}	157	21 %	−8	−163	−2 %	−9 %	7.81	7.83	1566	1534	271	295
TA 500–1000 μmol L^{-1}	99	13 %	+1307	+1900	+28 %	+72 %	6.97	7.11	651	697	304	318
TA < 500 μmol L^{-1}	395	52 %	+2070	+20090	+64 %	+350 %	6.30	6.24	222	232	400	538
TA < 100 μmol L^{-1}	82	11 %	+6840	+60560	+230 %	+1040 %	5.50	5.35	59	56	603	988
Ranked by DOC												
DOC < 200 μmol L^{-1}	179	24 %	+40	+776	+5 %	+62 %	7.89	7.92	1579	4807	163	149
DOC 200–300 μmol L^{-1}	167	22 %	+102	+2755	+5 %	+69 %	7.56	7.37	1132	1259	258	252
DOC 300–400 μmol L^{-1}	165	22 %	+887	+4473	+25 %	+101 %	6.90	6.93	499	866	341	344
DOC > 400 μmol L^{-1}	250	33 %	+3070	+27197	+59 %	+434 %	6.15	6.14	200	415	555	765
DOC > 800 μmol L^{-1}	79	10 %	+4995	+62784	+92 %	+886 %	5.80	5.62	94	180	1099	1438

acid anions. Hunt et al. (2011) also showed that in the absence of direct titration of NCA, which is labour-intensive and whose precision may be poor, this parameter could be calculated as the difference between the measured TA and the alkalinity calculated from measurements of pH and DIC and the dissociation constants of carbonic acid. Using the latter approach, Wang et al. (2013) obtained a positive correlation between NCA and DOC concentrations in the Congo River, evidencing the predominant role of organic acids in DIC speciation and pH in such acidic system. Because we did not directly measure DIC in this study, we could not calculate NCA with the same procedure as these studies. We attempted to calculate TA from our measured pH and pCO_2 with the CO2SYS program. However, TA values calculated this way were inconsistent with other measured variables (with sometimes negative values). Indeed, because pH and pCO_2 are too interdependent in the carbonate system, very small analytical errors on these variables lead to large uncertainties in the calculated TA (Cullison Gray et al., 2011). A second attempt to correct our TA data from NCA consisted in calculating organic alkalinity using pH and DOC as input parameters.

We compared the model of Driscoll et al. (1989), which assumes a single pK value for all organic acids, and the triprotic model of Hruska et al. (2003), which assumes three apparent pK values for organic acids. These two models applied to our pH and DOC gave very similar organic alkalinity values, which could be subtracted from the measured TA. In the most acidic samples (e.g. some sites from the Congo basin), modelled organic alkalinities were larger than measured TA and the difference was thus negative. Nevertheless, we then recalculated pCO_2 from the measured pH and the TA corrected from organic alkalinity. Calculated pCO_2 values corrected with that method were, however, still very different from those measured in the field, being sometimes higher and sometimes lower than the measured pCO_2, without any meaningful pattern (indeed, corrected pCO_2 was negatively correlated ($p < 0.001$) with measured pCO_2). Consequently, we were unable to derive any empirical relationship to correct for the bias in pCO_2 calculation from pH and TA. Nevertheless, the negative correlation between pH and DOC and positive correlation between pH and TA (Fig. 3) confirm a

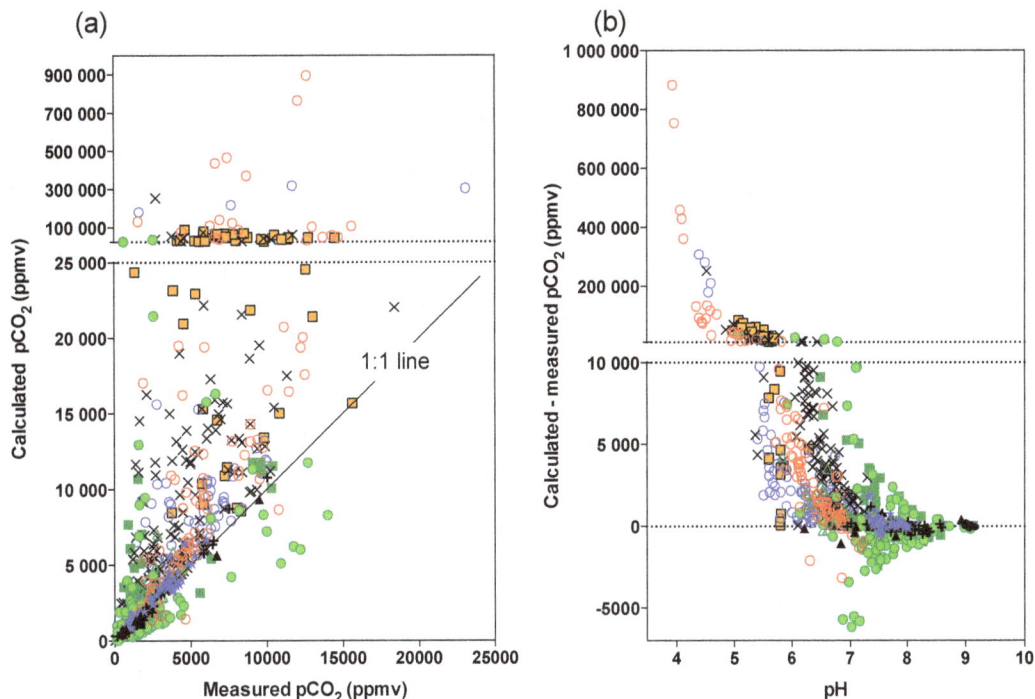

Figure 4. Comparison between measured and calculated pCO_2 for the whole data set: **(a)** calculated vs. measured pCO_2, the line shows when measured pCO_2 equals calculated pCO_2; **(b)** the difference between calculated and measured pCO_2 as a function of pH; same symbols as in Fig. 3.

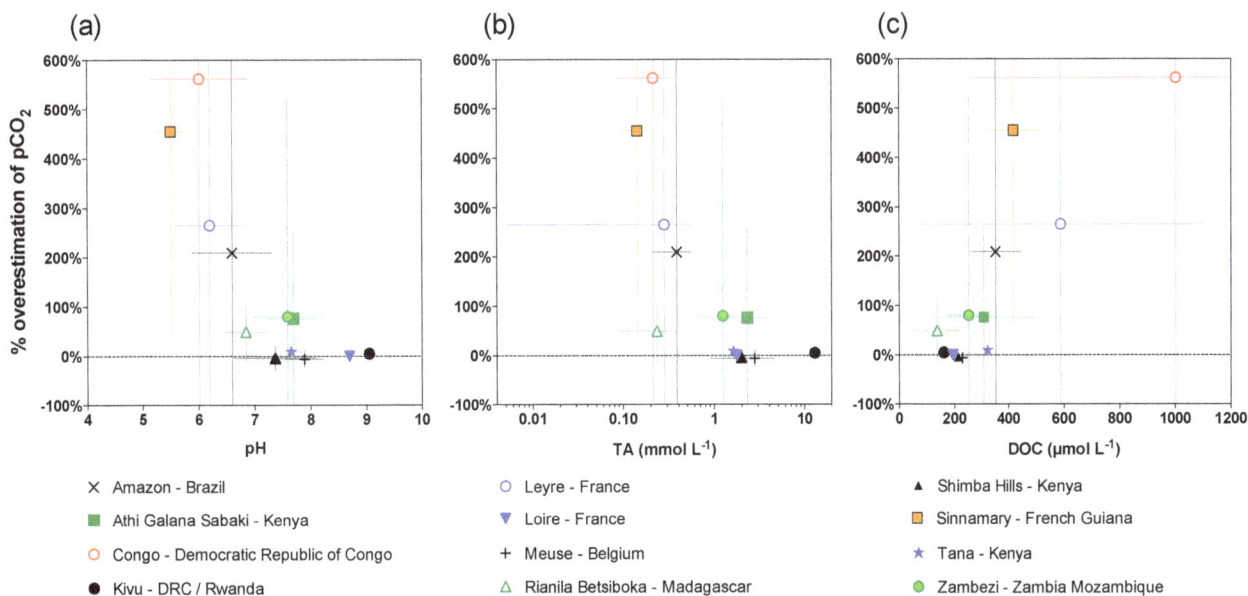

Figure 5. Average percentages of pCO_2 overestimation, calculated as $100 \times$ (calculated pCO_2 – measured pCO_2)/measured pCO_2, as a function of **(a)** pH, **(b)** TA, and **(c)** DOC, for the 12 studied sites. Error bars indicate the standard deviation from the mean for each freshwater system.

strong control of organic acids on pH and DIC speciation across the entire data set.

As discussed by Hunt et al. (2011), a significant contribution of organic acids to TA leads to an overestimation of cal-culated pCO_2 with the CO2SYS program, or with any pro-gram that accounts only for the inorganic species that con-tribute to TA. It is thus obvious that the observed increase in pCO_2 overestimation when pH decreases (Figs. 4b and

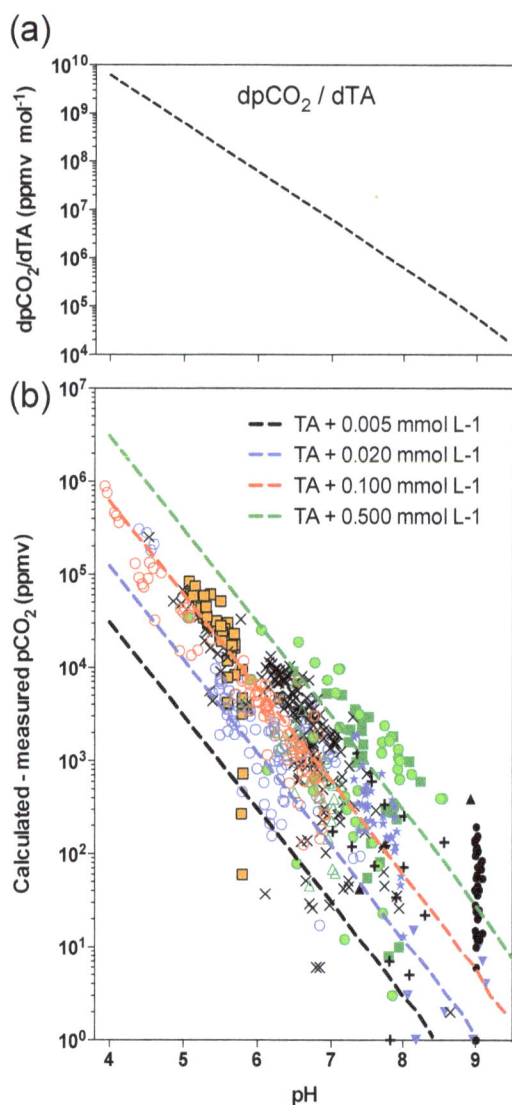

Figure 6. Sensitivity of pCO_2 overestimation to pH: (**a**) theoretical factor $dpCO_2 / dTA$, which describes the sensitivity of calculated pCO_2 to the TA value; (**b**) the solid lines show the increase in calculated pCO_2 induced by various increases in TA, as functions of pH; these lines mimic the overestimation of calculated pCO_2 generated by increasing contributions of organic alkalinity to the TA; field data (as calculated − measured pCO_2) have been plotted for comparison; same symbols as in Fig. 3. Note that negative values do not appear in the logarithmic scale.

5; Table 2) is due to an increasing contribution of organic acid anions to TA. However, this effect is not the only driver of the observed overestimation of pCO_2, which is also due to a decrease in the buffering capacity of the carbonate system at acidic pH. To investigate the magnitude of this second effect, we calculated the factor $dpCO_2 / dTA$ (in ppmv mol^{-1}), which describes the change in calculated pCO_2 induced by a change in TA. This factor, which is the opposite of a buffer factor as it reflects the sensitivity of pCO_2

calculation to the TA, increases exponentially when pH decreases (Fig. 6a), i.e. it is proportional to the H$^+$ concentration. To go further in this theoretical analysis, we computed the difference between the pCO_2 calculated at a given TA value and the one calculated at a slightly higher TA value (TA$+X\,\mu$mol L^{-1}). These calculations reveal an extreme sensitivity of calculated pCO_2 to TA at acidic pH (Fig. 6b). For instance, increasing TA by 5 μmol L^{-1} (a value close to the precision of TA titrations) increases the calculated pCO_2 by 31 ppmv at pH 7, by 307 ppmv at pH 6 and by 3070 at pH 5. Increasing TA by 100 μmol L^{-1} (a typical value of NCA found in freshwaters; Driscoll et al., 1994; Cai et al., 1998; Hunt et al., 2011), increases the calculated pCO_2 by 615 ppmv at pH 7, by 6156 ppmv at pH 6 and by 61560 ppmv at pH 5. Note that this increase in calculated pCO_2 is independent of the chosen initial TA value. The difference between calculated and measured pCO_2 from our data set shows that an NCA contribution around 100 μmol L^{-1} is sufficient to explain the overestimation of calculated pCO_2 of most samples at pH < 6, whereas an NCA contribution higher than 500 μmol L^{-1} would be necessary for several samples at circumneutral and slightly basic pH (Fig. 5b). Samples requiring this high NCA contribution are from the Athi-Galana-Sabaki and Zambezi watersheds, and correspond to TA values well above 1000 μmol L^{-1}. An NCA value of 500 μmol L^{-1} in these samples is thus plausible.

We have no definitive explanation for lower calculated than measured pCO_2, which is observed mainly at neutral to slightly basic pH, for example in the Zambezi River (Fig. 4). In most of these samples, owing to the relatively high TA value, an overestimation of pH of less than 0.2 units is sufficient to account for the low calculated pCO_2 compared to measured values. In general, it is not easy to judge the accuracy of pH measurements, especially when data come from environmental agencies. Thus, one factor of variability throughout the data set as well as in literature data is the accuracy of pH measurements – despite the care taken (e.g. calibrations with NBS buffers for each day of measurements), we cannot rule out that drift or malfunction of pH electrodes contributes to the observed variability, constituting an additional disadvantage compared to direct pCO_2 measurements with very stable gas analysers.

4.2 Impact on estimates of CO$_2$ emissions from freshwaters

According to our analysis, overestimation of calculated pCO_2 is largest in acidic, poorly buffered and organic-rich waters. Consequently, the overestimation of regional and global CO$_2$ emissions computed from calculated pCO_2 depends on the relative contribution of these types of waters worldwide. In their analysis, Raymond et al. (2013) have discarded all calculated pCO_2 values with a pH value of less than 5.4, as well as all pCO_2 values above 100 000 ppmv. These criteria would exclude only 8 % of samples from our data set. Indeed, from our analysis, it appears that overestimation of calculated $p\text{CO}_2$ occurs at pH much higher than 5.4 (Figs. 4, 5 and 6; Table 2). The two techniques were consistent at $\pm 10\%$ on average in only 5 of the 12 studied systems, which combine a circumneutral to basic pH with a TA concentration well above $1000\,\mu\text{mol}\,\text{L}^{-1}$ (Fig. 5). Although it would not be sufficient for the cases of the Zambezi and Athi-Galana-Sabaki rivers, where overestimation is still significant, a TA value above $1000\,\mu\text{mol}\,\text{L}^{-1}$ appears as a more robust criterion than a pH threshold to separate calculated $p\text{CO}_2$ values affected by bias from those consistent with measured $p\text{CO}_2$ (Table 2). In fact, $p\text{CO}_2$ calculation from pH and TA in freshwaters historically relies on theoretical background and validation data in high-alkalinity waters (Neal et al., 1998), including karstic waters (Kempe, 1975). At the global scale, high TA typically occurs in rivers draining watersheds with a significant proportion of carbonate rocks, typically $> 30\%$ of their surface area if the criterion of $\text{TA} > 1000\,\mu\text{mol}\,\text{L}^{-1}$ is chosen and the normalized weathering rates of Meybeck (1987) are applied. According to Meybeck (1987), the average and discharge-weighted TA is around $900\,\mu\text{mol}\,\text{L}^{-1}$ for world rivers and around $600\,\mu\text{mol}\,\text{L}^{-1}$ for tropical rivers. Among the 25 largest rivers in the world, 15 have a $\text{TA} > 1000\,\mu\text{mol}\,\text{L}^{-1}$ according to Cai et al. (2008). The two largest rivers in the world in terms of discharge, the Amazon and the Congo, are also well below this limit of $1000\,\mu\text{mol}\,\text{L}^{-1}$ and have large overestimation in calculated $p\text{CO}_2$ (on average 200 and 360 %, respectively). Very low TA and pH and high DOC values have also been reported in boreal streams and rivers (Humborg et al., 2010; Dinsmore et al., 2012; Wallin et al., 2014).

In lakes, the highest $p\text{CO}_2$ values in the literature come from tropical black water lakes and were also calculated rather than directly measured (Sobek et al., 2005). Calculated $p\text{CO}_2$ was 65 250 ppmv in Lago Tupé in the Brazilian Amazon, a Ria lake connected to the Rio Negro, where, according to our own data set, pH is below 5 and TA is around $70\,\mu\text{mol}\,\text{L}^{-1}$. It was 18 950 ppmv in Kambanain Lake in Papua New Guinea, corresponding to a pH value of 6.1 and a TA value of $350\,\mu\text{mol}\,\text{L}^{-1}$ (Vyverman, 1994). This suggests a widespread overestimation of calculated $p\text{CO}_2$ that significantly impacts the estimation of global CO_2 emissions from inland waters. However, a precise analysis based on exact quantitative information on the relative contribution of acidic and high- and low-alkalinity waters to the total surface area of inland waters is necessary in order to evaluate the exact magnitude of the overestimation.

5 Conclusions

From our analysis, it appears that the validity of calculating $p\text{CO}_2$ from pH, TA and temperature is most robust in freshwaters with circumneutral to basic pH and with TA exceeding $1000\,\mu\text{mol}\,\text{L}^{-1}$. At lower TA and pH, however, cal-

culated $p\text{CO}_2$ (and hence, CO_2 degassing rates) are overestimated by 50 to 300 % relative to direct, in situ $p\text{CO}_2$ measurements. Since a large majority of freshwater systems globally have characteristics outside the range of applicability of $p\text{CO}_2$ calculation, it appears reasonable to assume that recent estimates of global CO_2 emission from lakes and rivers, which are based exclusively on calculated $p\text{CO}_2$ data, are too high. We propose that while TA and pH measurements remain useful to describe the aquatic chemistry, data on $p\text{CO}_2$ should in the future rely on direct measurements of $p\text{CO}_2$. Even if some studies report relatively robust calculation of $p\text{CO}_2$ from pH and DIC measurements (Raymond et al., 1997; Kratz et al., 1997; Aberg and Wallin, 2014), direct $p\text{CO}_2$ values in the field are stable, precise and straightforward and do not depend on the quality of pH measurements, which are often uncertain. Further, high-quality DIC measurements are very time-consuming, fairly complicated to set up and do not allow continuous measurements to be carried out in a simple and straightforward fashion. Although there are some practical limitations to their use in the field, submerged IR sensors, which allow high temporal resolution, are also promising (Johnson et al., 2010). Long-term instrument stability and accuracy based on newly developed off-axis integrated cavity output spectroscopy and cavity ring-down spectroscopy technologies seem to improve in comparison to traditional IR instruments, although the latter are more affordable, more compact and have lower power requirements. Joint international efforts are necessary to define the most appropriate protocols for the measurement of DIC parameters in freshwaters.

Acknowledgements. The data set used in this study was collected in the framework of projects funded by the Cluster of Excellence COTE at the Université de Bordeaux (ANR-10-LABX-45, CNP-Leyre project), the European Research Council (ERC-StG 240002, AFRIVAL: African river basins: Catchment-scale carbon fluxes and transformations, http://ees.kuleuven.be/project/afrival/), the Fonds National de la Recherche Scientifique (FNRS, CAKI, 2.4.598.07, TransCongo, 14711103), the French national agency for research (ANR 08-BLAN-0221, CARBAMA project http://carbama.epoc.u-bordeaux1.fr/), the Research Foundation Flanders (FWO-Vlaanderen), the Belgian Federal Science Policy (BELSPO-SSD projects COBAFISH and EAGLES), the Research Council of the KU Leuven, and the Institut de Radioprotection et Sureté Nucléaire, France (FLORE project). We thank the Hydreco Laboratory in French Guiana, and Patrick Albéric (ISTO Orléans) who analysed some of the data reported here, Aurore Beulen (ULg) for collection of Meuse data set, Marc-Vincent Commarieu (ULg) for analytical support, two anonymous reviewers and C. W. Hunt (reviewer) for constructive comments on the previous version of the paper. AVB is a senior research associate at the FNRS.

Edited by: J. Middelburg

References

Åberg, J. and Wallin M. B.: Evaluating a fast headspace method for measuring DIC and subsequent calculation of pCO$_2$ in freshwater systems, Inland Wat., 4, 157–166, 2014.

Abril, G., Guérin, F., Richard, S., Delmas, R., Galy-Lacaux, C., Gosse, P., Tremblay, A., Varfalvy, L., Dos Santos, M. A., and Matvienko, B.: Carbon dioxide and methane emissions and the carbon budget of a 10-year old tropical reservoir (Petit-Saut. French Guiana), Global Biogeochem. Cy., 19, GB4007, doi:10.1029/2005GB002457, 2005.

Abril, G., Richard, S., and Guérin, F.: In-Situ measurements of dissolved gases (CO$_2$ and CH$_4$) in a wide range of concentrations in a tropical reservoir using an equilibrator, Sc. Total Envir., 354, 246–251, 2006.

Abril, G., Martinez, J.-M., Artigas, L. F., Moreira-Turcq, P., Benedetti, M. F., Vidal L., Meziane, T., Kim, J.-H., Bernardes, M. C., Savoye, N., Deborde, J., Albéric, P., Souza, M. F. L., Souza, E. L., and Roland, F.: Amazon River Carbon Dioxide Outgassing fuelled by Wetlands, Nature, 505, 395–398, 2014.

Barros, N., Cole, J. J., Tranvik L. J., Prairie Y. T., Bastviken D., Huszar V. L. M., del Giorgio P., and Roland F.: Carbon emission from hydroelectric reservoirs linked to reservoir age and latitude, Nat. Geosci.,4, 593–596, doi:10.1038/NGEO1211, 2011.

Borges, A. V., Bouillon, S., Abril, G., Delille, B., Poirier, D., Commarieu, M.-V., Lepoint, G., Morana, C., Servais, P., Descy, J.-P., and Darchambeau, F.: Variability of carbon dioxide and methane in the epilimnion of Lake Kivu, in: Lake Kivu: Limnology and biogeochemistry of a tropical great lake, edited by: Descy, J.-P., Darchambeau, F., and Schmid, M., Aquatic Ecology Series 5, Springer, 47–66, 2012.

Borges, A. V., Morana, C., Bouillon, S., Servais, P., Descy, J.-P., and Darchambeau, F.: Carbon cycling of Lake Kivu (East Africa): net autotrophy in the epilimnion and emission of CO$_2$ to the atmosphere sustained by geogenic inputs, PLoS ONE, 9, e109500, doi:10.1371/journal.pone.0109500, 2014.

Butman, D. and Raymond, P.A.: Significant efflux of carbon dioxide from streams and rivers in the United States, Nature Geosci., 4, 839–842, 2011.

Cai, W.-J., Wang, Y., and Hodson, R. E.: Acid-base properties of dissolved organic matter in the estuarine waters of Georgia, USA, Geochim. Cosmochim. Ac., 62, 473–483, 1998.

Cai, W.-J., Guo, X., Chen, C. T. A., Dai, M., Zhang, L., Zhai, W., Lohrenz, S. E., Yin, K., Harrison, P. J., and Wang, Y.: A comparative overview of weathering intensity and HCO$_3^-$ flux in the world's major rivers with emphasis on the Changjiang, Huanghe, Zhujiang (Pearl) and Mississippi Rivers, Cont. Shelf Res., 28, 1538–1549, 2008.

Cole, J. J., Caraco, N., Kling, G. W., and Kratz, T. K.: Carbon dioxide supersaturation in the surface waters of lakes, Science, 265, 1568–1570, 1994.

Cullison Gray, S. E., DeGranpre, M. E., Moore, T. S., Martz, T. R., Friedrich, G. E., and Johnson, K. S.: Applications of in situ pH measurements for inorganic carbon calculations, Mar. Chem., 125, 82–90, 2011.

Dinsmore, K. J., Wallin M. B., Johnson, M. S., Billett M. F., Bishop, K., Pumpanen, J., and Ojala, A.: Contrasting CO$_2$ concentration discharge dynamics in headwater streams: A multi-catchment comparison, J. Geophys. Res. Biogeosci., 118, 445–461, doi:10.1002/jgrg.20047, 2012.

Driscoll, C. T., Fuller, R., D., and Schecher, W. D.: The role of organic acids in the acidification of surface waters in the eastern US, Water Air Soil Pollut., 43, 21–40, 1989.

Frankignoulle, M. and Borges, A. V.: Direct and indirect pCO2 measurements in a wide range of pCO$_2$ and salinity values, Aquat. Geochem., 7, 267–273, 2001.

Frankignoulle, M., Borges, A. V., and Biondo, R.: A new design of equilibrator to monitor carbon dioxide in highly dynamic and turbid environments, Water Res., 35, 1344–1347, 2001.

Gran, G.: Determination of the equivalence point in potentiometric titrations of seawater with hydrochloric acid, Oceanol. Acta, 5, 209–218, 1952.

Guérin, F., Abril, G., Serça, D., Delon, C., Richard, S., Delmas, R., Tremblay, A., and Varfalvy, L.: Gas transfer velocities of CO$_2$ and CH$_4$ in a tropical reservoir and its river downstream, J. Mar. Syst., 66, 161–172, 2007.

Harned, H. S. and Scholes, S. R.: The ionization constant of HCO$_3^-$ from 0 to 50 °C, J. Am. Chem. Soc., 63, 1706–1709, 1941.

Harned, H. S. and Davis, R. D.: The ionization constant of carbonic acid in water and the solubility of carbon dioxide in water and aqueous salt solutions from 0 to 50 °C, J. Am. Chem. Soc., 65, 2030–2037, 1943.

Hemond, H. F.: Acid neutralizing capacity, alkalinity, and acid-base status of natural waters containing organic acids, Environ. Sci. Technol., 24, 1486–1489, 1990.

Hope, D., Dawson, J. J. C., Cresser, M. S., and Billett, M. F.: A method for measuring free CO$_2$ in upland streamwater using headspace analysis, J. Hydrol., 166, 1–14, 1995.

Hruska, J., Köhler, S., Laudon, H., and Bishop, K.: Is a universal model of organic acidity possible: Comparison of the acid/base properties of dissolved organic carbon in the boreal and temperate zones, Environ. Sci. Technol., 37, 1726–1730, 2003.

Humborg, C., Mörth, C. M., Sundbom, M., Borg, H., Blenckner, T., Giesler, R., and Ittekkot, V.: CO$_2$ supersaturation along the aquatic conduit in Swedish watersheds as constrained by terrestrial respiration, aquatic respiration and weathering, Glob. Change Biol., 16, 1966–1978, 2010.

Hunt, C. W., Salisbury, J. E., and Vandemark, D.: Contribution of non-carbonate anions to total alkalinity and overestimation of pCO$_2$ in New England and New Brunswick rivers, Biogeosciences, 8, 3069–3076, 2011,
http://www.biogeosciences.net/8/3069/2011/.

Johnson, M. J., Billett, M. F., Dinsmore, K. J., Wallin, M., Dyson, K. E., and Jassal, R. S.: Direct and continuous measurement of dissolved carbon dioxide in freshwater aquatic systems-method and applications, Ecohydrol., 3, 68–78, 2010.

Kempe, S.: A computer program for hydrochemical problems in karstic water. Annales de Spéléologie 30, 699–702, 1975.

Kempe, S.: Sinks of the anthropogenically enhanced carbon cycle in surface freshwaters, J. Geophys. Res., 89, 4657–4676, 1984.

Kratz, T. K., Schindler, J., Hope, D., Riera, J. L., and Bowser, C. J.: Average annual carbon dioxide concentrations in eight neighboring lakes in northern Wisconsin, USA. Verh. Internat. Verein. Limnol., 26, 335–338, 1997.

Liss, P. S. and Slater P. G,: Flux of gases across the air-sea interface. Nature, 233, 327–329, 1974.

Marwick, T. R., Tamooh, F., Ogwoka, B., Teodoru, C., Borges, A. V., Darchambeau, F., and Bouillon S.: Dynamic seasonal nitrogen cycling in response to anthropogenic N loading in a tropical catchment, Athi–Galana–Sabaki River, Kenya, Biogeosciences, 11, 1–18, doi:10.5194/bg-11-1-2014, 2014a

Marwick, T. R., Borges A. V., Van Acker K., Darchambeau F., and Bouillon S.: Disproportionate contribution of riparian inputs to organic carbon pools in freshwater systems, Ecosystems, 17, 974–989, 2014b.

Meybeck, M.: Global chemical weathering of surficial rocks estimated from river dissolved loads, American J. Science, 287, 401–428, 1987.

Millero, F. J.: The thermodynamics of the carbonic acid system in seawater, Geochim. Cosmochim. Ac., 43, 1651–1661, 1979.

Neal, C., House, W. A., and Down, K.: An assessment of excess carbon dioxide partial pressures in natural waters based on pH and alkalinity measurements, Sc. Total Envir., 210/211, 173–185, 1998.

Park, P. K.: Oceanic CO_2 system: An evaluation of ten methods of investigation, Limnol. Oceanogr., 14, 179–186, 1969.

Parkhurst, D. L. and Appelo, C. A. J.: User's guide to PHREEQC (version 2) – A computer program for speciation, batch-reaction, one-dimensional transport, and inverse geochemical calculations: US Geol. Surv. Water-Resour. Investigat. Report, 99–4259, 312 pp., 1999.

Polsenaere, P., Savoye, N., Etcheber, H., Canton, M., Poirier, D., Bouillon, S., and Abril ,G.: Export and degassing of terrestrial carbon through watercourses draining a temperate podsolised catchment, Aquatic Sciences, 75, 299–319, 2013.

Raymond, P. A., Caraco, N. F., and Cole J. J.: Carbon dioxide concentration and atmospheric flux in the Hudson River, Estuaries, 20, 381–390, 1997.

Raymond, P. A., Hartmann, J., Lauerwald R., Sobek, S., McDonald, C., Hoover, M., Butman, D., Striegl R., Mayorga, E., Humborg, C., Kortelainen, P., Dürr, H., Meybeck, M., Ciais, P., and Guth, P.: Global carbon dioxide emissions from inland waters, Nature, 503, 355–359, 2013.

Sobek, S., Tranvik L. J., and Cole, J. J.: Temperature independence of carbon dioxide supersaturation in global lakes, Global Biogeochem. Cy., 19, GB2003, doi:10.1029/2004GB002264, 2005.

Stumm, W. and Morgan, J. J.: Aquatic Chemistry, Wiley-Interscience, New York, 1996.

Tamooh, F., Borges, A. V., Meysman, F. J. R., Van Den Meersche, K., Dehairs, F., Merckx, R., and Bouillon, S.: Dynamics of dissolved inorganic carbon and aquatic metabolism in the Tana River basin, Kenya, Biogeosciences, 10, 6911–6928, doi:10.5194/bg-10-6911-2013, 2013.

Teodoru, C. R., del Giorgio P. A., Prairie Y. T., and Camire M., Patterns in pCO_2 in boreal streams and rivers of northern Quebec, Canada, Global Biogeochem. Cy., 23, GB2012, doi:10.1029/2008GB003404, 2009.

Teodoru, C. R., Nyoni, F. C., Borges, A. V., Darchambeau, F., Nyambe, I., and Bouillon, S.: Spatial variability and temporal dynamics of greenhouse gas (CO_2, CH_4, N_2O) concentrations and fluxes along the Zambezi River mainstem and major tributaries, Biogeosciences Discuss., 11, 16391–16445, doi:10.5194/bgd-11-16391-2014, 2014.

Vyverman, W.: Limnological Features of Lakes on the Sepik-Ramu Floodplain, Papua New Guinea Aust, J. Mar. Freshwater Res., 45, 1209–1224, 1994.

Wallin, M. B., Löfgren, S., Erlandsson, M., and Bishop, K.: Representative regional sampling of carbon dioxide and methane concentrations in hemiboreal headwater streams reveal underestimates in less systematic approaches, Glob. Biogeochem. Cy., 28, 465–479, 2014.

Wang, Z. A., Bienvenu, D. J., Mann, P. J., Hoering, K. A., Poulsen, J. R., Spencer, R. G. M., and Holmes, R. M. Inorganic carbon speciation and fluxes in the Congo River. Geophys. Res. Lett., 40, 511–516, 2013.

Weiss, R. F.: Carbon dioxide in water and seawater: the solubility of a non-ideal gas, Mar. Chem., 2, 203–215, 1974.

Weiss, R. F.: Determinations of carbon dioxide and methane by dual catalyst flame ionization chromatography and nitrous oxide by electron capture chromatography, J. Chromatogr. Sci., 19, 611–616, 1981.

5

Technical Note: Maximising accuracy and minimising cost of a potentiometrically regulated ocean acidification simulation system

C. D. MacLeod[1]**, H. L. Doyle**[2]**, and K. I. Currie**[2,3]

[1]Department of Zoology, University of Otago, Dunedin, New Zealand
[2]Department of Chemistry, University of Otago, Dunedin, New Zealand
[3]National Institute of Water and Atmospheric Research (NIWA), Dunedin, New Zealand

Correspondence to: C. D. MacLeod (colin.macleod@postgrad.otago.ac.nz)

Abstract. This article describes a potentiometric ocean acidification simulation system which automatically regulates pH through the injection of 100% CO_2 gas into temperature-controlled seawater. The system is ideally suited to long-term experimental studies of the effect of acidification on biological processes involving small-bodied (10–20 mm) calcifying or non-calcifying organisms. Using hobbyist-grade equipment, the system was constructed for approximately USD 1200 per treatment unit (tank, pH regulation apparatus, chiller, pump/filter unit). An overall tolerance of ± 0.05 pH_T units (SD) was achieved over 90 days in two acidified treatments (7.60 and 7.40) at $12\,^{\circ}C$ using glass electrodes calibrated with synthetic seawater buffers, thereby preventing liquid junction error. The performance of the system was validated through the independent calculation of pH_T ($12\,^{\circ}C$) using dissolved inorganic carbon and total alkalinity data taken from discrete acidified seawater samples. The system was used to compare the shell growth of the marine gastropod *Zeacumantus subcarinatus* infected with the trematode parasite *Maritrema novaezealandensis* with that of uninfected snails at pH levels of 7.4, 7.6, and 8.1.

1 Introduction

The carbon dioxide (CO_2) produced by human activity since 1850 has reduced average surface oceanic pH from approximately 8.2 to 8.1, while current CO_2 emission projections predict that oceanic pH will reach 8.06–7.77 by 2100 and approximately 7.41 by 2300 (IPCC, 2014). The mechanism responsible for this process is the sequestration of atmospheric CO_2 by the global ocean and a subsequent increase in hydrogen ion activity caused by a series of chemical reactions initiated by the dissolution of CO_2 into seawater:

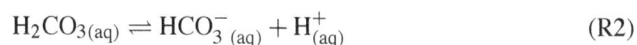

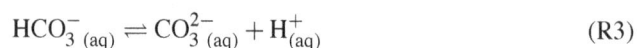

$$CO_{2(aq)} + H_2O_{(l)} \rightleftharpoons H_2CO_{3(aq)} \tag{R1}$$

$$H_2CO_{3(aq)} \rightleftharpoons HCO_3^-{}_{(aq)} + H^+_{(aq)} \tag{R2}$$

$$HCO_3^-{}_{(aq)} \rightleftharpoons CO_3^{2-}{}_{(aq)} + H^+_{(aq)} \tag{R3}$$

$$CO_3^{2-}{}_{(aq)} + H^+_{(aq)} \rightleftharpoons HCO_3^-{}_{(aq)} \tag{R4}$$

where H_2CO_3 is carbonic acid and HCO_3^- and CO_3^{2-} are the bicarbonate and carbonate ions, respectively. The global reduction of ocean pH has become known as ocean acidification (OA), although the term also refers to changes in the concentration of carbonic acid and bicarbonate and carbonate ions, in addition to increased hydrogen ion activity (Reactions R1–R4).

The altered chemical speciation of seawater caused by OA poses a variety of challenges to all marine species, e.g. the maintenance of intra- and extra-cellular acid–base homeostasis in a more acidic environment (Pörtner et al., 2004) or the synthesis and dissolution of calcium carbonate ($CaCO_3$) structures in seawater undersaturated with regard to component ions (Weiner and Dove, 2003). A meta-analysis conducted by Kröeker et al. (2013) showed that OA will likely have a varied yet negative effect on many marine organisms in future, while negative effects on calcifying species found in areas of naturally elevated acidity have already been reported (e.g. Gruber et al., 2012). To date, the majority of experimental research into the effects of OA has focussed on single marine species in an attempt to identify those with or without the ability to adapt to acidified conditions within a single generation. The identification of such phenotypic plasticity in response to stressors associated with OA is vital, as evolutionary adaptation may not occur at a sufficient rate to protect some species from changing marine conditions (Bell and Collins, 2008). However, it is now accepted that OA research must move beyond single-species experiments and begin investigating the effects of combined abiotic factors, such as pH and temperature (Boyd, 2011), and the potential effects of OA on biological interactions such as competition (Hoffman et al., 2012), predation (Dixon et al., 2010; Allan et al., 2013), and parasitism (MacLeod and Poulin, 2012). This paradigm does not negate the importance of single-species/single-factor experiments but rather broadens the scope of OA research. A thorough investigation of a species' response to novel abiotic stressors should begin with single-factor manipulations and then introduce increasing levels of complexity to fully document potential synergistic reactions between parameters. Given the current rate of ocean acidification (~ 0.0018 pH units yr^{-1}, Feely et al., 2009), the identification of species and species' interactions that are vulnerable to OA, alone or in combination with other abiotic factors, should be urgently addressed; lab-based simulations will play an important role in achieving this goal (Widdecombe et al., 2010).

This article provides a detailed description of a low-cost, easy set-up OA simulation system that reliably mimics the effects of elevated atmospheric CO_2 on seawater chemistry by controlling temperature, salinity, pH, and total alkalinity (A_T). In addition, we suggest goal tolerances, i.e. the variability around target parameter values expressed as standard deviations, for the control of these parameters: temperature ($\pm 0.5\,^{\circ}C$), salinity (± 0.6), pH (± 0.05), and A_T ($\pm 10\,\mu mol\,kg^{-1}$). We believe these tolerance values represent realistic and achievable goals for OA simulation systems, as they can be met with relatively inexpensive apparatus and cause minimal changes to calculated carbonate parameters (Table 3).

2 OA simulation systems

2.1 Review

OA simulation systems must be able to reliably manipulate the carbonate chemistry of seawater, which is characterised by seven parameters: (1) temperature ($^{\circ}C$); (2) salinity (reported on the Practical Salinity Scale); (3) depth (metres); (4) pH:

$$pH = -\log[H^+] \qquad (1)$$

(5) total alkalinity ($A_T\,\mu mol\,kg^{-1}$):

$$A_T = [HCO_3^-] + 2[CO_3^{2-}] + [B(OH)_4^-] + [OH^-] \qquad (2)$$
$$+ [HPO_4^{2-}] + 2[PO_4^{3-}] + [SiO(OH)_3^-] + [NH_3]$$
$$+ [HS^-] - [H^+] - [HSO_4^-] - [HF] - [H_3PO_4]\ldots$$

(6) dissolved inorganic carbon concentration (DIC-$\mu mol\,kg^{-1}$):

$$DIC = [CO_2] + [H_2CO_3] + [HCO_3^-] + [CO_3^{2-}] \qquad (3)$$

(7) partial pressure of seawater CO_2 (pCO_2-μatm):

$$p(CO_2) = x(CO_2)P \qquad (4)$$

where $x(CO_2)$ represents the mole fraction of CO_2 in the gas phase in equilibrium with seawater and P represents the total pressure. For detailed definitions of the analytical parameters used to characterise seawater carbonate chemistry, please see Dickson et al. (2007). Of the seven variables listed above, temperature, salinity, depth (if applicable), and two of the four analytical parameters must be known, in addition to appropriate equilibrium constants, to fully characterise the carbonate chemistry of the modified seawater and quantify variables central to the effects of OA, e.g. saturation states of calcium carbonate polymorphs or concentrations of HCO_3^- and CO_3^{2-}. Accordingly, one must *control* salinity, temperature, and two of the four analytical parameters described above to manipulate the carbonate chemistry of seawater in experimental OA simulation systems.

Riebesell et al. (2010) compiled a detailed guide for the standardisation of methodology used in the manipulation and measurement of carbonate chemistry (*The Guide to Best Practices for Ocean Acidification Research and Data Reporting*). Since the publication of the guide, there have been several published descriptions of OA simulation systems which use a variety of techniques to acidify seawater: gas injection ($CO_2 / O_2 / N_2$ – Bockmon et al. (2013); $100\,\%$ CO_2 – Wilcox-Freeburg et al., 2013), the addition of CO_2-enriched seawater (McGraw et al., 2010), and the addition of HCl (Riebesell et al., 2000). Despite the many differences between experimental approaches, almost all simulation systems are regulated through the measurement of pH as a master variable.

Monitoring pH in an OA simulation system by the automated spectrophotometric analysis of seawater samples integrated into a software-based regulation system (e.g. McGraw et al., 2010) provides a high degree of precision (± 0.0004, Carter et al., 2013; Clayton and Byrne, 1993) compared to potentiometric techniques (± 0.002–0.001, Dickson et al., 2007) and has been used to regulate OA simulation systems with minimal variation around target pH values (± 0.02, McGraw et al., 2010). However, spectrophotometric pH regulation can prove extremely expensive, as these systems must be custom-designed (Wilcox-Freeburg et al., 2013). Despite the reduced degree of precision, potentiometric measurement of pH is the central component of most OA simulation systems designed to explore the effects of reduced pH on biological organisms (Easley and Byrne, 2012). Indeed, in the 2013 special OA issue of the journal *Marine Biology* (August, Volume 160, Issue 8), 31 out of 32 (97 %) experimental articles used manipulation techniques controlled by, or monitored through, the potentiometric measurement of pH.

The regulation of temperature, salinity, and A_T is often not discussed in detail in the OA literature, despite the central role of these variables in the control of carbonate chemistry. Temperature is typically controlled by actively heating or cooling the acidified seawater to a target value using a variety of commonly available lab equipment, e.g. chiller units, temperature-controlled rooms, or heating coils. Salinity is often monitored but not controlled, as many simulation systems are supplied with seawater from a large reservoir or permanent connection to the ocean or passively controlled through the regular replacement of seawater. The A_T of an OA simulation system can be altered by the biological activity of experimental organisms. Consequently, A_T is often also regulated through the replacement of seawater or with a flow-through system. Possibly as a consequence of the commonplace (temperature) or passive (salinity and A_T) methods of regulation, tolerances of these parameters are often not reported in OA literature. In the 2013 special OA issue of the journal *Marine Biology*, 14 studies used temperature, salinity, pH, and A_T to control and describe seawater carbonate chemistry. Six of these studies reported no measure of temperature variance, 8 reported no salinity variance, and 5 reported no A_T variance. In addition, some articles gave parameter tolerances as standard error (SE), with or without the corresponding sample size, making comparisons of tolerance levels between studies difficult. As the measurement of pH is subject to many sources of uncertainty, the tolerances of temperature, salinity, and A_T should be stated explicitly and clearly in the description of OA simulation systems.

2.2 Described system

2.2.1 Overview

The described system manipulates the carbonate chemistry of seawater through the pH-controlled injection of 100 % CO_2 gas. The use of pH as a controlling variable and CO_2 gas as an acidifying agent has two key advantages over other acidification techniques. First, the addition of CO_2 gas more realistically mimics the effects of increased atmospheric CO_2 on seawater chemistry than the addition of an acid (Hurd et al., 2009, Schulz et al., 2009). Second, the pH-controlled addition of CO_2 gas reduces pH variation when compared to the injection of gas–air mixes at a fixed rate; the latter can result in unwanted fluctuations in pH caused by biological activity, changes in temperature, or increases in ambient atmospheric CO_2 (Wilcox-Freeburg et al., 2013). In this system, seawa-

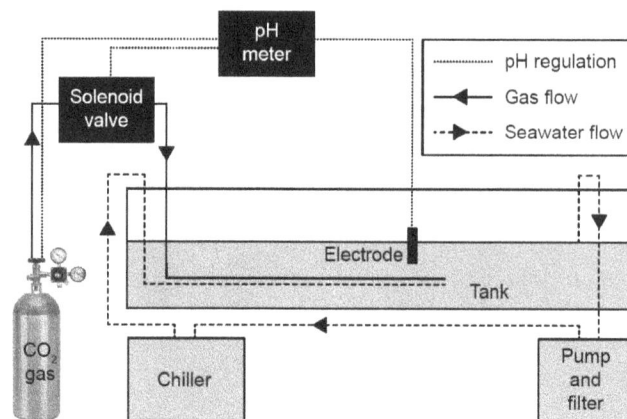

Figure 1. Schematic of one OA simulation unit. Solid lines indicate gas flow, dashed lines indicate seawater flow, and dotted lines indicate electrical connections between components of pH regulation apparatus.

ter temperature was actively maintained at $12.6 \pm 0.5\,°C$, while salinity (31.6 ± 0.6) and A_T ($2375 \pm 10\,\mu\text{mol kg}^{-1}$) were passively controlled through the regular replacement of seawater.

2.2.2 Apparatus

The described experimental apparatus consists of three identical units (Fig. 1), each capable of independently mimicking the effects of increased atmospheric CO_2 on seawater, i.e. elevated pCO_2 and DIC and reduced pH. The pH of culture tank seawater was constantly monitored potentiometrically and automatically regulated through the injection of 100 % food-grade CO_2 gas. In each tank, 80 L of seawater was contained in a 120 L open-top tank (870 mm (L) × 600 mm (W) × 295 mm (H), food-grade low-density polyethylene, Stowers Containment Solutions, NZ). Unamended seawater was supplied by the Portobello Marine Research Station, Dunedin, New Zealand, and was high-pressure filtered through sand prior to use. The unamended seawater had a total alkalinity of $2354 \times 10\,\mu\text{mol kg}^{-1}$ ($n = 6$), and a salinity of 31.5 ± 0.5. pH in each culture tank was regulated using TUNZE™ pH / CO_2 controller systems (glass electrodes, pH meter, solenoid switch unit, and a pressure reducer) connected to 33 kg gas cylinders containing 100 % food grade CO_2 (BOC). The TUNZE™ system automatically allowed CO_2 gas to flow from the pressurised cylinders through the solenoid switch unit into the culture tank when the pH of acidified seawater rose above target values. Carbon dioxide gas diffused into the acidified seawater through a perforated 4 mm plastic tube which was wrapped around the water inflow pipe. This allowed for a maximum rate of dispersal of dissolved gas through the culture tank, minimising any pH gradient relative to the gas input point. To ensure that ambient temperature variations did not alter pH (TUNZE™ pH meters have no automatic temperature compensation function), seawater was pumped through a 1/5 hp refrigeration unit (Hailea HC-150A) using an aquarium pump–filter system (Aqua One®, Aquis700) at a rate of approximately $400 \, L \, h^{-1}$. To minimise changes in salinity and A_T caused by evaporation, calcification, shell dissolution, or respiration, 20 L of seawater was removed from each tank every 48 h and gradually ($30 \, L \, h^{-1}$) replaced with unamended seawater. Each culture tank was also aerated with ambient air by an aquarium bubbler (AquaOne 9500), and oxygen saturation (measured daily with a YSI ProODO) was greater than 95 % for the duration of the experimental period.

2.2.3 Measurement of analytical parameters

As noted in Easley and Byrne (2012), there are a number of challenges inherent in the potentiometric measurement of pH: calibration buffers must be of similar ionic strength to samples to avoid liquid junction error (see the Discussion section for a complete description of liquid junction error) (Millero et al., 1993; Waters, 2012); preparing saltwater buffers in the lab can lead to pH variation due to human error; post-preparation, the pH of buffers can be altered through contact with ambient atmospheric CO_2; electrode function can degrade over time and result in a deviation from the ideal Nernstian slope required to convert volts to pH units; and all electrodes are subject to a certain degree of drift over time (Dickson et al., 2007).

In the described system, pH meters were calibrated using homemade saltwater buffers (*2-amino-2-hydroxy-1,3-propanediol* (TRIS) and *2-aminopyridine* (AMP)) prepared in accordance with Dickson et al. (2007). Buffer salinity was slightly higher than that of seawater in the culture tanks (35 vs. ~ 32); however, the consequent error was assumed to be less than 0.005 pH units (Dickson et al., 2007). In case of small deviations of buffer pH caused by human error during preparation, buffers were analysed with an Agilent 8453 spectrophotometer using pure meta-Cresol Purple (mCP) (provided by the laboratory of Professor Robert H. Byrne, University of South Florida) at 25 °C, and pH_T was calculated from a measured mCP spectrum using the calibration of Liu et al. (2011). After preparation, saltwater buffers were aliquoted into 100 mL borosilicate Schott bottles in front of an air pump modified to produce CO_2-depleted air, thus minimising the effect of ambient CO_2 on buffer pH. With appropriate storage protocols, saltwater buffers prepared in this way have proved stable for up to one year, and subsequent degradation is approximately 0.0005 pH units per year (Nemzer and Dickson, 2005).

Table 1. Average values ($\pm$SD, $n = 64$) for pH_T, temperature, and salinity, recorded over a 90-day period in three pH treatment tanks during the culture of *Z. subcarinatus*.

	pH_T (Measured)	Temp. (°C)	Salinity
8.1 treatment	8.09 ± 0.03	12.5 ± 0.3	31.7 ± 0.6
7.6 treatment	7.60 ± 0.03	12.6 ± 0.6	31.9 ± 0.6
7.4 treatment	7.40 ± 0.03	12.6 ± 0.5	31.3 ± 0.6

In addition to frequent calibration of pH electrodes to compensate for drift, TRIS and AMP buffers were used to ensure that all electrode responses were within 0.2–0.3 % of the ideal Nernst value (0.05916 V) at 25 °C (Dickson et al., 2007; Millero et al., 1993):

$$\text{Electrode response} = EMF_{AMP} - EMF_{TRIS}/pH_{TRIS} - pH_{AMP} \quad (5)$$

where EMF refers to electromotive force, measured in volts. Variability in culture tank pH was minimised through a two-stage monitoring process. Seawater pH in each tank was constantly measured with electrodes connected to the CO_2 delivery system (TUNZE™, two-point calibration, $\pm$0.01 pH units). As individual electrodes are prone to drift even with frequent calibration (Dickson et al., 2007), an independent, hand-held pH meter (Denver Instrument Company AP50, two-point calibration, $\pm$0.002 pH units) was also used to measure culture tank pH daily. If the Denver pH meter detected deviations from the target pH, the TUNZE™ apparatus was adjusted, allowing for centralised control of pH using the most precise meter available.

The performance of the potentiometric apparatus was also validated with the calculation of pH_T (12 °C) based on A_T and DIC data taken from culture tank seawater, using SWCO2 Software (Hunter, 2007) and the dissociation constants of Mehrbach et al. (1973) refit by Dickson and Millero (1987). Total alkalinity was measured with a closed-cell potentiometric apparatus, based on the system described by Dickson et al. (2007), while DIC was measured using infrared analyses of CO_2 evolved from an acidified sample (AIR-ICA DIC analyser (Automated Infra Red Inorganic Carbon Analyzer), by MARIANDA). Measurements of A_T and DIC were calibrated using certified reference materials (CRM) from the lab of Professor Andrew Dickson, University of California San Diego. Seawater taken from culture tanks was stored in 1000 mL borosilicate Schott bottles and fixed with a saturated solution of mercuric chloride prior to A_T and DIC analysis (per recommendations of Riebesell et al., 2010).

3 Assessment

3.1 Carbonate parameters

Carbonate parameters were monitored throughout a 90-day experiment to culture the New Zealand mud snail

(*Zeacumantus subcarinatus*), collected from Otago Harbour, Dunedin, New Zealand. During the experimental period, temperature, salinity, and pH were measured daily (Table 1), while A_T and DIC were analysed from samples taken approximately every 18 days (Table 2). Table 2 also lists other relevant carbonate parameters calculated using DIC and A_T as measured variables.

pH$_T$ (12 °C), measured both potentiometrically and calculated from DIC and A_T data, varied by $\pm$0.03–0.04 units (SD) in all three culture tanks over the 90-day period (measured: 7.40 $\pm$ 0.03, 7.60 $\pm$ 0.04; calculated: 7.45 $\pm$ 0.04, 7.64 $\pm$ 0.04) (Fig. 2). While the calibration of all electrodes occurred weekly, there was very little drift in the electrodes connected to the CO_2 regulation apparatus. Temperature, controlled by the chiller units, was also stable across all culture tanks, while salinity and A_T showed minimal variation (Table 1). However, there was a greater relative uncertainty in salinity (approximately 2 %) than A_T (< 0.5 %) over the experimental period. We assume that this was due to a greater variability in salinity over the entire 90-day period, detected by more frequent sampling ($n = 64$) compared to A_T ($n = 6$). As expected, DIC (measured) and pCO$_2$ (calculated) increased in all culture tanks after the injection of CO_2 gas (Hansen et al., 2013; Campbell and Fourqueran, 2011; Findlay et al., 2008), while A_T remained unchanged in all treatments (Table 2).

Sources of error in our measurement of pH include the following: spectrophotometric measurement of buffer pH ($\pm$0.004, Carter et al., 2013); differences between buffer salinity and seawater salinity (< 0.005, Dickson et al., 2007); and the potentiometric measurement of seawater pH ($\pm$0.01–0.002, pH meter specifications).

In addition, while the variability of temperature, salinity, and A_T was relatively minor, measurement errors or incorrect calibrations ("offsets") in these parameters will result in offsets in the calculated parameters central to the study of the effects of OA on marine organisms. Table 3 contains examples of the offsets in calculated carbonate parameters caused by values of uncertainty found in this study. The uncertainty in calculated pH resulting from uncertainties in measured A_T (10 μmol kg^{-1}) and DIC (10 μmol kg^{-1}), and uncertainty in the dissociation constants (pK) of H_2CO_3 (0.01) and HCO_3^- (0.02), gives an uncertainty in calculated pH$_T$ of approximately 0.05 pH (Dickson and Riley, 1978). Thus, this error estimate in pH is in good agreement with the difference between our measured and calculated values for seawater pH; measured pH was between 0.03 and 0.05 lower than calculated pH in all pH treatments.

3.2 Culture of biological organisms

To investigate the potential interaction of infection stress and stressors associated with OA on the growth of *Z. subcarinatus*, 180 snails (average length, 14.4 $\pm$ 1.3 mm; average mass, 0.22 $\pm$ 0.05 g) were distributed evenly between three

pH treatments: 8.1, 7.6, and 7.4. Of the 60 snails in each treatment, 30 were infected with the marine trematode parasite *Maritrema novaezealandensis* and 30 had no parasitic infection. Each group of 30 snails was further subdivided into groups of 5 and placed in mesh chambers which allowed the flow-through of seawater. Prior to exposure to acidified seawater, all snails were soaked for 24 h in a saltwater solution of calcein, a soluble fluorochrome which is incorporated into growing calcified structures and produces a fluorescent band which can be treated as a baseline for subsequent growth (Riascos et al., 2007). The snails were maintained in the three pH treatments for a total of 90 days, although during that time each tank was assigned a particular pH for only 30 days. During the reassignment of tank pH, snails from the control (8.1 pH) culture tank were first removed and placed in a second aerated container. The now vacant tank was then acidified to 7.6 pH and snails transferred from the tank previously assigned that treatment. This process was repeated for the snails in the 7.4 pH treatment, and the tank originally assigned 7.4 pH was allowed to re-equilibrate with atmospheric CO_2 before the "control" snails were replaced. This stepwise changeover removed the potential for tank effect to bias experimental data and reduced any variation in pH conditions experienced by the snails.

After 90 days, all snails were removed from the culture tanks and the growing edge of their shell imaged under UV light (Leica camera (DFC320) and dissecting scope (MZFL11), 6.4$\times$ magnification). New shell growth, visible beyond the fluorescent band, was measured with ImageJ software, and these data were analysed with a two-factor ANOVA (analysis of variance) to test the effects of pH and infection on shell growth. Analysis of variance showed that there was significantly reduced growth under acidified conditions in infected and uninfected snails (Fig. 3) and that infected snails grew more than uninfected individuals in all pH treatments. The complete details of this study and the biological interpretations of the findings will be published elsewhere.

4 Discussion and recommendations

4.1 Overview

This article describes an OA simulation system that maintained temperature, salinity, pH, and A_T within goal tolerances in three 80 L seawater culture tanks over 90 days. The system was used to culture the New Zealand mud snail, *Zeacumantus subcarinatus*, to investigate the effects of reduced pH on individuals infected with the marine trematode *M. novaezealandensis* relative to uninfected conspecifics. All apparatus used in the construction of the described system was purchased through aquarium suppliers at a cost of approximately USD 3600, i.e. USD 1200 per unit.

Table 2. Average values ($\pm$SD, $n = 6$) for A_T and DIC (measured) and pH_T and pCO_2 (calculated) recorded over a 90-day period in three pH_T treatments during the culture of *Z. subcarinatus*.

	Alkalinity (μmol kg^{-1})	DIC (μmol kg^{-1})	pH_T (calculated)	pCO_2 (calculated)
8.1 treatment	2361 ± 10	2138 ± 11	8.12 ± 0.03	365 ± 30
7.6 treatment	2389 ± 7	2351 ± 16	7.64 ± 0.04	1304 ± 115
7.4 treatment	2375 ± 12	2397 ± 13	7.45 ± 0.04	1980 ± 110

Table 3. A comparison of the offsets resulting in calculated carbonate parameters by offsets or calibration errors in measured variables. The top line shows calculated values for DIC, pCO_2, Ωa, and Ωc, calculated based on the average oceanic values for temperature, salinity, pH, and A_T reported in Riebesell et al. (2010). Text in bold indicates the parameter that was varied.

	Measured parameters				Calculated parameters			
	Temperature (°C)	Salinity	pH_T	A_T (μmol kg^{-1})	DIC (μmol kg^{-1})	pCO_2 (μatm)	Ωa	Ωc
Oceanic average (2010)	18.7	34.8	8.062	2305	2050	384	2.83	4.38
Temperature ($\pm$0.5 °C)	**18.2–19.2**	34.8	8.062	2305	2054–2045	384–384	2.79–2.88	4.31–4.45
Salinity ($\pm$0.6)	18.7	**34.2–35.4**	8.062	2305	2054–2046	386–382	2.81–2.86	4.35–4.41
pH_T ($\pm$0.05)	18.7	34.8	**8.012–8.112**	2305	2075–2022	440–334	2.58–3.11	3.99–4.80
A_T ($\pm$10 μmol kg^{-1})	18.7	34.8	8.062	**2295–2315**	2040–2058	381–384	2.83–2.85	4.37–4.41
Temp. and salinity	**18.2–19.2**	**34.2–35.4**	8.062	2305	2057–2041	385–381	2.77–2.91	4.29–4.48
Temp., salinity, and A_T	**18.2–19.2**	**34.2–35.4**	8.062	**2295–2315**	2048–2050	383–382	2.76–2.92	4.27–4.50
Temp., salinity, A_T, and pH_T	**18.2–19.2**	**34.2–35.4**	**8.012–8.112**	**2295–2315**	2074–2023	440–334	2.51–3.19	3.88–4.92
Liquid junction error ($\pm$0.065 pH)	18.7	34.8	**7.997–8.127**	2305	2083–2014	458–320	2.51–3.19	3.88–4.93

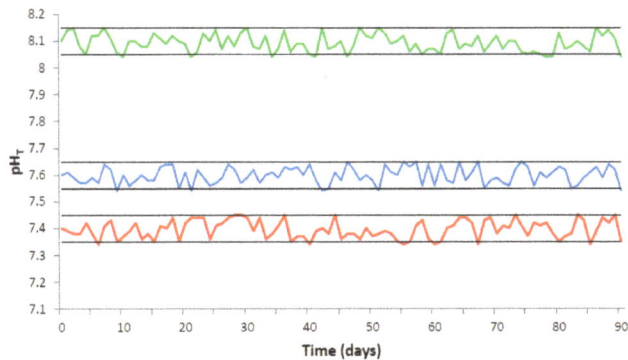

Figure 2. pH_T recorded over the course of a 90-day experiment in which snails were maintained in three culture tanks: 8.1 (green), 7.6 (blue), 7.4 (red) pH_T. Coloured lines represent pH_T data recorded on Denver AP50 hand-held pH meter, and black lines represent $\pm$0.05 error around target pH_T values.

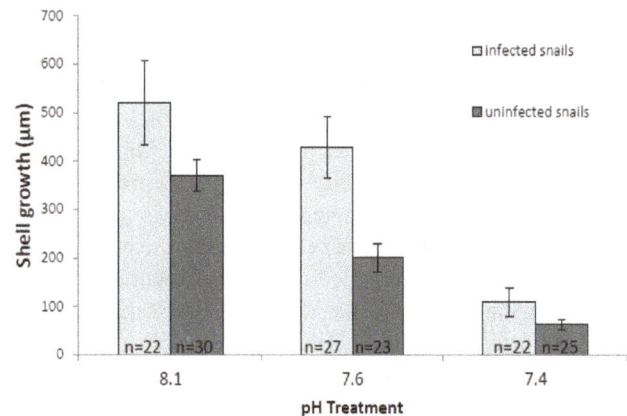

Figure 3. Average shell growth ($\pm$SE, sample size as indicated) of infected and uninfected snails in three pH treatment: 7.4, 7.6, 8.1.

The design of OA simulation systems is under constant development and review (e.g. Findlay et al., 2008; McGraw et al., 2010; Wilcox-Freeburg et al., 2013). The system described here improves the tolerance and repeatability of the potentiometric measurement and regulation of pH in an OA simulation system by (a) using two synthetic seawater buffers to calibrate glass electrodes and report pH on the total hydrogen ion scale (pH_T) and (b) measuring two additional, non-pH, carbonate parameters to independently validate pH and monitor changes to seawater chemistry caused by the culture of calcifying organisms. This article also includes an evaluation of offsets in calculated carbonate parameters caused by potential offsets and calibration errors in our measurement of temperature, salinity, pH_T, and A_T (Table 3). We recommend that this type of assessment is carried out by all researchers working with OA simulation systems.

4.2 Calibration buffers

To date, the most commonly used buffers for the calibration of electrodes used in OA simulation systems are defined by the National Bureau of Standards (NBS), now known as the National Institute of Standards and Technology (NIST), and report pH on the NBS scale (pH_{NBS}). NBS buffers are inexpensive, commonly available in most labs, and have pH values which are typically pre-programmed into pH meters to facilitate ease of electrode calibration. In the 2013 special OA issue of the journal *Marine Biology*, 18 out of 32 (56 %) experimental articles used these buffers and reported pH on the NBS scale. However, NBS/NIST buffers have a low ionic strength compared to seawater (0.1 vs. 0.7 M, Waters, 2012; Hurd et al., 2009) and are not recommended for the measurement of seawater pH (Zeebe and Gladrow, 2001; Dickson, 1984; Millero, 1986).

When measuring pH with potentiometric apparatus, the use of calibration buffers with a different ionic strength from sampled media leads to an error based on a fundamental assumption of potentiometric theory, i.e. that the difference in electric potential between the electrode solution and buffer solution is the same as that between the electrode solution and sample solution (Covington, 1985). This error is referred to as liquid junction error and has been discussed in several articles describing the potentiometric measurement of pH (Dickson et al., 2007; Illingworth, 1981; Easley and Byrne, 2012). The pH scale is essentially a quantification of the difference in electric potential between an ion-selective electrode and a sample solution. If the difference in ionic strength between the calibration buffer and sample is great, the electrode will not accurately report the difference in electric potential or provide repeatable measurements (Zeebe and Gladrow, 2001; Wedborg et al., 2009). Liquid junction error has been reported to cause uncertainties of ± 0.01–0.14 units in the measurement of seawater pH when using electrodes calibrated with low ionic strength buffers (Dickson, 1993; Easley and Byrne, 2012). The use of NBS buffers not only compromises the repeatability of potentiometrically regulated OA simulation experiments, this error is also propagated through calculations of other important seawater characteristics commonly reported in the OA literature, e.g. the saturation states of aragonite (Ωa) and calcite (Ωc). If we apply an error of ± 0.065 pH units (the median of reported liquid junction error values) to Ωa and Ωc in the software program SWCO2, we generate errors of 19 and 15 % respectively (Table. 3). The saturation states of aragonite and calcite are particularly vulnerable to this degree of error, as the current range of these variables is 1.2–5.4 (Ωa) and 1.9–9.2 (Ωc) (Riebesell et al., 2010), and Ω values less than 1.0, commonly achieved in OA simulation systems, indicate that the dissolution of these $CaCO_3$ polymorphs is thermodynamically favoured (Andersson et al., 2007). This type of error could prevent the correct interpretation of data sets generated in OA experimental studies, as they may indicate a dissolution of calcified structures at saturation states greater than 1.0.

An additional consideration when reporting data generated by an OA simulation system is the choice of pH scale. Measurement of seawater pH can be reported on three scales: the free proton scale (pH_F), the total hydrogen ion scale (pH_T), and the seawater scale (pH_{SWS}). There has been considerable debate over which scale is the most appropriate for reporting seawater pH in OA experiments (e.g. Waters and Millero,

2013), although the total hydrogen ion scale (pH_T) is most commonly reported in published data. In the 2013 special OA issue of the journal *Marine Biology*, pH_T was reported in 14 out of 32 (44 %) of experimental articles, while pH_F and pH_{SWS} were not used at all. One reason for this trend is that pH_T is generated directly by pH meters calibrated with saltwater buffers without additional calculation or conversion, as with the free proton and seawater scales. With the increasing availability of these buffers and the importance of establishing comparability between data sets, it seems appropriate that pH_T should be adopted as the default scale in OA research.

4.3   DIC and A_T analysis

Throughout the 90-day trial of this system, seawater samples were periodically taken from each culture tank and used to measure A_T and DIC. The primary purpose of this analysis was to validate the performance of the described system, with respect to the regulation of pH, by using DIC and A_T data to independently calculate the pH of culture tank seawater using the SWCO2 software. As previously discussed, the calculated pH was in good agreement with the potentiometrically measured pH, and it is advisable that this additional validation process should be standard procedure after the initial construction of a potentiometrically regulated OA simulation system. A secondary function of measuring A_T and DIC is the identification of alterations to seawater chemistry caused by the culture of calcifying organisms in acidified seawater. As discussed in Hurd et al. (2009), the addition of 100 % CO_2 to seawater is expected to cause an increase in DIC but not affect A_T. However, the culture of marine organisms in OA simulation systems can alter the concentration of carbon species in seawater through photosynthesis (decreased CO_2), respiration (increased CO_2), or dissolution of calcified structures (increased HCO_3^-). During an earlier trial of this system, when acidified treatments were 7.1 and 7.4 pH_T (12 °C), A_T greatly exceeded the expected value of ~ 2300 μmol kg^{-1} (2938.04 $\pm$ 1.29 μmol kg^{-1} (7.1 pH), 2564.16 $\pm$ 3.50 μmol kg^{-1} (7.4 pH)) and DIC was also unusually high compared to data generated by other systems that used CO_2 gas to reduce pH (3098.54 $\pm$ 5.14 μmol kg^{-1} (7.1 pH) and 2614.34 $\pm$ 2.61 μmol kg^{-1} (7.4 pH)). We assumed that the observed changes in seawater chemistry were caused by the release of HCO_3^- through the dissolution of calcified structures, as the snail shells had visibly dissolved, and therefore we increased the replacement rate of seawater from 20 L week^{-1} to 20 L 48 h^{-1}. As reported earlier in this paper, further analysis of A_T and DIC showed that these parameters had returned to expected levels, supporting the assumption that the dissolution of calcified structures had altered seawater chemistry. It is important to note that the replacement rate of seawater used in this simulation system may be specific to the size and number of snails in culture and the volume of the culture tanks. These observations illustrate the importance of measuring both A_T and DIC during the culture of calcifying organisms in acidified seawater, especially in closed or partially closed systems. If only DIC had been measured and A_T assumed to be constant, elevated DIC could have been solely attributed to the addition of CO_2 and could have resulted in the introduction of an unknown, additional abiotic factor to the experimental design.

5   Conclusions

The described system increases the accessibility of reliable OA simulation apparatus by using relatively inexpensive equipment that is readily available from aquarium suppliers. With careful calibration and the use of appropriate buffers, it is possible to generate high-quality and repeatable data. Incorporating DIC and A_T analysis in the validation of this system also provides a greater degree of reliability with regard to pH manipulation and a more complete understanding of the complex nature of seawater chemistry. Additional stressors such as temperature, salinity, and UV radiation could also be easily incorporated into experimental design due to the modular design of this system. Consequently, this system will facilitate the increase in research effort required to identify species, and species' interactions, vulnerable to novel stressors associated with OA, alone or in combination with other abiotic factors.

Acknowledgements. This research was supported by funding from the University of Otago Doctoral Scholarship Program (C. D. MacLeod), the University of Otago Research Centre for Oceanography (H. L. Doyle), the National Institute of Water and Atmospheric Research (K. I. Currie and H. L. Doyle), and the Departments of Zoology and Chemistry, University of Otago. The authors also thank Robert Poulin, Andrew Dickson, an anonymous reviewer, and members of the Evolutionary and Ecological Parasitology Research Group, University of Otago, for constructive comments on an earlier draft of this manuscript, and Lisa Bucke, University of Otago, for the preparation of a schematic included in this article.

Edited by: K. Fennel

References

Allan, B. J. M., Domenici, P., McCormick, M. I., Watson, S., and Munday, P. L.: Elevated CO_2 affects predator-prey interactions through altered performance, PLOS ONE, 8, 1–7, doi:10.1371/journal.pone.0058520, 2013.

Andersson, A. J., Bates, N. R., and Mackenzie, F. T.: Dissolution of carbonate sediments under rising $p$$CO_2$ and ocean acidification: observations from Devil's Hole, Bermuda, Aquat. Geochem., 13, 237–264, doi:10.1007/s10498-007-9018-8, 2007.

Bell, G. and Collins, S.: Adaptation, extinction and global change, Evol. Appl., 1, 3–16, doi:10.1111/j.1752-4571.2007.00011.x, 2008.

Bockmon, E. E., Frieder, C. A., Navarro, M. O., White-Kershek, L. A., and Dickson, A. G.: Technical Note: Controlled experimental aquarium system for multi-stressor investigation of carbonate chemistry, oxygen saturation, and temperature, Biogeosciences, 10, 5967–5975, doi:10.5194/bg-10-5967-2013, 2013.

Boyd, P. W.: Beyond ocean acidification, Nat. Geosci., 4, 273–274, doi:10.1038/ngeo1150, 2011.

Campbell, J. E. and Fourqurean, J. W.: Novel methodology for in situ carbon dioxide enrichment of benthic ecosystems, Limnol. Oceanogr-Meth., 9, 97–109, doi:10.4319/lom.2011.9.97, 2011.

Carter, B. R., Radich, J. A., Doyle, H. L., and Dickson, A. G.: An automated system for spectrophotometric seawater pH measurements, Limnol. Oceanogr-Meth., 11, 16–27, doi:10.4319/lom.2013.11.16, 2013.

Clayton, T. D. and Byrne, R. H.: Spectrophotometric seawater pH measurements: total hydrogen ion concentration scale calibration of *m*-cresol purple and at-sea results, Deep-Sea Res., 40, 2115–2129, doi:10.1016/0967-0637(93)90048-8, 1993.

Covington, A. K., Bates, R. B., and Durst, R. A.: Definition of pH scales, standard reference values, measurement of pH and related terminology, Pure Appl. Chem., 57, 531–542, doi:10.1351/pac198557030531, 1985.

Dickson, A. G.: pH scales and proton-transfer reactions in saline media such as sea water, Geochim. Cosmochim. Ac., 48, 2299–2308, 1984.

Dickson, A. G.: The measurement of sea water pH, Mar. Chem., 44, 131–142, doi:10.1016/0304-4203(93)90198-W, 1993.

Dickson, A. G. and Millero, F. J.: A comparison of the equilibrium constants for the dissociation of carbonic acid in seawater media, Deep-Sea Res., 34, 1733–1743, doi:10.1016/0198-0149(87)90021-5, 1987.

Dickson, A. G. and Riley, J. P.: The effect of analytical error on the evaluation of the components of the aquatic carbon-dioxide system, Mar. Chem., 6, 77–85, doi:10.1016/0304-4203(78)90008-7, 1978.

Dickson, A. G., Sabine, C. L., and Christian, J. R.: Guide to best practices for ocean CO_2 measurements, PICES Special Publication 3, 191, 2007.

Dixson, D. L., Munday, P. L., and Jones, G. P.: Ocean acidification disrupts the innate ability of fish to detect predator olfactory cues, Ecol. Lett., 13, 68–75, doi:10.1111/j.1461-0248.2009.01400.x, 2010.

Easley, R. A. and Byrne, R. H.: Spectrophotometric Calibration of pH Electrodes in Seawater Using Purified m-Cresol Purple, Environ. Sci. Technol., 46, 5018–5024, doi:10.1021/es300491s, 2012.

Feely, R. A., Doney, S. C., and Cooley, S. R.: Ocean acidification: present conditions and future changes, Oceanogr., 22, 36–47, doi:10.5670/oceanog.2009.95, 2009.

Findlay, H. S., Kendall, M. A., Spicer, J. I., Turley, C., and Widdicombe, S.: Novel microcosm system for investigating the effects of elevated carbon dioxide and temperature on intertidal organisms, Aquat. Biol., 3, 51–62, doi:10.3354/ab00061, 2008.

Gruber, N., Hauri, C., Lachkar, Z., Loher, D., Folicher, T. L., and Plattner, G.: Rapid Progression of Ocean Acidification in the California Current System, Science, 337, 220, doi:10.1126/science.1216773, 2012.

Hansen, T., Gardeler, B., and Matthiessen, B.: Technical Note: Precise quantitative measurements of total dissolved inorganic carbon from small amounts of seawater using a gas chromatographic system, Biogeosciences, 10, 6601–6608, doi:10.5194/bg-10-6601-2013, 2013.

Hofmann, L. C., Straub, S., and Bischof, K.: Competition between calcifying and noncalcifying temperate marine macroalgae under elevated CO_2 levels. Mar. Ecol-Prog. Ser., 464, 89–105, doi:10.3354/meps09892, 2012.

Hunter, K. A. SWCO2, http://neon.otago.ac.nz/research/mfc/people/keith_hunter/software/swco2, last access: 1 March 2014.

Hurd, C. L., Hepburn, C. D., Currie, K. I., Raven, J. A., and Hunter, K. A.: Testing the effects of ocean acidification on algal metabolism: considerations for experimental designs, J. Phycol., 45, 1236–1251, doi:10.1111/j.1529-8817.2009.00768.x, 2009.

Illingworth, J. A.: A common source of error in pH measurements, Biochem. J., 195, 259–262, 1981.

IPCC: Climate Change 2014: Impacts, Adaptation, and Vulnerability. Part A: Global and Sectoral Aspects. Contribution of Working Group II to the Fifth Assessment Report of the Intergovernmental Panel on Climate Change, edited by: Field, C. B., Barros, V. R., Dokken, D. J., Mach, K. J., Mastrandrea, M. D., Bilir, T. E., Chatterjee, M., Ebi, K. L., Estrada, Y. O., Genova, R. C., Girma, B., Kissel, E. S., Levy, A. N., MacCracken, S., Mastrandrea, P. R., and White, L. L., Cambridge University Press, Cambridge, United Kingdom and New York, NY, USA, 1132 pp., 2014.

Kroeker, K. J., Kordas, R. L., Crim, R., Hendriks, I. E., Ramajo, L., Sihngh, G. S., Duarte, C. M., and Gattuso, J.: Impacts of ocean acidification on marine organisms: quantifying sensitivities and interaction with warming, Glob. Change Biol., 19, 1884–1889, doi:10.1111/gcb.12179, 2013.

Liu, X., Patsavas, M. C., and Byrne, R. H.: Purification and Characterization of meta-Cresol Purple for Spectrophotometric Seawater pH Measurements, Environ. Sci. Technol., 45, 4862–4868, doi:10.1021/es200665d, 2011.

MacLeod, C. D. and Poulin, R.: Host–parasite interactions: a litmus test for ocean acidification?, Trends Parasitol., 28, 365–369, doi:10.1016/j.pt.2012.06.007, 2012.

McGraw, C. M., Cornwall, C. E., Reid, M. R., Currie, K. I., Hepburn, C. D., Boyd, P., Hurd, C. L., and Hunter, K. A.: An automated pH-controlled culture system for laboratory-based ocean acidification experiments, Limnol. Oceanogr-Meth., 8, 686–694, doi:10.1016/j.marchem.2011.04.002, 2010.

Mehrbach, C., Culberson, C. H., Hawley, J. E., and Pytkowicz, R. M.: Measurements of the apparent dissociation constants of carbonic acid in seawater at atmospheric pressure, Limnol. Oceanogr., 18, 897–907, 1973.

Millero, F. J.: The pH of estuarine waters, Limnol. Oceanogr., 31, 839–847, 1986.

Millero, F. J., Zhang, J.-Z., Fiol, S., Sotolongo, S., Roy, R. N., Lee, K., and Mane, S.: The use of buffers to measure the pH of seawater, Mar. Chem., 44, 143–152, doi:10.1016/0304-4203(93)90199-X, 1993.

Nemzer, B. V. and Dickson, A. G.: The stability and reproducibility of Tris buffers in synthetic seawater, Mar. Chem., 96, 237–242, doi:10.1016/j.marchem.2005.01.004, 2005.

Pörtner, H. O., Langenbuch, M., and Reipschlager, A.: Biological Impact of Elevated Ocean CO_2 Concentrations: Lessons from Animal Physiology and Earth History, J. Oceanogr., 60, 705–718, doi:10.1007/s10872-004-5763-0, 2004.

Riascos, J., Guzman, N., Laudien, J., Heilmayer, O, and Oliva, M.: Suitability of three stains to mark shells of *Concholepas concholepas* (Gastropoda) and *Mesodesma donacium*, J. Shellfish Res., 26, 43–49, doi:10.2983/0730-8000(2007)26[43:SOTSTM]2.0.CO;2, 2007.

Riebesell, U., Zondervan, I., Rost, B., Tortell, P. D. Zeebe, R. E., and Morel, F. M. M.: Reduced calcification of marine plankton in response to increased atmospheric CO_2, Nature, 407, 364–367, doi:10.1038/35030078, 2000.

Riebesell, U., Fabry, V. J., Hansson, L., and Gattuso, J. P.: Guide to best practices for ocean acidification research and data reporting, Luxembourg, Publications Office of the European Union, 260 pp., 2010.

Schulz, K. G., Barcelos e Ramos, J., Zeebe, R. E., and Riebesell, U.: CO_2 perturbation experiments: similarities and differences between dissolved inorganic carbon and total alkalinity manipulations, Biogeosciences, 6, 2145–2153, doi:10.5194/bg-6-2145-2009, 2009.

Waters, J. F.: Measurement of seawater pH: a theoretical and analytical investigation, PhD Thesis, University of Miami, Miami, FL, 199 pp., 2012.

Waters, J. F. and Millero, F. J.: The free proton concentration scale for seawater pH, Mar. Chem., 149, 8–22, doi:10.1016/j.marchem.2012.11.003, 2013.

Wedborg, M., Turner, D. R., Anderson, L. G., and Dyrssen, D.: Determination of pH, in: Methods of seawater analysis, edited by: Grasshoff, K., Kremling, K., and Ehrhardt, M., New York, Wiley-VCH, 109–125, 2009.

Weiner, S. and Dove, P. M.: An Overview of Biomineralization Processes and the Problem of the Vital Effect, Rev. Mineral. Geochem., 54, 1–29, doi:10.2113/0540001, 2003.

Widdecombe, S., Dupont, S., and Thorndyke, M.: Laboratory experiments and benthic mesocosm studies, in: Guide to best practices for ocean acidification research and data reporting, edited by: Riebesell, U., Fabry, V. J., Hansson, L., and Gattuso, J. P., Luxembourg, Publications Office of the European Union, 260 pp., 2010.

Wilcox-Freeburg, E., Rhyne, A., Robinson, W. E., Tlusty, M., Bourque, B., and Hannigan, R. E.: A comparison of two pH-stat carbon dioxide dosing systems for ocean acidification experiments, Limnol. Oceanogr-Meth., 11, 485–494, doi:10.4319/lom.2013.11.485, 2013.

Zeebe, R. E. and Wolf-Gladrow, D.: CO_2 in Seawater Equilibrium, Kinetics, Isotopes, Burlington: Elsevier Science, 361 pp., 2001.

6

Lena Delta hydrology and geochemistry: long-term hydrological data and recent field observations

I. Fedorova[1,2]**, A. Chetverova**[2,1]**, D. Bolshiyanov**[2,1]**, A. Makarov**[1,2]**, J. Boike**[3]**, B. Heim**[3]**, A. Morgenstern**[3]**,
P. P. Overduin**[3]**, C. Wegner**[4]**, V. Kashina**[2]**, A. Eulenburg**[3]**, E. Dobrotina**[1]**, and I. Sidorina**[2]

[1]Arctic and Antarctic Research Institute, St. Petersburg, Russia
[2]Hydrology Department, Institute of Earth Science, Saint Petersburg State University, St. Petersburg, Russia
[3]The Alfred Wegener Institute, Helmholtz Centre for Polar and Marine Research, Potsdam, Germany
[4]Helmholtz Centre for Ocean Research, Kiel, Germany

Correspondence to: I. Fedorova (umnichka@mail.ru)

Abstract. The Lena River forms one of the largest deltas in the Arctic. We compare two sets of data to reveal new insights into the hydrological, hydrochemical, and geochemical processes within the delta: (i) long-term hydrometric observations at the Khabarova station at the head of the delta from 1951 to 2005; (ii) field hydrological and geochemical observations carried out within the delta since 2002. Periods with differing relative discharge and intensity of fluvial processes were identified from the long-term record of water and sediment discharge. Ice events during spring melt (high water) reconfigured branch channels and probably influenced sediment transport within the delta. Based on summer field measurements during 2005–2012 of discharge and sediment fluxes along main delta channels, both are increased between the apex and the front of the delta. This increase is to a great extent connected with an additional influx of water from tributaries, as well as an increase of suspended and dissolved material released from the ice complex. Summer concentrations of major ion and biogenic substances along the delta branches are partly explained by water sources within the delta, such as thawing ice complex waters, small Lena River branches and estuarine areas.

1 Introduction

1.1 The Lena River delta study area

The Lena River, which flows into the Arctic Ocean, is one of the biggest rivers in Russia: 4400 km long from its source near Lake Baikal to its mouth. The mean annual Lena River discharge rate in 2007 was $16\,800\,\mathrm{m}^3\,\mathrm{s}^{-1}$, and the mean annual sediment flux was $680\,\mathrm{kg}\,\mathrm{s}^{-1}$ for suspended and $170\,\mathrm{kg}\,\mathrm{s}^{-1}$ for bottom sediments (Alekseevsky, 2007). Accompanying these fluxes are mean flux rates for major ions ($1460\,\mathrm{kg}\,\mathrm{s}^{-1}$), plankton ($12\,\mathrm{kg}\,\mathrm{s}^{-1}$), and heat discharge ($0.49 \times 10^{12}\,\mathrm{J}\,\mathrm{s}^{-1}$). The Lena can be divided into several areas, differing in the gradient of water surface elevation, fluvial forms, hydraulics, and transporting capacity. As it passes through its estuarine area, the main Lena flow is divided into numerous arms and transverse branches, creating the largest delta in the Russian Arctic. The Lena Delta area also comprises two large regions of late Pleistocene accumulation plains that are mostly untouched by modern active deltaic processes (Schwamborn et al., 2002). The total area of the Lena River delta, if Stolb Island is assumed to be its upstream limit, is over $25\,000\,\mathrm{km}^2$ and includes more than 1500 islands, about 60 000 lakes, and many branches of the Lena River (Antonov, 1967). If the delta's upstream limit is defined as including the Bulkurskaya Lena River branch to Tit-Ary Island, the delta area exceeds $32\,000\,\mathrm{km}^2$ (Walker, 1983). The Lena River delta is a complex of more than 800 branches with a total length of 6500 km. River branches flow in different directions, some diverging, others converging. The biggest branch is the Trofimovskaya branch; from this branch the Sardakhskaya branch diverges after Sardakh Island (Fig. 1). The second largest branch by volume is the branch that turns sharply to the east after Sardakh Island and flows into Buor Khaya Gulf. The next two largest

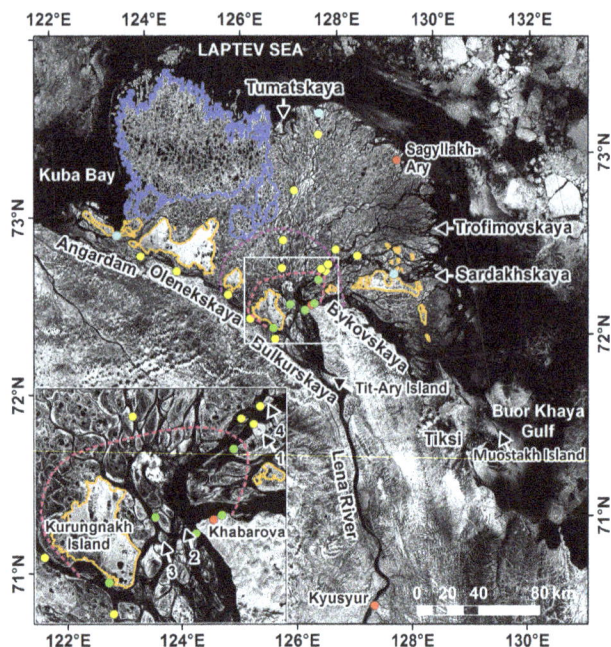

Figure 1. Place names in the Lena River delta and the locations of measurement profiles during expeditions from 2002 to 2012. Red circles: polar stations of Russian Federal Service for Hydrometeorology and Environmental Monitoring (Roshydromet) (Hydrometeoizdat, 2002–2012); green circles: standard hydrometeorological cross-sections of Roshydromet; yellow circles: additional cross-sections along the branches; light-blue circles: outlet (estuarine) cross-sections. Units with non-deltaic deposits within the Lena Delta: orange outline: ice complex deposits (third terrace); blue outline: late Pleistocene fluvial sands (second terrace); pink dashed line: outer contour of a central delta: purple dashed line: outer contour of a middle delta; 1: Sardakh Island; 2: Stolb Island; 3: Samoylov Island; 4: Gogolevsky Island.

branches are Olenekskaya branch, which flows west into the Kuba Gulf, and the Tumatskaya branch. Recently, a decrease in discharge has been observed in the Olenekskaya and Tumatskaya branches (Fedorova et al., 2009a). The quantity of eroded material carried by the river and the processes that occur where the river water and sea water come into contact have led to the formation of a broad, shallow shelf surrounding the Lena Delta below the Laptev Sea.

1.2 Review of existing literature

1.2.1 Hydrology of the Lena River delta

We investigate changes in water discharge and sediment fluxes channel cross-section morphology and sandbank extent that occur in the delta branches.

Observations of the principal Lena River delta hydrological features have been carried out since 1951, when the Khabarova station was established (Fig. 1). Hydrographic studies of the Arctic and Antarctic Research Institute (Ma-

rine transport, 1956), Moscow State University (Korotaev, 1984a), Tiksi hydrological party (Seleznev, 1986; Atlas, 1948), and others have been conducted in the delta. Data collected by the beginning of the 21st century described the long-term change of river water volume and the redistribution of water and sediment discharge in the delta branches. Publications since around 2000 have dealt either with assessments carried out on the basis of previously published hydrological data (Berezovskaya et al., 2005; Ivanov, 1963; Ivanov et al., 1983: Ivanov and Piskun, 1999; Rawlins et al., 2009; Shiklomanov and Lammers, 2009) or with new data from the Lena River catchment area discharging at the Kyusyur gauging section, upstream of the delta (Fig. 1; Ye et al., 2003, 2009). In contrast to the Lena River, where the magnitude of fluxes is dominated by the lateral river discharge, vertical fluxes (precipitation and evapotranspiration) dominate the summer water budget on the low-gradient polygonal tundra of the first terrace of the delta (Boike et al., 2013). Though redistribution of storage water due to lateral fluxes takes place within the microtopography of polygonal tundra (Helbig et al., 2013), the water balance here was controlled by the vertical fluxes. The long-term water budget modeled using precipitation–evapotranspiration, on the basis of ERA reanalysis data, was roughly balanced, tending towards positive values (precipitation > evapotranspiration; Boike et al., 2013).

1.2.2 Hydromorphology of the Lena River delta

A few researchers studied the long-term change in the supply of suspended materials and the characteristics of fluvial processes that are related to the cryolithic zone, but they have not investigated features of related hydrological processes within the delta itself. Syvitski (2003) modeled an increase of Lena River sediments due to water discharge increase and found that an increase in temperature in a river basin increases runoff more than does increased precipitation in the catchment area. Although the model was not validated using independent data and calculations for the delta itself were not carried out, the study concluded that erosion and runoff of sediments is intensified in places where the ice cover of the catchment area is degraded.

Lena River delta coastal zone erosion and accumulation in the delta front and inner shelf of the Laptev Sea have been studied (Are and Reimnitz, 2000; Grigoriev, 1993; Korotaev, 1984a, b, c; Korotaev and Chistyakov, 2002; Stein and Fahl, 2004; Wegner et al., 2013). Rachold et al. (2000) assumes that most sediment entering the Lena River delta is transported through the delta to the sea, an assumption contradicted by the findings of Charkin et al. (2011). However, Rachold et al. (2000) and Are and Reimnitz (2000) showed that most Laptev Sea sediments are composed of material from thawing coastal ice complex deposits, which are near-surface syngenetic permafrost deposits with high ice content (Schirrmeister et al., 2013). Through thermo-erosion, they contribute sediment volume almost 2.5 times as great as the fluvial sediment fluxes (Rachold et al., 2000). Charkin et al. (2011) showed that whereas the old particulate organic carbon (POC) in the Laptev Sea shelf waters originates from the ice complex–coastal systems, the younger to modern POC and lignin tracers originate from the fluvial discharges and are widely distributed on the inner and the middle Laptev Sea shelf. The main part of sediment supply by the Lena to the Siberian shelf is transported in the bottom nepheloid layer in submarine channels (Wegner et al., 2013).

Semiletov et al. (2011), Charkin et al. (2011), Heim et al. (2014), and Gordeev (2006) have analyzed the geochemical composition of material transported by the Lena River. However, sediment fluxes and composition and their distribution among the branches of the Lena Delta are not analyzed. We show that the spatial distribution of water discharge, sediment load, geochemistry, and river-bed morphology changes within the delta. Changes to the hydromorphology of delta channels provide a possible explanation for the observed changes in discharge and suspended load along the delta arms.

1.2.3 Hydrochemistry and geochemistry of the Lena River delta

It is difficult to access the Arctic zone throughout much of the year; therefore, data describing Arctic river hydrochemistry and geochemistry are poorly reported in the literature. The first expeditions to collect Arctic river hydrochemical data describing the chemical composition of Arctic river waters were conducted by the Omsk and Yakutsk territorial department offices of the Federal Service for Hydrometeorology and Environmental Monitoring of Russia (Roshydromet; e.g., Hydrological yearbook, 1974). Lena River hydrochemistry at the Kyusyur gauging section (Fig. 1) and seasonal hydrochemistry in the main channel have been studied (Alekseevskiy, 2007; Gordeev et al., 1999; Hoelemann et al., 2005; Izrael et al., 2004, 2012; Schpakova, 1999; Zubakina, 1979). Studies of the geochemistry of suspended matter are presented by Gordeev (2009), Hoelemann et al. (2005), and Savenko (2006). We present the hydrochemistry and geochemical composition of suspended material of the delta branches during the summer (July, August) 2005 and 2010–2012.

2 Materials and methods

2.1 Long-term hydrological data

Five standard hydrometric cross-sections are located within the Lena River delta, one on the main channel (4.7 km upriver from Khabarova Station) and the others on the Bykovskaya, Trofimovskaya, Tumatskaya, and Olenekskaya main delta branches (Fig. 1). Observations began in 1951 at the Bykovskaya and Trofimovskaya cross-sections and in

1977 at the other three. Long-term observations on water discharge and sediment loads were used (Hydrological yearbooks, 1951–2007). At the Khabarova water gauge located on the Bykovskaya branch, water levels (H, m) were measured visually using a depth gauge installed in the branch and according to standard Roshydromet methods; H values on hydrological cross-sections of other delta branches are calculated from rating curves. Daily water discharges (Q, $m^3 s^{-1}$) are deselected from water discharge curves according to Instruction for Hydrometeorological stations and posts (1958) and Guidance document (1989). The cross-section at Kyusyur, which began operating in 1936, is used as the last hydrological cross-section for assessing Lena River runoff before water is diverted into the delta branches near Tit-Ary Island (Fig. 1). Measurements of water and sediment discharge at Kyusyur were carried out until 2007. From 1951 to 2005 depth and water and sediment discharge measurements were also conducted at the Khabarova cross-section, after which only water levels were recorded.

Long-term data are presented on the basis of monthly mean discharge for the period of record for each of the available stations, permitting visualization of intra-annual and interannual variability. These fluctuations can be revealed on the basis of difference-integral flow curves analysis. The method of plotting the difference-integral curve for assessing the fluctuations was proposed by Glushkov (1934) and has found wide use in hydrology. To determine time periods with differing discharges and to compare average annual values of Lena River water discharge with average long-term runoff, difference-integral curves (residual mass curve, integral storage curve) were plotted (Reshet'ko and Shvarzeva, 2010; Rozhdestvenskiy and Chebotarev, 1974). Andreyanov (1960) was the first to conduct a comparative analysis of data based on standardized difference-integral curves of discharge rates.

$$\sum_{i=1}^{t} \frac{(K_i - 1)}{C_v} = F(t), \quad (1)$$

where K_i is modular ratio, $K_i = \frac{Q_i}{Q_0}$, Q_i is water discharge for i observation, Q_0 is average water discharge value for the observation period, C_v is the coefficient of variation, $F(t)$ is the curve of flow accumulation, t is a period of time. If the difference $\frac{(K_i-1)}{C_v}$ is equal to zero for some period then the average value of flow in this period coincides with average water discharge during the whole observation period. The sum of positive values of difference $\frac{(K_i-1)}{C_v}$ corresponds to heightened water flow, and the sum of negative values conforms to low water flow. The list of parameters are in Table 7.

However, at the centennial scale, the difference-integral curve leads to inaccurate higher and lower phases of intracentury intervals or does not reproduce them at all. Therefore, an analysis of average monthly sediment discharge and of the total runoff was conducted over the long-term period (as described above for each station) to identify shorter intervals with differing water discharge and therefore erosive power.

The discharge of water or sediment load through a channel cross-section is plotted against time starting at some initial time (Shiklomanov, 1979). The average water content of observation periods was determined by

$$K_{av} = 1 + \frac{F_l - F_i}{n}, \tag{2}$$

where n is number of years in the interval, F_l and F_i are last and initial ordinates of the difference-integral curve. Curves were constructed for the head of delta and for the main channel for various periods.

2.2 Field research

In order to analyze the current hydrological regime and the characteristics of water and sediment flux distributions in the delta branches, annual summer expeditions to the Lena Delta were undertaken from 2002 to 2012. Water and sediment discharges were measured and suspended and bottom sediments were sampled for geochemical and grain-size composition. Hydrological measurements were carried out every year at the standard hydrometric cross-sections; in some years other sections were added. Figure 1 illustrates all measured cross-sections. All data describing discharge of water and suspended sediments – dates of measurements, coordinates of each cross-section, and channel parameters – are presented in the PANGAEA database (Fedorova et al., 2013).

Hydrological measurements were made along several branch lengths over 2–3-day periods when no sizable water-level fluctuations occurred. Measurements along the Olenekskaya branch were realized in 2005 and 2012, and more briefly (with fewer cross-sections) in 2008, 2010, and 2011. While measurements were made along the Olenekskaya branch in 2005, for example, water level at the Khabarova water gauge varied by only 20 cm during 2–3 days. Measurements along the Tumatskaya branch, from Samoylovsky Island to the mouth, were taken in 2006. Detailed Sardakhskaya branch measurements were carried out in 2002 and in 2005. Discharges recorded by the Bykovskaya branch water gauge at Khabarova showed differences of $\leq 3\%$, allowing values to be compared with no need to introduce additional adjustments, with the exception of diurnal measurements at estuarine stations. Water discharges at the standard hydrometric cross-sections were calculated to the water level at the Khabarova water gauge, allowing those data to be used for long-term comparisons (Instruction for Hydrometeorological stations and posts, 1978).

Hydrometric observations included water depth, current velocity, and total suspended solids (TSS) content in water. Depth was measured twice using Garmin GPSmap 178C and GPSmap 421s echo sounders on board a motor boat or a river transport vessel. Some positions were determined using a Garmin GPSMap76CSX navigator. Vertical profiles of

at least three measurements of current velocity were measured at characteristic points of bottom relief on each cross-section. Current velocity measurements on each vertical were carried out on standard horizons, i.e., surface, 0.2, 0.6, and 0.8 h, and bottom (detailed five-point method, points given as fraction of total water depth, h). Truncated velocity measurements were frequently made: (a) 0.6 h (single-point); (b) 0.2 and 0.8 h (standard two-point), and (c) 0.2, 0.6 and 0.8 h (three-point). Current velocity measurements at the selected hydrometric cross-sections and surveying work at the cross-sections followed Instructions on Hydrometric Stations and Posts (1978).

Current velocities from 2002 to 2010 were measured with a GR-21M precalibrated velocity meter; in 2011 and 2012 measurements were carried out with a 2D-ACM multiparametric probe of Falmouth Scientific, Inc. (FSI). To ensure that data collected using two different devices were equivalent, measurements were conducted using both devices simultaneously. Maximum discrepancies were $\pm 0.01\,\mathrm{m\,s^{-1}}$ and measurements from the two devices were treated as equivalent. Water discharge was calculated according to

$$Q = 0.7v_1 f_0 + \left(\frac{v_1 + v_2}{2}\right) f_1 + \ldots + \left(\frac{v_{n-1} + v_n}{2}\right) f_{n-1} + 0.7v_n f_n, \tag{3}$$

where Q, $\mathrm{m^3\,s^{-1}}$ is water discharge; v_{1-n} is average current velocity ($\mathrm{m\,s^{-1}}$) on the first–n velocity verticals; f_0, $\mathrm{m^2}$ is water-section area between the bank and the first velocity vertical; f_1 is water-section area ($\mathrm{m^2}$) between the first and second velocity verticals, etc.; f_n is water-section area between the last vertical n and the bank. Velocity V_m averaged over the first–n velocity verticals was calculated according to

for the five-point method:

$$V_m = 0.1 \cdot (V_s + 3 \cdot V_{0.2h} + 3 \cdot V_{0.6h} + 2 \cdot V_{0.8h} + V_b), \tag{4}$$

for the three-point method:

$$V_m = 0.25 \cdot (V_{0.2h} + 2 \cdot V_{0.6h} + V_{0.8h}), \tag{5}$$

for the standard two-point method:

$$V_m = 0.5 \cdot (V_{0.2h} + V_{0.8h}). \tag{6}$$

When measuring velocity at one point, the mean velocity (V_m) was taken to be equal to the velocity at the 0.6 h horizon. Areas between velocity verticals were calculated according to

$$f_0 = 2/3h_1 b_0, \tag{7}$$

$$f_1 = \left(\frac{h_1 + h_2}{2}\right) b_1 + \left(\frac{h_2 + h_3}{2}\right) b_2 + \ldots$$
$$+ \left(\frac{h_{n-1} + h_n}{2}\right) b_n, \tag{8}$$

$$f_n = 2/3h_n b_n, \tag{9}$$

where h_{1-n} is the water depth of the measured verticals; $b_1, b_2, \ldots, b_{n-1}$ are the distances between the measured verticals; b_0, b_n are the distances between the outer measured verticals and the bank. Depth measurements were adjusted for vessel draft where necessary and averaged where duplicate values were available (the usual case).

Calculations were carried out using 102 water discharge measurements from all cross-sections. Water discharge measurements by a GR-21M velocity meter have an expected error of 3–5 % (Zheleznov and Danilevich, 1966). The velocity meter measurement systematic error is $\sigma_{sys} = 0.02 \, \text{m s}^{-1}$ and the random experimental error $\sigma_{ran} = 1.23 \, \text{m s}^{-1}$. The summarized field observations error S_Q is also $1.23 \, \text{m s}^{-1}$ following

$$S_Q = \sqrt{\sigma_{sys}^2 + \sigma_{ran}^2}. \qquad (10)$$

The critical measured velocity u_{cr} can be calculated by (Zheleznov and Danilevich, 1966)

$$u_{cr} = 7.1 \frac{u_0}{\sqrt{\beta}}, \qquad (11)$$

$$\beta = 6.9u_0 - 0.06 + \sqrt{(2.3u_0 - 0.055)^2 + 0.00058}, \qquad (12)$$

where u_0 is an initial velocity of the GR-21M velocity meter and is $0.01 \, \text{m s}^{-1}$. For our measurements u_{cr} is $0.32 \, \text{m s}^{-1}$ less than the summarized field observations error S_Q and can be accepted as satisfactory; measured water discharged can be used for analyses.

Water discharge calculated from field measurements differs from long-term discharge records, which are calculated using the discharge curve $Q = f(H)$, by up to 30–40 % (Fedorova et al., 2009a). This is due to the fact that the required adjustments of correlation coefficients between water levels and water discharge volumes are not carried out at hydrometric stations. In recent years the water gauge altitude elevations also appear to be in doubt. Starting in 2007, water-level and runoff data have been checked for such errors at the Arctic and Antarctic Research Institute (AARI, St Petersburg, Russia) in order to prepare them for publication in Hydrological Yearbooks.

To calculate sediment fluxes, SPM samples were selected from the same horizons where current velocities were measured. Vertical profiles for suspended matter determination were sometimes reduced to one or two-points as detailed above because it took a long time to collect the water in a vacuum bathometer. For TSS measurements, samples were filtered through ashless filters of 11 cm diameter and 5–8 µm pore size using a GR-60 vacuum pump. For geochemical analyses polycarbonate filters 0.45 mm diameter, 0.7 µm pore size (PC; Sartorius AG) were used for major and trace element content. Filters were dried at 60 °C for paper and PC filters and weighted before filtration.

Suspended sediment supply, R, was calculated using

$$R = \sum_{i=1}^{n} s_i q_i, \qquad (13)$$

where q_i is water discharge ($\text{m}^3 \text{s}^{-1}$) between verticals and s_i is mean value of TSS (mg L^{-1}) between verticals.

Bottom sediments were collected using either a UWITEC gravity corer with a 60 cm long, 6 cm diameter PVC liner or a Hydrobios Van-Veen grab sampler and stored in plastic bags, which were transported, frozen, to the laboratory.

Water samples were taken at the same points as suspended particulate matter (SPM) samples. Water samples for main and trace elements were collected in 60 mL plastic bottles and samples were kept cool. Water samples for nutrients were collected into plastic 40 mL plastic bottles and frozen. All samples were transported to St Petersburg for processing in the Russian–German Otto-Schmidt Laboratory for Polar and Marine Research (OSL)of the AARI laboratory.

2.3 Methods of laboratory sample processing

Suspended and bottom sediment samples collected in the field were analyzed in OSL and at the Alfred Wegener Institute (AWI, Potsdam, Germany). In keeping with the Russian literature, we designate species as major dissolved ions (Ca^{2+}, K^+, Mg^{2+}, Na^+, Cl^-, SO_4^{2-}, HCO_3^-), aqueous trace elements (Al, Fe, Si, Li, Ba, Sr, Ni, Pb), and nutrients (silicate, phosphate, nitrite, and nitrate) (Alekin, 1970). The bulk dissolved species parameter – salinity ("mineralization" in the Russian literature) – is determined by summing of major ions' concentrations.

Geochemical analysis of water and sediment samples (determination of major and trace element concentrations) was carried out via atomic emission spectrometry using an inductively coupled plasma optical emission spectrometer (ICP-OES; CIROS VISION). Solid samples of bottom sediments as well as SPM samples collected on PC filters were dissolved prior to analysis in Teflon weighing bottles and heated in a mixture of acids: nitric (HNO_3) – 3 mL, hydrofluoric (HF) – 4 mL, and perchloric ($HClO_4$) – 3 mL. A sodium hydroxide (NaOH) solution was used to neutralize the solution, then the rest of the prepared solution was diluted with deionized water to 25 mg. The final solution is measured on the ICP-OES. The methods of sample preparation and laboratory analyses are described in detail by Wetterich et al. (2009).

2.4 Hydromorphological analysis

Long-term studies of changes in the morphometric parameters of lakes and delta branches require the use of cartographic methods to display the spatial, temporal, and quantitative relationships between geomorphological, hydrological, and river-bed processes. For this purpose we employ change detection based on aerial and satellite images from different years (Snischenko, 1988; Usachyov, 1985). This method makes it possible to assess the rate of macro-form changes (Kondratyev et al., 1982). Changes in river-bed morphology were analyzed across the Trofimovskaya branch at Sardakh Island (Fig. 1). The obtained spatial change

detections were compared with field measurements of Trofimovskaya branch depths on the Sardakh Island cross-section
(Bolshiyanov et al., 2003; Korotaev, 1984a; Atlas, 1948).

Twenty-six aerial images of the studied Lena River delta
area from 1951 were used as baseline data and several 1 :
200 000 topographic maps were also included. Three Landsat satellite images from 26 July 1973, 5 August 2000,
and 26 June 2009, with a resolution of 60 m in 1973 and
15 m (for the panchromatic band) in 2000 and 2009 (http:
//glovis.usgs.gov) were also used to investigate hydromorphological changes. The aerial images were georeferenced
and mosaiced in Photomod Lite 5 software. The program
is intended for photogrammetric processing of the remote
sensing data. The Landsat satellite image data were georeferenced and matched to one another using MapInfo Profes-
sional 9.0.2 software. Changes in vector layers of river-bank
line contours between years revealed bank cave-ins and areas
of scouring or sedimentation. Average maximal rates of shifting were calculated (in meters per year) by dividing the obtained distance by which the bank had shifted (in meters) by
the time interval between images (in years). The spatial res-
olutions of the aerial images and the Landsat images differ.
Pixel resolution of the Landsat 1973 image is 60 m, and of the
2000 and 2009 Landsat images is 15 m (panchromatic band)
(Usachyov, 1985; Riordan et al., 2006). The lower boundary
of areal changes that can be detected in the case of the Landsat MSS baseline data (1973) is 0.014 km^2 (± 60 m $\times \pm 60$ m
mixed pixel error). For changes between 2000 and 2009 the
lower boundary of change detection accuracy is 0.0009 km^2
(± 15 m $\times \pm 15$ m).

Images were made between 26 July and 7 August, during
the descending phase of the water regime. The water level
changed from 250 to 270 cm relative to the height mark of the
nearest water gauge at Sagyllakh-Ary. The area of braided
bars was digitized and measured, and calculated in MapInfo
software.

The volume of deposited or eroded sediments was calculated by representing those sediments as a regular geomet-
ric figure, in the case of this study as a truncated pyramid.
The calculation involves determining the volumes of different truncated pyramids. The area that existed during the most
recent year of a period of interest, for example 1973, was
taken as the upper plane of the pyramid; the area that existed
during the first year of the period of interest, for example
1951, was taken as the lower plane. The selection of periods is limited by image availability. Volumes were calculated
from digitized areas via

$$W = \frac{1}{3} \Delta H \left(f_0 + f_1 + \sqrt{f_0 f_1} \right), \tag{14}$$

where f_0 and f_1 are areas of sandbanks that existed on the
dates when the images were captured, bounded by water surface; ΔH is the difference between water levels in the years
under investigation.

3 Results

3.1 Long-term discharge changes

3.1.1 Data from the hydrometeorological network: 1951–2007

Analysis of long-term Lena River hydrological data from
Kyusyur showed that, from the middle of the last century until the end of the record, average annual water discharge and
suspended sediment flux show a positive linear trend (Fig. 2),
yet the average annual water discharge remains below the
long-term average value (Fig. 3). This is typical both for the
outlet cross-section of the Lena River at Kyusyur and for the
cross-section 4.7 km upriver at Khabarova, on the main principal delta area channel. Figure 3 shows a decrease in water
discharge before the beginning of the 1970s and then a slight
increase. In 1983 there was a sharp drop in water discharge
which continued until the end of the 1980s, when the delta
area water discharge decrease fell to its lowest recorded level
(Fig. 2). From the late 1980s until today water discharge has
continued to increase.

A long period of observing the intra-annual water discharge distribution shows that the largest increase of water
discharge is observed during high water in May–June. Suspended sediments load is lower during high water (June) and
higher during winter low water (February). More than 50 %
of the suspended sediment discharge from the Olenekskaya
and Tumatskaya branches occurs in June (Fig. 4).

The rate of increase of cumulative suspended sediment
discharge from the main delta branches shows variability
over time (Fig. 5). Several points are evident at which the
rate of increase changes, indicating hydromorphological processes of erosion and accumulation in the delta. The timing
of the critical points is different for each branch. One can
clearly see a critical point on the Olenekskaya branch during
high water in 1983–1984. In August (middle of summer low
water) this critical point on the Olenekskaya, Trofimovskaya,
and, to a greater extent, Tumatsksya branches is typical for
1985–1986.

One can also observe a difference in angles of positive
trend slopes during high water and low water. It is illustrated
on Fig. 5: the same augmentation of the suspended supply cumulative curve carried out for different periods. Since about
1987, the June water content and sediment runoff have increased slightly in comparison with previous years. An even
greater increase has been observed since the end of the 1990s
for all branches. At the same time there has been a slight decrease of water volume during the low-water period.

3.1.2 Field hydrological observations: 2002–2012

Discharges at the main branches measured at the standard
hydrometric cross-sections, and calculated to the one water

Table 1. Measured discharge Q ($m^3 s^{-1}$) for the main branches. All discharges have been calculated normalized to one water level, equal to 365 cm at the Bykovskaya branch water gauge at Khabarova.

The Lena Delta main branches	2002	2004	2005	2006	2007	2008	2010	2011	2012
Main Lena channel	18 854		29 897		26 171		23 776* (1 Aug)	31 998* (20 Aug)	25 380* (29 Aug)
Olenekskaya branch	2023	2021	1693	2335	1700	1778	1406* (2 Aug)	1180	1696
Bykovskaya branch	4007		5641		6140			4353	
Trofimovskaya branch	12 824		15 038		14 800		20 800		15 299
Tumatskaya branch	2023	1746	1462	1730	1690	1037	2800	1225	643

* Measured water discharges without normalization.

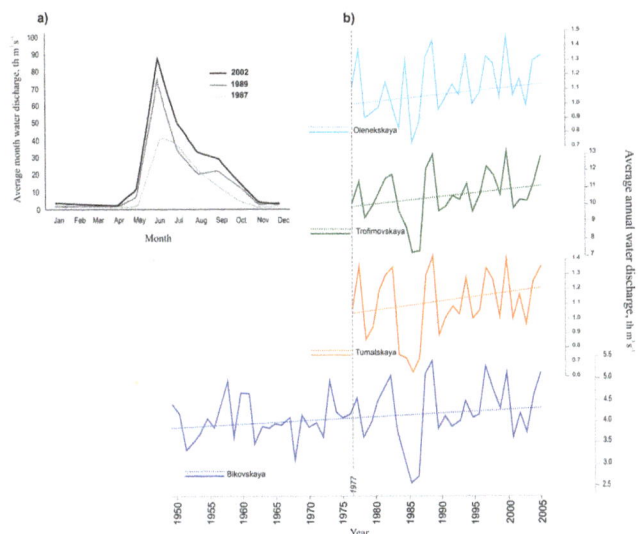

Figure 2. (a) Typical hydrograph for the Lena River delta (on the main channel section): average monthly water discharges for years with minimal (1987), maximal (1989), and average (2002) runoff; **(b)** average annual water discharge from 1950 to 2005 on branches of the Lena River delta according to data from Roshydromet (Hydrological yearbooks, 2002–2005).

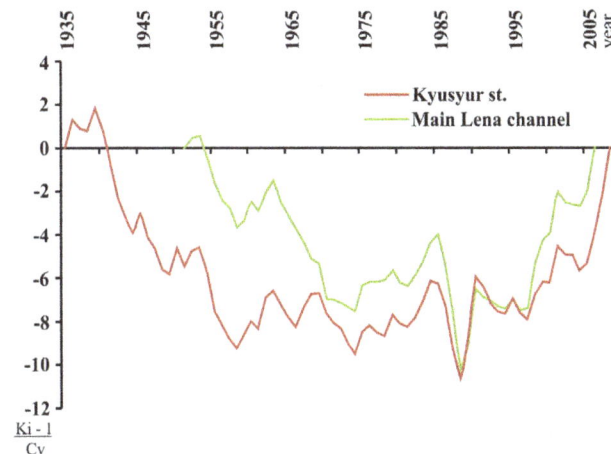

Figure 3. Difference-integral curve of average annual water discharge from 1935 to 2007 in the Lena River at Kyusyur, and from 1951 to 2005 in the Lena River main channel, at a cross-section 4.7 km upriver from Khabarova. "0" is the rate of water discharge.

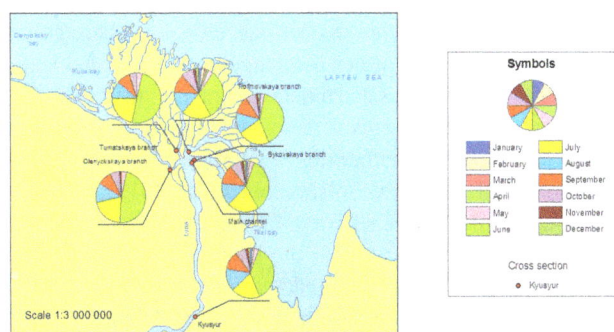

Figure 4. Intra-annual distribution of average annual suspended sediment supply in the Lena River delta.

level of 365 cm at the Bykovskaya branch water gauge at Khabarova, are specified in Table 1.

Our own field observation measurements between 2002 and 2012 showed that during the summer low-water period (August) discharge volumes from the main delta branches were in the ratio of 1 : 1 : 7 : 21 for the Olenekskaya : Tumatskaya : Bykovskaya : Trofimovskaya channels, respectively. The data also show that discharge from the main Lena River channel before it branches near Stolb Island at the time of summer low water sometimes exceeded 30 000 $m^3 s^{-1}$.

From the central delta to the sea there is, in general, a two-fold decrease in branch water discharge and suspended sediment supply (Fig. 6). But on some branches, the Sardakhskaya for example, water discharge can decrease

from 7942 $m^3 s^{-1}$ near Gogolevky Island to 11 $m^3 s^{-1}$ at the mouth. The discharge of sediments shows a similar change over the same distance, from 183 to 0.03 $kg s^{-1}$. Because there are particular areas of channel scour and sediment accumulation within the delta itself, the discharge decrease along

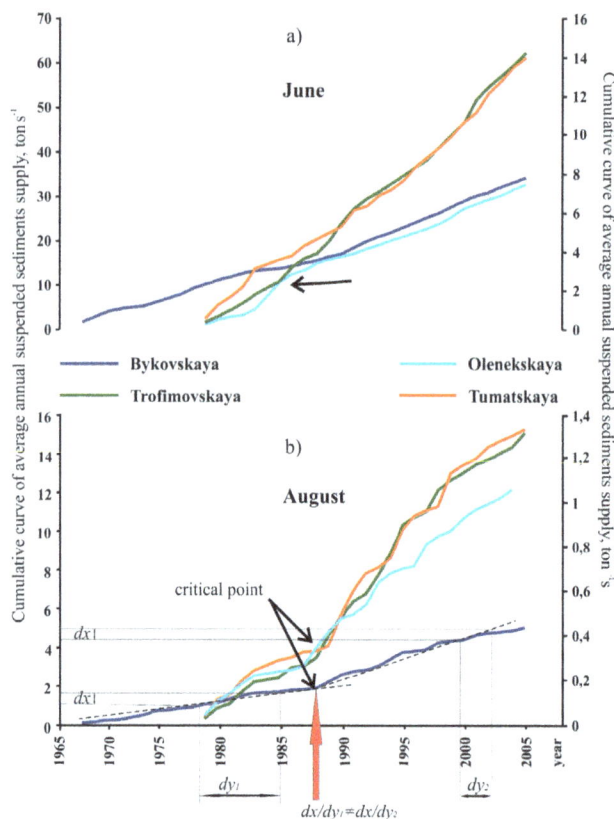

Figure 5. Cumulative average monthly suspended sediment supply from the main delta branches based on data from the Russian hydrometeorological network for June (**a**) and August (**b**). The right y-axis is for the Olenekskaya and Tumatskaya branches; the left y-axis is for the Bykovskaya and Trofimovskaya branches. The arrows indicate points at which the rate of increase of cumulative suspended sediment discharge shifts; dashed lines are trend lines.

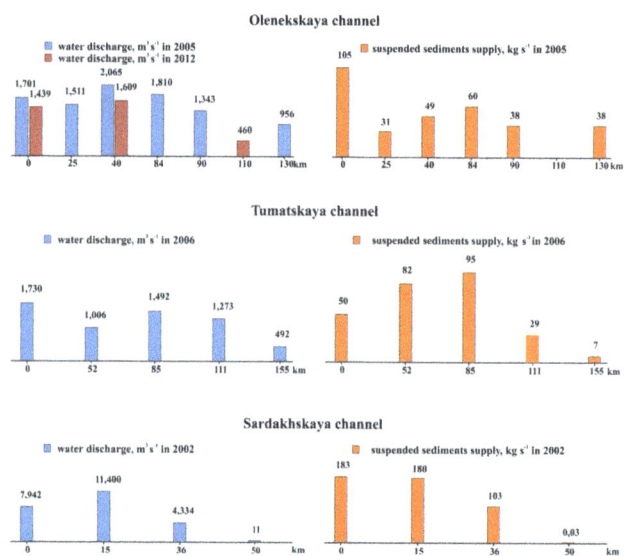

Figure 6. Discharge of water and sediments along the branches: on the Olenekskaya branch during 2005 and 2012, on the Tumatskaya branch during 2006, and on the Sardakhskaya branch during 2002. The distance of the measurement cross-section from the standard hydrometeorological cross-sections of Roshydromet is shown on the x-axis. The position of the standard hydrometeorological cross-sections of Roshydromet on the branch (cross-section near Gogolevsky Island for the Sardakhskaya branch) is taken as zero.

the length of the branches occurs unevenly, i.e., there could be a local increase of water discharge and suspended sediment supply in one area and a decrease in another.

This heterogeneity reflects the complex hydrographic layout of the delta and peculiarities of delta geological and geomorphological structure (Bolshiyanov et al., 2013). Thus, in 2005 on the middle Olenekskaya branch the measured water discharge was 2065 m^3 s^{-1}, at the beginning of this branch (after the influx of the Bulkurskaya branch) discharge was 1701 m^3 s^{-1}, and at the mouth it was only 956 m^3 s^{-1}. In 2012 discharges at the same cross-sections were 1609 and 1439 m^3 s^{-1}, respectively. The same situation can be seen on the Tumatskaya and Sardakhskaya branches (Fig. 6).

It is also typical for TSS to change along the length of a branch. In general, TSS decreases from 50–100 mg L^{-1} in the head of the delta to 3–5 mg L^{-1} on the sea edge. A major part of suspended sediments brought by the Lena River from the water catchment has already been deposited before the Lena reaches Lenskaya Truba (Lena's Tube), where TSS of more than 250 mg L^{-1} was observed during low water (Fe-

dorova et al., 2009b). TSS also varies within the delta: in the center and at the edge it can vary by a factor of 2–10, while it remains more or less the same between the two locations. Thus, during low water, values of TSS in the central delta (Fig. 1) vary from 20 to 45 mg L^{-1}, in the middle reach of the branch they remain around 20–25 mg L^{-1}, and at the edge they vary from 3 to 30 mg L^{-1}.

3.2 River-bed hydromorphology changes in the Trofimovskaya branch area and change of water discharge near Sardakh Island

Using data from previous studies (Antonov, 1967; Korotaev, 1984b, c) and from field observations carried out within the framework of a Russian–German Lena River delta expedition in the area of Sardakh Island and on the Trofimovskaya branch at the Sardakh-Khaya–Trofim-Kumaga cross-section made it possible to analyze the velocity and direction of river-bed morphology changes. Cross-section profiles of the branch channel were obtained for various years during the low-water period clearly demonstrate erosion in this profile, indicating an accumulation of alluvial deposits on the left bank of the Trofimovskaya branch; the main watercourse shifted to the right river bank, i.e., near Sardakh Island, which is a rocky island resistant to scouring (Fig. 7). Over the period from 1948 to 1981 the width of the Trofimovskaya branch channel decreased by more than half, while the depth increased from 10 to 22 m. Over the next 20 years there were no fundamental channel changes, but from 2001 to 2010 sediments accumulated in the cross-section and the channel width increased, i.e., lateral erosion increased.

These changes were also traceable in comparing the different image acquisitions. Figure 8 shows the state of the Trofimovskaya channel close to Sardakh Island in summer 1951 (aerial image) and in summer 2000 (Landsat satellite image). These images show where sediment accumulated and where erosion occurred. The area of the Trofim-Kumaga sands significantly increased from 1951 to 2000.

In general, the Trofim-Kumaga sands opposite Sardakh Island are constantly changing. The braided sandbar area increased from 1951 to 2000, but by 2009 began to decrease again. Table 2 presents the results of aerial and volume changes according to Eq. (14) between 1951 and 2000.

During the period from 1951 to 1973 the area of Trofim-Kumaga sands increased by $4.13\,\text{km}^2$, while at the same time the sand volume increased by $2.45\,\text{km}^3$. During the period from 1973 to 2000 the area increased by just $1.5\,\text{km}^2$, but the volume increased by $6.09\,\text{km}^3$. Roshydromet long-term data of water discharge and suspended sediment supply for Trofimovskaya branch confirm an increase and are presented in Fig. 9. Measurements carried out from 1977 to 2005 show a positive trend and mostly overlap the period during which changes in the Trofim-Kumaga sands' morphometric characteristics were observed.

3.3 Geochemical results: ion sinks and the composition of suspended sediments

Studies conducted in the Lena River estuarine area (Zubakina, 1979), i.e., on the main delta branches, Tiksi Bay, Olenek Bay, the Buor-Khaya Gulf, and the Laptev Sea coast, established that water salinity of the Lena River varies throughout the year. In the area of Stolb on the principal channel salinity ranges from 84 to $613\,\text{mg}\,\text{L}^{-1}$, while in the Bykovskaya branch it ranges from 55 to $561\,\text{mg}\,\text{L}^{-1}$. Salinity of the Lena River delta varies inversely with water discharge. Dissolved major ion concentration is practically unchanged throughout its depth, as well as downstream. In winter low water occurs near Stolb Island and chloride minerals are prevalent with higher salinity. When the high water recedes, Ca^{2+} and Mg^{2+} ions dominate with higher salinity (up to $540\,\text{mg}\,\text{L}^{-1}$). From then until the freeze-up period, low salinity with dominance of carbonate and calcium ions prevails. Estuarine water pH fluctuates within narrow limits, from 7.27 to 7.82, reaching its minimum value during spring high water.

Considerable attention is paid in the modern literature to estimating the amounts of dissolved mineral and organic substances carried by Arctic rivers to the Arctic Ocean. According to Alekseevsky (2007), the average long-term annual major ion delivery at the Lena River closing cross-section equals 48.4–59.8×10^6 tons per year, including 37–104×10^6 tons per year sulphate, 6.3–11.3×10^6 tons per year chloride, 16.5–26.0×10^6 tons per year hydrocarbonate, 7.6–

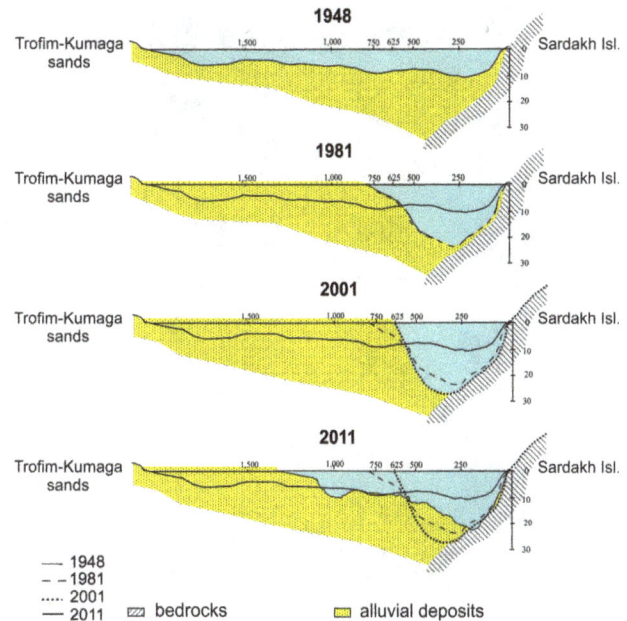

Figure 7. River-bed deformation changes of the Trofimovskaya branch at the Sardakh-Khaya–Trofim-Kumaga cross-section near Sardakh Island for the years 1948, 1981, 2001, and 2011 (33-, 20-, and 10-year intervals, respectively). The x-axis shows distance (m) from Sardakh Island; the y-axis shows depth (m) of the Trofimovskaya branch.

Table 2. The measured areas and volumes of Trofim-Kumaga sands on the Trofimovskaya branch close to Sardakh Island over various time periods.

Period	Changes of volume for each period, km^3	Changes of area for each period, km^2	Mean changes of volume, $\text{km}^3\,\text{year}^{-1}$
1951–1973	2.45	4.13	0.11
1973–2000	6.09	1.50	0.23
1951–2000	7.73	5.63	0.16

24.7×10^6 tons per year calcium, 2.4–5.8×10^6 tons per year magnesium, and 7.0–9.5×10^6 tons per year sodium.

Intra-annually, the maximum major ions' flux occurs in the spring, due to the larger water volume carried by Arctic rivers and the high concentrations during this period. Silicon, iron, and ammonium nitrogen have the highest concentrations (Yearbook, 1989–2012). Annual transport is about 44.6×10^3 tons per year ammonium nitrogen, 2.6×10^3 tons per year nitrite nitrogen, 32.5×10^3 tons per year nitrates, 3.7×10^3 tons per year phosphates, and 7.3×10^3 tons per year total phosphorus (Gordeev et al., 1999).

The intra-annual Lena River ion runoff distribution varies considerably: up to 47 % of ion runoff occurs during the high-water period and up to 34 % in the ice-covered period.

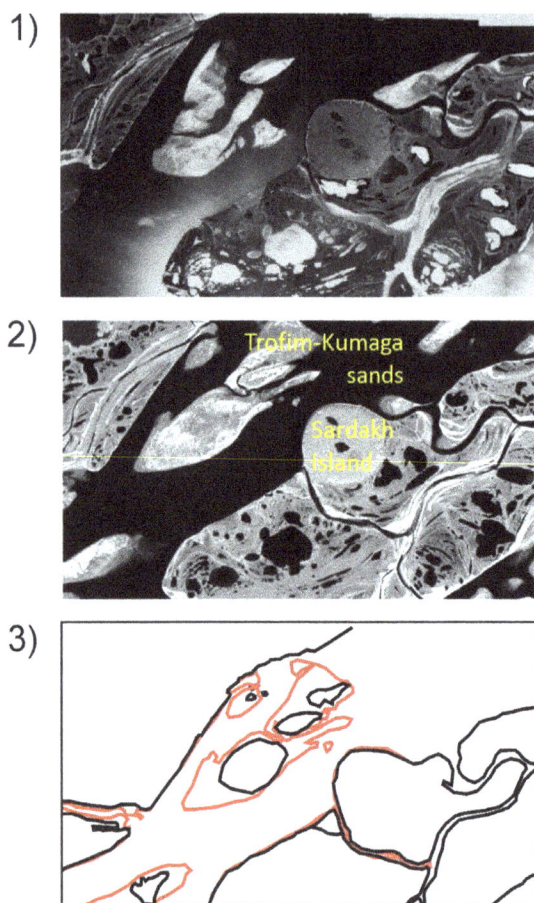

Table 3. The range of dissolved element content concentrations of main ions and trace elements in water of Lena River delta large branches from the summer period (July–August), 2010 to 2011.

Type of element	Element	Range of concentration, $mg\,L^{-1}$	Mean value, $mg\,L^{-1}$
Major ions	Ca^{2+}	15.2–18.9	16.8
	K^+	0.5–1.1	0.6
	Mg^{2+}	3.6–4.5	4.0
	Na^+	4.1–8.8	5.5
	Cl^-	4.7–13.5	7.1
	SO_4^{2-}	8.8–18.1	10.6
	HCO_3^-	12.0–50.8	27.8
Salinity		63.8–83.9	71.8
Trace elements	Al_{aq}	0.009–0.07	0.017
	Fe_{aq}	0.012–0.042	0.023
	Si_{aq}	1.6–2.1	1.8
	Li_{aq}	0.010	
	Ba_{aq}	0.007–0.016	0.013
	Sr_{aq}	0.124–0.148	0.13
	Ni_{aq}	0.020	
	Pb_{aq}	0.050	
Nutrients	Silicates SiO_2	1.4–2.4	1.8
	Phosphates PO_4	0.003–0.026	0.005
	Nitrites NO_2	0.003–0.011	0.006
	Nitrates NO_3	0.003–0.035	0.02

Figure 8. Changes of areal extent of the Trofim-Kumaga sandbanks in the Trofimovskaya branch close to Sardakh Island area: (1) on 10 July 1951 (aerial image), (2) on 5 August 2000 (Landsat TM satellite image), (3) digitized contours of islands, watercourses, and water basins (red contour line for 1951, black for 2000).

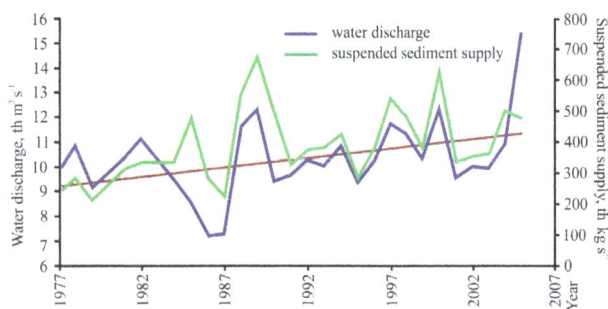

Figure 9. Average annual water discharge (blue line) with trend (red line) and suspended sediment supply (green line) on the Trofimovskaya branch over the period of instrumental observations according to data from Roshydromet (Hydrometeoizdat, 2002–2005).

The highest discharge of ammonium, nitrates, and iron occurs during the spring high-water period, when from 64 to 84 % of the annual nutrients' runoff occurs. Data from laboratory analyses of water sampled during summer campaigns and field measurements are presented in Tables 3 and 4, respectively.

The hydrochemistry of the Lena within the delta is similar to data published by Zubakina (1979) and Alekseevsky (2007). Water from the major branches (Table 3) is characterized by low salinity in the summer ($\leq 84\,mg\,L^{-1}$), low trace elements ($< 0.05\,mg\,L^{-1}$) and nutrients ($< 2.5\,mg\,L^{-1}$) contents, and high silicate concentration ($\leq 2.4\,mg\,L^{-1}$). The delta's small branches (Table 4) and streams had high salinity ($\leq 285\,mg\,L^{-1}$). Geochemical characteristics of suspended sediments, i.e., the major petrogenic elements and trace elements content, were determined for large branches of the Lena Delta. These values are in good compliance with published data (Gordeev, 2009; Hoelemann et al., 2005; Savenko, 2006) for the Lena River. Mean data and their range for the main delta channels are presented in Table 5.

4 Discussion

4.1 Hydrology of the Lena River delta and river-bed morphology

Long-term water discharge and sediment fluxes showed a positive trend. This has been observed before: Berezovskaya et al. (2004), Bolshiyanov et al. (2004), and Fedorova et

Table 4. Water salinity, turbidity, and temperature in the Lena River delta smaller branches and streams.

Stream/channel	Temperature, °C	Turbidity, gL^{-1}	Salinity, mgL^{-1}
Ysy-Khaya-Tyobyulege branch	10.6	–	80
Stream 1 from Kurungnakh Island	6	495	285
Stream 2 from Kurungnakh Island	6	102	227
Sistyakh-Aryi-Uesya branch	11.2	–	53
Krestyakhskaya branch	10.2*	0.03	56
Stream 3 from Arga-Bilir-Aryita Island	–	0.01	162

* Measurement conducted on 22 August 2012; other data gathered collected on 8 August 2012.

Table 5. Chemical elements found in suspended material from the Lena River delta in 2002–2012 in comparison with data from Savenko (2006), Hoelemann et al. (2005), and Gordeev (2009).

Component	Range of concentration	Mean value	Element concentration in SPM		
			Hoelemann et al. (2005)	Savenko (2006)	Gordeev (2009)
Al_2O_3, mgL^{-1}	11.9–15.9	14.2		13.91	
CaO, mgL^{-1}	0.8–1.5	1.2		5.43	
Fe_2O_3, mgL^{-1}	4.9–6.4	5.7		2.25	
K_2O, mgL^{-1}	2.2–3.0	2.6		1.57	
MgO, mgL^{-1}	1.5–2.1	1.8		2.15	
Na_2O, mgL^{-1}	1.4–1.9	1.7		2.82	
SiO_2, mgL^{-1}	66–88	70		71.87	
Li, μgL^{-1}	49–61	53		42	
Ba, μgL^{-1}	535–944	618		734	
Pb, μgL^{-1}	57–347	157	38	102	28
Sr, μgL^{-1}	147–221	182	–	195	194
Ni, μgL^{-1}	43–64	53	28	52	47
V, μgL^{-1}	97–127	113	–	84	97

al. (2009a) reported statistically significant positive trends (by Student's t test and F ratio) in water discharge of an average of $35\,m^3\,s^{-1}\,year^{-1}$ (0.22 % of the average long-term discharge for 1951–2005 period). Nevertheless, such an increase is slight, and until 2000 Lena River discharge (measured at Kyusyur) was lower than the rate of water discharge (Fig. 3). From 2000 on, discharge increase began to rise significantly.

However, the increased water discharge occurs mostly during high water (June). During summer low water (August) there is a slight water discharge decrease (Fig. 5). In our opinion, it is premature to draw any conclusions about winter low water and possibly more crucial discharge variations due to climate change. Model calculations (Fedorova et al., 2009b) show that hydrological systems require a long period of adaptation when parameters that control discharge formation change. Also, discharge measurements that have been carried out in the 21st century have been of varying quality, and sometimes do not meet the requirements of the hydrometeorological network; out-of-date devices and methods are often used. For example, winter measurements of

water depth and flow have not been made for a period of more than 10 years. Long interruptions, for example, in sediment discharge measurements in February are given in Table 6. The possibility of measurement inaccuracies at the Roshydromet stations has previously been noted by others (Berezovskaya et al., 2004). Previous measurements in the delta branches (Antonov, 1967; Bolshiyanov and Tretiakov, 2002; Gordeev, 2006; Ivanov et al., 1983; Ivanov and Piskun, 1999; had not shown long-term changing of water and sediment flow distribution between branches. Fedorova et al. (2009a) found the following distribution of the increase in flow, expressed as percentages of the observed discharge increase in the principal channel: 6.8 % in the Olenekskaya branch, 6.4 % in the Tumatskaya branch, 61.5 % in the Trofimovskaya branch, and 25.3 % in the Bykovskaya branch. At the Sardakh-Trofimovskaya branch point during the open water period, 20–26 % flows into the Trofimovskaya and 23–33 % into the Sardakhskaya branches. On this background of longer term increases in discharge, however, is superimposed spatial and temporal variability in flow, caused by ice. Ice jams have an impact on this distribution of discharge within

Table 6. Average monthly (for February) sediment discharges (kg L^{-1}) during all periods of long-term observations on the standard hydrometeorological cross-sections of Roshydromet on the branch.

Year of measurements	Kyusyur	Main channel	Bykovskaya	Olenekskaya	Tumatskaya	Trofimovskaya
1944	3.2					
1960	2.4					
1961	0.73					
1962	8.2					
1963	3.6					
1964	6.5					
1965	2.6					
1966	1.8					
1967	2.6					
1968	6.1					
1969	0.64	2.7	0.39			
1970	2.7	2.8	0.42			
1971	1.0					
1972	3.2					
1977		5.2	0.65			19
1978		5.0				19
1979		26	4.2	3.6	0.66	23
1980	9.6	19	3.4	0.67	0.20	12
average	3.66	11.1	2.34	2.13	0.43	18.25
max	9.6	26	5.0	3.6	0.66	23
min	0.64	2.7	0.39	0.67	0.2	12

the delta and its temporal variation (Izrael et al., 2012). They may, for example, cause a sharp increase of Bykovskaya branch water level, and can block the Olenekskaya and Tumatskaya branches entirely.

Ice events in the delta play a significant role in river-bed processes; for example, an ice jam can cause greater fluvial adjustments than a change of water runoff volume. During one flood caused by an ice jam 40 m of shoreline was washed away due to thermal erosion and banks being cut by ice (Are, 1983). In our opinion, catastrophic ice events were the primary cause of the dramatic increase of sediment runoff on the Olenekskaya branch in 1984 (Fig. 5), despite the fact that, from 1983 to 1984, no Olenekskaya branch jams were officially registered in the yearbooks. However, this cross-section is far from Khabarova, and visual observations are lacking because it is dangerous to access this area. In 1982 an ice jam was registered near Kyusyur on 1–6 June, and on 10–12 June a jam occurred near Tit-Ary Island, i.e., at the place where the delta begins to branch out. Here, the Bulkurskaya branch begins, which later enters the Olenekskaya branch further upriver from the hydrometric cross section. According to yearbook data, there was no runoff of water and sediments on the Olenekskaya branch in June (during high water) in 1983. This was apparently due to the branch channel being blocked by ice. Ice jams were also observed near Kyusyur and Khabarova on 3–9 and 10–12 June, respectively. There was a sharp increase of average suspended sed-

iments on the Olenekskaya branch, from 290 kg s^{-1} in 1982 up to 1400 kg s^{-1} in 1984.

The annual cutoff of river bank edges during high water (Figs. 10, 11; Supplement 1) produces an unmeasured quantity of suspended and bottom sediments which are carried into the delta and, as a consequence, ejected into the delta front. Costard et al. (2003) noted the important role of slope erosion during flood periods for the middle part of the Lena River due to thermal erosion. A change detection study for Kurungnakh Island in the central Lena Delta showed mean annual river bank erosion rates of 2.9 and 1.8 m year^{-1} for two different cliff sections over the period 1964-2006 (Günther, 2009). Such erosion can be a trigger for river-bed processes intensification and cause additional sediment runoff to streams (Morgenstern et al., 2011).

In the opinion of Charkin et al. (2011) and Heim et al. (2014), during the low water-level period in summer a relatively minor amount of suspended sediments is contributed to the sea from the Lena River branches delta. Heim et al. (2014) and Charkin et al. (2011) measured 3–5 mg L^{-1} in the estuarine parts of the delta and in the coastal waters of Buorkhaya Gulf in the surface water layer where velocities are reduced. This is confirmed by our field data. The main marine sediment transport functions with the bottom nepheloid layer (Wegner et al., 2013). However, the volume and, more importantly, the composition of sediments on the

Figure 10. Ice on the sands of Sistyakh-Ariyta Island after a flood (photo by I. Fedorova).

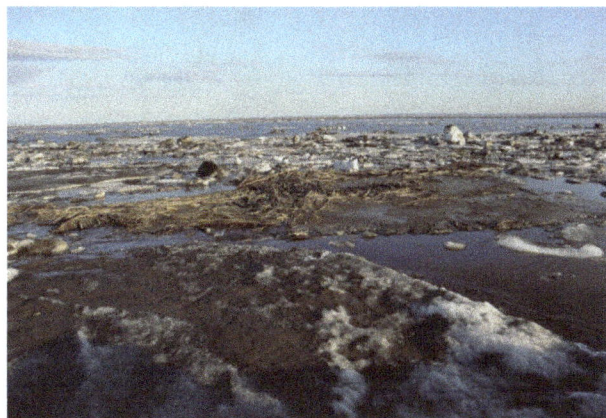

Figure 11. River ice during a flood on the Olenekskaya branch (photo by M. Grigoriev).

delta edge and the quantity of sediments ejected onto the inner shelf have not been identified.

Questions that remain to be studied include how dissolved and solid substances are provided by the erosion of ice complex and flood plain terrace material and its resedimentation (Bolshiyanov et al., 2013). An increase of water and sediment discharge occurs in the middle part of the Olenekskaya branch, where separate ice complex masses are exposed to the warming action of the water and active processes of bank thermoerosion and thermodenudation. The role of groundwater runoff from the thawed horizon in thermokarst areas is also mentioned in Woo et al. (2008). On Kurungnakh Island, water pools loaded with high amounts of sediments have been observed flowing down a thermoerosion valley and discharging into the Olenekskaya branch during summer 2008 (Supplement 2). When ice dams break up, water carrying particulate material flows down thermoerosion valleys and enters the delta branches. Increases in summer precipitation are often responsible for increased summer discharge (Kane et al., 2003). In a thermokarst landscape, such increases may also be due to catastrophic lake drainage following an ice jam.

Possible mechanisms underlying the observed increase of water and sediment discharge within the Lena Delta include neotectonic (isostatic) processes (Bolshiyanov et al., 2013) that can cause an increase of the water surface altitude gradient and, as a consequence, an increase of erosive power that exposes new sediment material to erosion. Such a hypothesis could be confirmed using geodesic benchmarks inside the delta, as well as by conducting additional branch water-level and sea-level measurements.

Any local decrease or increase of water and sediment discharge in the estuary zone is often constrained by the sea. However, our studies carried out in 2005 (Fedorova et al., 2007) and 2012 (Fedorova et al., 2013) in the Olenek-

skaya branch delta showed an absence of seawater influx 60 and 15 km deep into the branch; electroconductivity did not exceed $125\,\mu S\,cm^{-1}$. In addition, neither salt water nor a change of water current direction at the Angardam branch cross-section was observed. The influx of sea water could certainly have an impact by changing the inclination of the branch water surface, but confirming this hypothesis will require high-precision geodesic work.

The situation on the Bykovskaya branch is slightly different. In recent times, according to observation by the hydrometeorological network (at Muostakh and Bykovsky), the hydrological regime of this branch has been estuarine with a prevalent marine influence. Roshydromet is currently considering the possibility of reconfiguring the estuarine station with regard to the interface between river and sea, including additional measurements on Bykovsky Peninsula.

There are other possible explanations for the observed decrease in discharge within the delta. River water may infiltrate into the talik below the river bed of the Lena Delta. We have shown that water discharge from the estuarine areas of the measured branches decreases by orders of magnitude compared to the discharge of the middle delta. For example, flow in the Sardakhskaya branch decreased from more than $11\,000\,m^3\,s^{-1}$ near Gogolevsky Island (in the middle delta) to $11\,m^3\,s^{-1}$ at the branch outlet. Certainly runoff can decrease due to flow branching or because of a rise in sea level, but the existence of such a large difference in discharge requires additional study. A hydraulic connection between flow in the river and flow in the talik beneath it is possible and could include outflow to the talik in summer and inflow to the river in winter. Similar variations in water salinity might be explained by the same mechanism (Zubakina, 1979). Burdyikina (1951) provides another explanation for interdeltaic discharge decrease: infiltration of spring flood water from the Lena River through the Lena–Anabar depression to the Olenek and Anabar River basins.

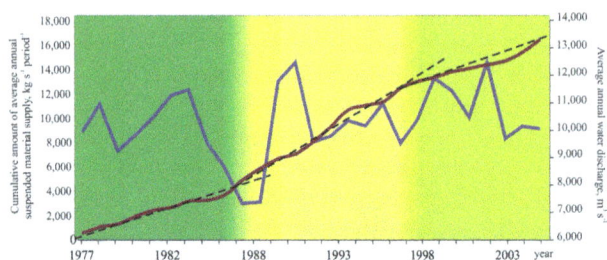

Figure 12. Long-term average annual discharge (blue line, right-hand axis) and the cumulative amount of average annual suspended sediments (red line, left-hand axis) over the period of instrumental observation. Dashed lines are trends of suspended sediments discharges for three periods with different annual water volumes (green tints).

4.2 Cyclicity of hydrological processes

Hydrological and river-bed processes in the Lena River delta are cyclic. Three large-scale periods characterized water flow fluctuations: low-water (1938–1957, $K_{av} = 0.93$); water-average (1958–1987, $K_{av} = 1.00$); high-water (1988–2006, $K_{av} = 1.06$). Within the large-scale phases of flow fluctuations there are periods of heightened and low water content which have shorter duration. For instance, water average period (1958–1987) includes three high-water, three low-water and four water-average phases of water content. The relationship between long-term average annual water discharge and increasing average annual suspended sediment supply over the period of instrumental observations is shown in Fig. 12.

Inflection points in the plot of average monthly sediment discharge can indicate critical points (Fig. 5). In spite of the three large-scale periods characterized by water flow fluctuations with different K_{av} values, water content of the river cannot be, in our opinion, the main criterion for highlighting certain periods because many delta processes may impact fluvial deformations and rearrangements of the delta shape. One can see from Fig. 12 that from 1977 to the middle of the 1980s a cycle existed that was characterized by low water volume and little fluvial deformation, as evidenced by sediment discharge. From the mid-1980s to the mid-1990s river water content increased and fluvial processes were active. Currently, with increased water content, the transport capacity of the Lena River delta has actually decreased slightly. Of course, a hydrological system does not immediately respond to changes in water volume or fluvial deformation; there is a certain lag time between change and effect.

Nevertheless, as has been mentioned above, at the beginning of the 1980s abrupt increases in Olenekskaya branch sediment runoff were observed. By the 1980s rapid increases had also occurred in the Trofimovskaya branch channel near Sardakh Island (Fig. 7). Trofim-Kumaga sands accumulated until 1973; in 1981 a process of active bottom erosion of

these sands began near Sardakh Island. Over the third interval (from the late 1990s until 2005) another decrease of fluvial process activity was observed, manifested by the gradual silting up of the Trofimovskaya branch channel, decreasing the channel depth and increasing the channel width.

4.3 Geochemistry of the delta

An assessment (Chetverova et al., 2011) of the amount of dissolved substances upstream of the Lena River over the period from 1960 to 1987 produced averages of annual dissolved substances at the outlet cross-section of the Lena River (Kyusyur). Results obtained by the authors are consistent with published assessments (Alekseevsky, 2007). Analysis of long-term dissolved substance runoff data has enabled conclusions to be drawn regarding the seasonal and long-term dynamics of Lena River runoff.

Analysis of the intra-annual variations of dissolved substance runoff showed that levels are highest when the Lena River water volume is high. From 1960 to 1987 the Lena River ion transport decreased almost 3-fold. A decreasing tendency is observed for all major ions, except for magnesium. The flux of calcium decreased by 54 %, sodium and potassium by 43 %, hydrocarbons by 44 %, sulphates by 7 %, and chlorides by 30 %. A decreased flux for nutrients was also observed, including a 2.2-fold decrease for nitrates and a 7 % decrease for phosphates, and more than a 2-fold decrease for silicon; in contrast, a 9-fold increase of iron was observed.

Geochemical processes in the delta are closely connected with the amount of river discharge and changes in that discharge due to division of the channel into smaller branches. This process is the basis of the postulated mechanism of a marginal filter that has been developed by Lisitzin (1988). However, a quantitative discussion of these linkages is lacking due to insufficient study of the delta watercourses. For river systems upstream from the delta, this question is being answered on the basis of linking runoff characteristics to stream order, as determined within the conceptual framework advanced by Horton (1945). The concept of conventional orders, proposed by Alekseevsky and Chalov (2009), quantifies stream order within the delta. As the main branch in the delta divides into smaller and smaller watercourses, all other characteristics of river discharge change, including the ability of the water to transport dissolved substances including ions, trace elements, and nutrients.

Dissolved and suspended chemical compounds and element concentrations vary spatially between the delta apex and the coastline. Major dissolved elements and most elements of suspended material are transported through the delta without significant concentration changes at the branch bifurcations, as shown by the small concentration ranges observed in suspended material from differently sized branches. Some dissolved nutrient and trace element concentrations changed at channel bifurcations. Concentrations of nitrites and phosphates are conservative (no changes of concentration due to bifurcation). Nitrate concentration increases towards the delta coastline due to sedimentation of silt particles, which adsorb nitrate, and the biochemical transformation of this nitrogen. Dissolved barium concentration exhibits the opposite behavior; it decreases closer to the coastline because silt particles and nutrient compounds are incorporated into trophic chains and biogenic processes. The opposite situation occurs for salinity. It increased upstream of the river–seawater mixing zone, which could be the result of more mineralized underground water flowing into the river or the influence of marine sources of dissolved solids.

Dissolved hydrochemical components also showed differences in spatial variability within the delta. For conservative components, bifurcation of delta channels does not cause large changes in concentration, whereas physicochemical and biochemical processes can change non-conservative component concentrations along channels (Nikanorov, 2001).

Calculations using field measurement data showed that the following components were conservative: (i) dissolved components – main ions (Ca^{2+}, K^+, Mg^{2+}, Na^+, SO_4^{2-}, HCO_3^-, Cl^-), iron (Fe), aluminum (Al), barium (Ba), strontium (Sr), silicon (Si), nitrite (NO_2), phosphate (PO_4), and (ii) suspended components – petrogenic (Al_2O_3, CaO, Fe_2O_3, K_2O, MgO, Na_2O, SiO_2), trace elements (lithium (Li), vanadium (V), and strontium (Sr)). A non-conservative concentration change along the delta branches was observed for nitrate (NO_3), dissolved barium (Ba), and Ba in suspension, which decreased along channels in comparison with the delta apex. Sedimentation of finely dispersed particles and inclusion of nutrient compounds into trophic chains, i.e., involvement of compounds in biochemical processes, could explain these greater changes. The reverse situation is typical for salinity. Its increase is registered long before the mixing zone is reached.

However, observations in the Lena River delta showed that there is no direct dependence of water discharge and material content on stream order. Changes in the concentration of individual substances are either conservative or non-conservative. From our point of view, observations have failed to reveal dependence on river order for two main reasons: first, the mechanism of the marginal filtration of nutrients in the Lena River delta has been under-studied; second, field data have shown an influx of dissolved and suspended substances into the delta itself (Table 4). Streams of melting water from ice complex carry cold (about 4–6 °C), turbid (up to $500\,\mathrm{g\,L^{-1}}$) water with salinity (up to $285\,\mathrm{mg\,L^{-1}}$) into the channels influenced by ice complex. Additional studies will be required to elucidate the role of these substances in hydrochemical and geochemical processes. This is not addressed in the current hypothesis of Alekseevsky and Chalov (2009) about geochemical processes in the delta.

Table 7. List of parameters.

H, m	–	water levels
Q, $\mathrm{m^3\,s^{-1}}$	–	daily water discharges
R, $\mathrm{kg\,s^{-1}}$	–	sediment discharges
K_i	–	value of a single element in the series
Q_0	–	water discharge for i observation
Q_0	–	average water discharge of all the observations
C_v	–	coefficient of variation
$F(t)$	–	difference-integral curve
K_{av}	–	average water content of observation periods
F_l	–	last ordinates of difference-integral curve
F_i	–	initial ordinates of difference-integral curve
t	–	time period
h, m	–	depth on the vertical
v_{1-n}, $\mathrm{m\,s^{-1}}$	–	average current velocity on the first–n velocity verticals
f_0, $\mathrm{m^2}$	–	water-section area between the bank and the first velocity vertical
f_1, $\mathrm{m^2}$	–	water-section area between the first and second velocity vertical, etc.
f_n, $\mathrm{m^2}$	–	water-section area between the last vertical n and the bank
V_m	–	averaged velocity over the first–n velocity verticals
V_s	–	velocity on the surface of a vertical
V_{02h}	–	velocity on the horizontal, 0.2 h
V_{06h}	–	velocity on the horizontal, 0.6 h
V_{08h}	–	velocity on the horizontal, 0.8 h
V_b	–	bottom velocity
h_{1-n}	–	depths of the measured verticals
$b_1, b_2, \ldots, b_{n-1}$	–	distance between the measured verticals
b_0, b_n	–	distances between outer measured verticals and encroachment lines
q_i, $\mathrm{m^3\,s^{-1}}$	–	water discharge between verticals
s_i, $\mathrm{mg\,L^{-1}}$	–	mean value of TSS content between verticals
σ_{sys}	–	systematic error of water discharge measurements
σ_{ran}	–	random experimental error of water discharge measurements
S_Q	–	summarized field observations error
u_{cr}	–	critical velocity for measurements by a GR-21M velocity meter
β	–	parameter of Zheleznov for measurements by a GR-21M velocity meter
u_0	–	initial velocity of a GR-21M velocity meter

5 Conclusions

Long-term Lena River delta field observations (2002–2012) combined with Roshydromet data and geoinformation technology have made it possible to obtain a number of new insights into the hydrological and geochemical peculiarities of the Lena River delta. The velocities of the fluvial processes that occur in the middle part of the delta were also documented. In summary, the following can be concluded:

1. Water discharge and suspended sediment supply in the delta over a long-term period was reviewed. According to Roshydromet data, a positive trend until 2007 was

confirmed, as well as a decrease of dissolved substance flux from 1960 to 1987.

2. Three periods were selected that are characterized by similarity of water volume and erosive power in the delta. From 1977 (from the beginning of instrumental measurements in all the delta branches) to the mid-1980s, low water volume and minor hydromorphological changes occurred in the delta. From the mid-1980s to the mid-1990s water volumes flowing through the delta increased and active fluvial processes were observed; after the mid-1990s, concomitant with increased discharge, the dissolved and suspended material transport of the Lena River delta decreased slightly.

3. New data were obtained from detailed field observations in the delta; the most valuable of these arose from along-branch hydrological measurements, which yielded new data about sources and sinks regions for discharge and fluvial transport. Between the head of the delta and its edge, an increase of water discharge and suspended sediment supply occurs. We hypothesize that it is caused by the degradation of the ice complex, erosion of river terraces, and river bank abrasion. A decrease of water and sediment discharge from the main branches on the delta edge is connected to channel branching; additional field measurements are required in this under-studied part of the delta, to investigate the possibility of a connection between the network of channels and a river talik.

4. New data were obtained on the geochemistry of main branch suspended sediments in the middle parts of the delta that confirm the ranges of previously published data on Lena River and estuarine coastal waters. The range of dissolved matter content changes for the main delta branches is small; the content is comparable to the long-term values. Such local factors as ice complex runoff water with higher TSS influence the hydrochemical characteristics of smaller branches.

The collection of long-term observational data described here not only has produced new results but also has demonstrated the necessity of carrying out more detailed observations of the hydrological, geochemical, and channel processes inside the Lena River delta, of studying the estuarine branch areas, and of developing an assessment of the sea's impact on the delta edge.

Acknowledgements. The results presented here were made possible by combining data from fieldwork undertaken in summer expeditions from 2002 to 2012 (Bolshiyanov et al., 2003; Bolshiyanov and Tretiakov, 2002; Fedorova et al., 2006, 2007; Morgenstern et al., 2012; Wagner et al., 2012) with the assistance of the Russian–German Samoylov Island Scientific Research Station project, the Russian Federal Target Program entitled *Development of methods and methodologies for assessment of geosystem changeability and environmental monitoring of the Laptev Sea region* in collaboration with German research institutions, grants from the Russian Foundation of Basic Research (No. 05-05-64419-at and No. 10-05-00727-a), grants from the German Federal Ministry of Education and Research (BMBF), OSL-11-02 *Substantial flow transformation processes in the Lena river delta* and OSL-12-01 *The study of geochemical processes of the Lena River delta*, and a grant from the German Research Foundation (DFG).

Edited by: T. Laurila

References

Alekin, O. A.: Basis of hydrochemistry, Hydrometeoizdat, Leningrad, 413 pp., 1970 (in Russian).

Alekseevsky, N. I. (Ed.): Geoecological state of Russian Arctic coast and their safety of nature management, GEOS Publ., Moscow, Russia, 586 pp., 2007 (in Russian).

Alekseevsky, N. I. and Chalov, S. R.: Hydrological impacts of braided channels, Publication of MSU, Moscow, Russia, 240 pp., 2009 (in Russian).

Andreyanov, V. G.: Inter annual runoff distribution (main particularities and their using in hydrological and water service estimation), edited by: Sokolov, D. L., Hydeometeoizdat, Leningrad, USSR, 327 pp., 1960 (in Russian).

Antonov, V. S.: The Lena river delta, Hydrometeoizdat, Leningrad, USSR, 108 pp., 1967.

Are, F. and Reimnitz, E.: An overview of the Lena River Delta settings: geology, tectonics, geomorphology, and hydrology, J. Coastal Res., 16, 1083–1093, 2000.

Are, F. E.: Thermal abrasion of coasts, in: Proceedings of the Fourth International Conference on Permafrost, Alaska USR, 18–22 July 1983, 24–28, 1983 (in Russian).

Atlas – Atlas of the Sardakhskaya channel, sailing direction, edited by: Yakutian steamship company, Russia, 1948 (in Russian).

Berezovskaya, S., Yang, D., and Kane, D. L.: Compatibility analysis of precipitation and runoff trends over the large Siberian watersheds, J. Geophys. Res., 31, L21502, doi:10.1029/2004GL021277, 2004.

Berezovskaya, S., Yang, D., and Hinzman, L.: Long-term annual water balance analysis of the Lena River, Global Planet. Change, 48, 84–95, 2005.

Boike, J., Kattenstroth, B., Abramova, K., Bornemann, N., Chetverova, A., Fedorova, I., Fröb, K., Grigoriev, M., Grüber, M., Kutzbach, L., Langer, M., Minke, M., Muster, S., Piel, K., Pfeiffer, E.-M., Stoof, G., Westermann, S., Wischnewski, K., Wille, C., and Hubberten, H.-W.: Baseline characteristics of climate, permafrost and land cover from a new permafrost observatory in the Lena River Delta, Siberia (1998–2011), Biogeosciences, 10, 2105–2128, doi:10.5194/bg-10-2105-2013, 2013.

Bolshiyanov, D. Yu. and Tretiakov, M. V.: Investigation of run-off in the Sardakh-Trofimovsky bifurcation point of the Lena River Delta, Reports on Polar and Marine Research, Alfred Wegener

Institute for Polar and Maine Research, Germany, 426, 57–62, 2002.

Bolshiyanov, D. Yu., Fedorova, I. V., and Tretiakov, M. V: Hydrological investigation in the Lena River Delta, Reports on Polar Research, Alfred Wegener Institute for Polar and Marine Research, Germany, 466, 82–91, 2003.

Bolshiyanov, D. Yu., Tret'yakov, M. V., and Fedorova, I. V.: Riverbed deformation in the Lena River delta channels, in: Proceeding of the VIth All-Union hydrological congress, Section 6: Problem of river-bed processes, erosion and sediments, St. Petersburg, Russia, 28 September–1 October, 257–260, 2004 (in Russian).

Bolshiyanov, D. Yu., Makarov, A. S., Schneider, V., and Stof, G.: Evolution and development of the Lena River delta, Publ. of AARI, St. Petersburg, Russia, 268 pp., 2013 (in Russian).

Burdyikina, A. P.: To a question of rivers freezing of the Lena-Anabar depression, Library holdings of the Arctic and Antarctic Research Institute, 2, 86–92, 1951 (in Russian).

Charkin, A. N., Dudarev, O. V., Semiletov, I. P., Kruhmalev, A. V., Vonk, J. E., Sánchez-García, L., Karlsson, E., and Gustafsson, Ö.: Seasonal and interannual variability of sedimentation and organic matter distribution in the Buor-Khaya Gulf: the primary recipient of input from Lena River and coastal erosion in the southeast Laptev Sea, Biogeosciences, 8, 2581–2594, doi:10.5194/bg-8-2581-2011, 2011.

Chetverova, A. A., Potapova, T. M., and Fedorova, I. V.: Geochemical runoff of Arctic rivers by the example of West Siberia Rivers and The Lena River, in: Proceeding of the IVth regional school-conference for young scientists/Aqueous medium and nature-territorial complex: study, using, protection, Petrozavodsk, Karelian scientific center of RAS, 26–28 August, 83–90, 2011 (in Russian).

Costard, F., Dupeyrat, L., Gautier, E., and Carey-Gailhardis, E.: Fluvial thermal erosion investigations along the rapidly eroding riverbank: application to Lena River (central Siberia), Earth Surf. Proc. Land., 28, 1349–1359, doi:10.1002/esp.592, 2003.

Fedorova, I., Fedorov, G., and Makarov, A.: Hydrological investigation in the Lena River Delta, Reports on Polar and Marine Research, Alfred Wegener Institute for Polar and Marine Research, Germany, 539, 48–57, 2006.

Fedorova, I., Bolshiyanov, D., Nikels, D., and Makarov, A.: Report on hydrological work in the Lena River Delta in August 2005, Reports on Polar and Marine Research, Alfred Wegener Institute for Polar and Maine Research, Germany, 550, 233–241, 2007.

Fedorova, I. V., Bolshiyanov, D. Yu., Makarov, A. S., Tret'yakov, M. V., and Chetverova, A. A.: Current hydrological state of the Lena delta, in: System of the Laptev Sea and adjacent seas of Arctic: current state and evolution, edited by: Kassens, H., Lisitzin, A. P., Tiede, J., Polyakova, E. I., Timokhov, L. A., MSU publishers, Moscow, Russia, 278–292, 2009a (in Russian).

Fedorova, I. V., Charkin, A. N., and Chetverova, A. A.: Changes of content and composition of suspended material in middle and lower course of the Lena River, in: Proceeding of the 25th plenary session on erosion, river-bed processes and estuary areas problems, Barnaul, Russia, 5–9 October 2009, 200–201, 2009b (in Russian).

Fedorova, I., Bolshiyanov, D. Yu, Chetverova, A., Makarov, A., and Tretiyakov, M.: Database of Alfred Wegener Institute of Polar and Maine Research, Germany, doi:10.1594/PANGAEA.808854, 2013.

Glushkov, V. G.: Perspectives and ways of Hydrology development in USR, Izvestiya of Russian State Hydrological Institute, 65, 23–30, 1934 (in Russian).

Gordeev, V. V.: Fluvial sediment flux to the Arctic Ocean, Geomorphology, 80, 94–104, doi:10.1016/j.geomorph.2005.09.008, 2006.

Gordeev, V. V.: Microelements in water, suspended material and sediments of Ob' bay, Enisey gulf and the Lena River delta and surrounded areas of the Kara and Laptev Seas, in: System of the Laptev Sea and adjacent seas of Arctic: current state and evolution, edited by: Kassens, H., Lisitzin, A. P., Tiede, J., Polyakova, E. I., and Timokhov, L. A., MSU publishers, Moscow, Russia, 202–225, 2009 (in Russian).

Gordeev, V. V. Zektser, I. S., Djamalov, R. G., Zhulidov, A. V., and Bryzgalo, V. A.: Score removal of nutrients of river runoff to the marginal seas of the Russian Arctic, Water Resour., 26, 206–211, 1999.

Grigoriev, M. N.: Cryomorphogenesis of the Lena River mouth area, Siberian Branch of Russian Academy of Sciences, Yakutsk, USSR, 176 pp., 1993 (in Russian).

Guidance document: Addition to instruction for hydrometeorlogical stations and posts, 6, part 1: Hydrological measurements and works on big and middle rivers, RD 52.08.163-88, Hydrometeoizdat, Leningrad, USSR, 90 pp., 1989 (in Russian).

Günther, F.: Investigation of thermokarst evolution in the southern Lena Delta using multitemporal remote sensing and field data, unpublished diploma thesis, Technical University of Dresden, Dresden, Germany, 96 pp., 2009.

Heim, B., Abramova, E., Doerffer, R., Günther, F., Hölemann, J., Kraberg, A., Lantuit, H., Loginova, A., Martynov, F., Overduin, P. P., and Wegner, C.: Ocean colour remote sensing in the southern Laptev Sea: evaluation and applications, Biogeosciences, 11, 4191–4210, doi:10.5194/bg-11-4191-2014, 2014.

Helbig, M., Boike, J., Langer, M., Schreiber, P., Runkle, B. R. K., and Kutzbach, L.: Spatial and seasonal variability of polygonal tundra water balance, Lena River Delta, northern Siberia, Hydrogeol. J., 21, 133–147, doi:10.1007/s10040-012-0933-4, 2013.

Hoelemann, J. A., Schirmaher, M., and Prange, A.: Seasonal variability of trace metals in the Lena River and the south-eastern Laptev Sea: impact of the spring freshet, Global Planet. Change, 48, 112–125, 2005.

Horton, R. E.: Erosional development of streams and their drainage basins; hydrophysical approach to quantitative morphology, Bull. Geol. Soc. Am., 56, 156 pp., 1945.

Hydrological yearbooks, basin of Laptev, East and Siberia and Chukchi Seas, Hydrometeoizdat, Leningrad, Russia, 8, 0–7, 1951–2007 (in Russian).

Instruction for Hydrometeorological stations and posts: Part III: Making and preparing for printing of a Hydrological yearbook, Hydrometeoizdat, Leningrad, Russia, 6, 291 pp., 1958 (in Russian).

Instruction for Hydrometeorological stations and posts: Part I: Hydrological measurements and work on big and middle rivers, Hydrometeoizdat, Leningrad, Russia, 6, 384 pp., 1978 (in Russian).

Ivanov, V. V.: Runoff and velocities of the main Lena River delta branches, in: Proceeding of Arctic and Antarctic Research Institute/Hydrology of rivers of the Soviet Arctic, Marine transport, Leningrad, USSR, 234, 76–86, 1963 (in Russian).

Ivanov, V. V. and Piskun, A. A.: Distribution of the river water and suspended sediment loads in the deltas of rivers in the basin of the Laptev and East-Siberian Seas, in: Land-Ocean Systems in the Siberian Arctic: Dynamics and History, edited by: Kassens, H., Bauch, H. A., Dmittrenko, I. A., Eicken, H., Hubberten, H.-W., Mellis, M., Thiede, J., and Timokhov, L. A., Springer-Verlag Berlin, 239–250, 1999.

Ivanov, V. V., Piskun, A. A., and Korabel, R. A.: Runoff distribution between main branches of the Lena River delta, in: Proceeding of Arctic and Antarctic Research Institute: Calculations and predictions of hydrological regime of estuary area of Arctic Rivers, edited by: Ivanov, V. V., Hydrometeoizdat Leningrad, USSR, 378, 59–71, 1983.

Izrael, Y. A., Chernogayev, G. M, Egorov, V. I., Peshkov, Y. V., and Kotlyakova, M. G.: Review of contamination in the Russian Federation in 2003: Coast of the Arctic seas: The Lena River basin, Roshydromet, Moscow, Russia, 142–144, 2004 (in Russian).

Izrael, Y. A., Chernogayev, G. M., Egorov, V. I., Peshkov, Y. V., and Kotlyakova, M. G.: Review of state and pollution of Russian Federation environment in 2011, Roshydromet, Moscow, Russia, 105–106, 2012 (in Russian).

Kane, D. L., McNamara, J. P., Yang, D., Olsson, P. Q., and Gieck, R. E.: An extreme rainfall/runoff event in Arctic Alaska, J. Hydrometeorol., 4, 1220–1228, 2003.

Kondratyev, N. E., Popov, I. V, and Snischenko, B. F.: Fundamentals of hydro-morphological theory of channel process, Hydrometeoizdat, Leningrad, USSR, 272 pp., 1982 (in Russian).

Korotaev, V. N.: Formation of the hydrographic net of the Lena delta in Holocene, Vestnik of MSU: Geography, 6, 39–44, 1984a (in Russian).

Korotaev, V. N.: Study of river-bed and hydrological regime of the Lena River estuary and feasibility of measure of deep-water route support in the navigate branches in the mouth area (results of study 1979–1981), in: Report of MSU, Moscow, USSR, 68 pp., 1984b (in Russian).

Korotaev, V. N. (Ed.): Investigation of river-bed and hydrological regime of the Lena estuary area and support of a deep route arrangement on navigation channel of the delta and coastline (results of work 1979–1981), Report of MSU, Moscow, USSR, 68 pp., 1984c (in Russian).

Korotaev, V. N. and Chistyakov, A. A.: Sedimentation processes in the estuaries areas of rivers, Vestnik of MSU: Geography, 5, 3–7, 2002 (in Russian).

Landsat satellite images source, available at: http://glovis.usgs.gov/, U.S. Geological Survey, last access: 14 January 2015.

Lisitzin, A. P.: Avalanche sedimentation and sedimentation disconformities in seas and oceans, Nauka, Moscow, USSR, 308 pp., 1988 (in Russian).

Marine transport: Hydrology of the Soviet Arctic Rivers, Proceeding of Arctic and Antarctic Research Institute, Leningrad, Russia, 1, 95 pp., 1956 (in Russian).

Morgenstern, A., Grosse, G., Günther, F., Fedorova, I., and Schirrmeister, L.: Spatial analyses of thermokarst lakes and basins in Yedoma landscapes of the Lena Delta, The Cryosphere, 5, 849–867, doi:10.5194/tc-5-849-2011, 2011.

Morgenstern, A., Fedorova, I., Roessler, S., and Ivlev, P.: Lake investigations. Reports on polar and marine research, Alfred Wegener Institute for Polar and Marine Research. Bremerhaven, Germany, 642, 68–70, hdl:10013/epic.38929, 2012.

Nikanorov, A. M.: Hydrochemistry: manual, Hydrometeoizdat, St. Petersburg, Russia, 444 pp., 2001 (in Russian).

Rachold, V., Grigoriev, M., Are, F., Solomon, S., Reimnitz, E., Kassens, H., and Antonov, M.: Coastal erosion in riverine sediments discharge in the Arctic Shelf seas, Int. J. Earth Sci., 89, 450–460, doi:10.1007/s005310000113, 2000.

Rawlins, M. A., Ye, H., Yang, D., Shiklomanov, A., and McDonald, K. C.: Divergence in seasonal hydrology across northern Eurasia: Emerging trends and water cycle linkages, J. Geophys. Res., 114, D18119, doi:10.1029/2009JD011747, 2009.

Reshet'ko, N. V. and Shvarzeva, N. M.: Basis of climatology and hydrology: methodic instruction for laboratory work realization for a course of lectures "Basis of climatology and hydrology" for student of 3d course, studying by specialization 280400 "Nature provision of the necessary facilities", Publisher by the Tomsk Polytechnic University, Tomsk, Russia, 39 pp., 2010 (in Russian).

Riordan, B., Verbyla, D., and McGuire, A. D.: Shrinking ponds in subarctic Alaska based on 1950–2002 remotely sensed images, J. Geophys. Res., 111, G04002, doi:10.1029/2005JG000150, 2006.

Rozhdestvenskiy, A. V. and Chebotarev, A.: Statistical methods in hydrology, Hydrometeoizdat, Leningrad, USSR, 424 pp., 1974 (in Russian).

Savenko, V. S.: Chemical composition of world River's suspended matter, GEOS, Moscow, Russia, 174 pp., 2006 (in Russian).

Schirrmeister, L., Froese, D., Tumskoy, V., Grosse, G., and Wetterich, S.: Yedoma: Late Pleistocene Ice-Rich Syngenetic Permafrost of Beringia, in: The Encyclopedia of Quaternary Science, edited by: Elias, S. A., Elsevier, 3, 542–552, 2013.

Schpakova, R. N.: Formation of water quality in the Lena River nowadays, in: Abstracts of PhD thesis of Geographical science, Moscow, Russia, 175 pp., 1999 (in Russian).

Schwamborn, G., Rachold, V., and Grigoriev, M. N.: Late Quaternary sedimentation history of the Lena Delta, Quaternary Int., 89, 119–134, 2002.

Seleznev, P. V. (Ed.): Report of hydrological work in the Lena River delta in 1985, TTUGMiKPS, Hydrographic party, Tiksi, USSR, 88 pp., 1986.

Semiletov, I. P., Pipko, I. I., Shakhova, N. E., Dudarev, O. V., Pugach, S. P., Charkin, A. N., McRoy, C. P., Kosmach, D., and Gustafsson, Ö.: Carbon transport by the Lena River from its headwaters to the Arctic Ocean, with emphasis on fluvial input of terrestrial particulate organic carbon vs. carbon transport by coastal erosion, Biogeosciences, 8, 2407–2426, doi:10.5194/bg-8-2407-2011, 2011.

Shiklomanov, A. I. and Lammers, R. B.: Record Russian river discharge in 2007 and the limits of analysis, Environ. Res. Lett., 4, 045015, doi:10.1088/1748-9326/4/4/045015, 2009.

Shiklomanov, I. A.: Anthropogenic changes of water content of rivers, Hydrometeoizdat, Leningrad, USSR, 345 pp., 1979 (in Russian).

Snischenko, D. V.: Using of aero cosmic information using in river-bed processes estimation, in: Proceeding of the Vth All-Union hydrological congress/River-bed processes and sediments, Hydrometeoizdat, Leningrad, USSR, 10, 55–59, 1988 (in Russian).

Stein, R. and Fahl, K.: The Laptev Sea: Distribution, sources, variability and burial of organic carbon, in: The Organic Carbon Cycle in the Arctic Ocean, edited by: Stein, R. and Macdonald, R. W., Springer, 213–237, 2004.

Syvitski, J. P. M.: Supply and flux of sediment along hydrological pathways: research for the 21st century, Global Planet. Change, 39, 1–11, 2003.

Usachyov, V. F. (Ed.): Recommendations about aero cosmic information using in river-bed processes investigation. Hydrometeoizdat, Leningrad, USSR, 50 pp., 1985 (in Russian).

Wagner, D., Overduin, P., Grigoriev, M. N., Knoblauch, C., Bolshiyanov, D. Yu. (Eds.): Russian-German Cooperation SYSTEM LAPTEV SEA: The Expedition LENA 2008, Reports on Polar and Marine Research, Alfred Wegener Institute for Polar and Marine Research, Germany, 642, 132 pp., hdl:10013/epic.38929, 2012.

Walker, H. J.: Arctic deltas, J. Coastal Res., 14, 718–738, 1983.

Wegner, C., Bauch, D., Hölemann, J. A., Janout, M. A., Heim, B., Novikhin, A., Kassens, H., and Timokhov, L.: Interannual variability of surface and bottom sediment transport on the Laptev Sea shelf during summer, Biogeosciences, 10, 1117–1129, doi:10.5194/bg-10-1117-2013, 2013.

Wetterich, S., Schirrmeister, L., Andreev, A., Pudenz, M., Plessen, B., Meyer, H., and Kunitsky, V. V.: Eemian and Late Glacial/Holocene palaeoenvironmental records from permafrost sequences at the Dmitry Laptev Strait (NE Siberia, Russia), Palaeogeogr. Palaeocl., 279, 73–95, doi:10.1016/j.palaeo.2009.05.002, 2009.

Woo, M.-K., Kane, D. L., Carey, S. K., and Yang, D.: Progress in Permafrost Hydrology in the New Millennium, Permafrost Periglac., 19, 237–254, 2008.

Ye, B., Yang, D., and Kane, D. L.: Changes in Lena River streamflow hydrology: human impacts versus natural variations, Water Resour. Res., 39, 1200, doi:10.1029/2003WR001991, 2003.

Ye, B., Yang, D., Zhang, Z., and Kane, D. L.: Variation of hydrological regime with permafrost coverage over Lena Basin in Siberia, J. Geophys. Res., 114, D07102, doi:10.1029/2008JD010537, 2009.

Yearbook – Yearbook of quality of surface water in the territory of the Omsk Department for Hydrometeorology and Environmental Monitoring, Hydrometeoizdat, Leningrad, USSR, 6, 4–9, 1989–2012 (in Russian).

Zheleznov, G. B. and Danilevich, B. B.: Accuracy of hydrological measurements and calculations, Hydrometeoizdat, Leningrad, USSR, 240 pp., 1966 (in Russian).

Zubakina, A. N.: Particularities of hydrochemical regime of the Lena estuary area, Proceeding of GOIN, 143, 69–76, 1979 (in Russian).

7

Historical TOC concentration minima during peak sulfur deposition in two Swedish lakes

P. Bragée[1], F. Mazier[2], A. B. Nielsen[5,1,6], P. Rosén[3], D. Fredh[1], A. Broström[1,*], W. Granéli[4], and D. Hammarlund[1]

[1]Quaternary Sciences, Department of Geology, Lund University, Sweden
[2]GEODE, UMR5602, Jean Jaures University, Toulouse-Le Mirail, France
[3]Department of Ecology and Environmental Science, Umeå University, Sweden
[4]Department of Biology, Aquatic Ecology, Lund University, Sweden
[5]Department of Physical Geography and Ecosystem Science, Lund University, Sweden
[6]Department of Biology and Environmental Science, Linnæus University, Sweden
[*]now at: Swedish National Heritage Board, Contract Archaeology Service, Sweden

Correspondence to: P. Bragée (petra.bragee@geol.lu.se)

Abstract. Decadal-scale variations in total organic carbon (TOC) concentration in lake water since AD 1200 in two small lakes in southern Sweden were reconstructed based on visible–near-infrared spectroscopy (VNIRS) of their recent sediment successions. In order to assess the impacts of local land-use changes, regional variations in sulfur, and nitrogen deposition and climate variations on the inferred changes in TOC concentration, the same sediment records were subjected to multi-proxy palaeolimnological analyses. Changes in lake-water pH were inferred from diatom analysis, whereas pollen-based land-use reconstructions (Landscape Reconstruction Algorithm) together with geochemical records provided information on catchment-scale environmental changes, and comparisons were made with available records of climate and population density. Our long-term reconstructions reveal that inferred lake-water TOC concentrations were generally high prior to AD 1900, with additional variability coupled mainly to changes in forest cover and agricultural land-use intensity. The last century showed significant changes, and unusually low TOC concentrations were inferred at AD 1930–1990, followed by a recent increase, largely consistent with monitoring data. Variations in sulfur emissions, with an increase in the early 1900s to a peak around AD 1980 and a subsequent decrease, were identified as an important driver of these dynamics at both sites, while processes related to the introduction of modern forestry and recent increases in precipitation and temperature may have contributed, but the effects differed between the sites. The increase in lake-water TOC concentration from around AD 1980 may therefore reflect a recovery process. Given that the effects of sulfur deposition now subside and that the recovery of lake-water TOC concentrations has reached pre-industrial levels, other forcing mechanisms related to land management and climate change may become the main drivers of TOC concentration changes in boreal lake waters in the future.

1 Introduction

Several studies have demonstrated increases in dissolved organic carbon (DOC) concentrations and colour in surface waters across large parts of Europe and North America over the last three decades (Stoddard et al., 2003; Hongve et al., 2004; Evans et al., 2005; Worrall and Burt, 2007; Erlandsson et al., 2008; Arvola et al., 2010). These trends have raised concerns about drinking water quality, as contaminants and toxic compounds may be associated with DOC (Ledesma et al., 2012). This may lead to increased demands for chemical pre-treatment in drinking water plants. Increased DOC export to surface waters may also have major consequences for aquatic ecosystems (Karlsson et al., 2009) and recreational values, as well as the role of lakes as carbon sources to the atmosphere (Cole et al., 2007).

A number of hypotheses have been put forward as explanations of the recent increase in DOC concentration. Several studies have proposed a link to declining atmospheric acid deposition (Evans et al., 2006; Vourenmaa et al., 2006; Monteith et al., 2007), while others have coupled enhanced leaching of DOC from soils to changes in climate (Freeman et al., 2001; Hongve et al., 2004; Worrall and Burt, 2007; Haaland et al., 2010) or nitrogen deposition (Findlay, 2005). Local-scale land-use and land management practices have also been demonstrated to influence DOC concentrations (Corell et al., 2001; Mattsson et al., 2005; Armstrong et al., 2010; Yallop et al., 2011). The lack of agreement on the mechanisms controlling DOC and colour variations in lake water during recent decades may partly reflect that many studies have been performed on catchment areas with heterogeneous types of land use, making it difficult to distinguish between co-existing forcing factors. Moreover, most studies have been based on monitoring data covering only a few decades, and have therefore failed to place the recent DOC trends in the perspective of the pronounced dynamics of anthropogenic atmospheric sulfur emissions that have occurred during the last century. Correspondingly, long-term changes in vegetation, land use and climate have also not been considered.

One way of gaining an increased understanding of this important environmental problem is to obtain long-term records of past changes in total organic carbon (TOC) concentration in lake water by using inference models derived from visible–near-infrared spectroscopy (VNIRS) of lake sediments (Rosén, 2005; Cunningham et al., 2011; Rosén et al., 2011). Following methodological development, this palae-olimnological approach has recently gained increased attention as a trustworthy proxy for ambient variations in lake-water DOC concentrations, building on the fact that the dominant fraction (> 95 %) of TOC in Scandinavian lake waters consists of DOC, usually defined as organic matter not retained by a filter of $0.45\,\mu\text{m}$ in nominal pore size (Wetzel, 2001). The remaining fraction is particulate organic carbon (POC), which consists of larger organic compounds. In boreal forested catchment areas, DOC is primarily allochthonous, originating from leaching of terrestrial soils. Additional autochthonous DOC may be produced in lakes by phytoplankton and aquatic macrophytes, although this part commonly constitutes only a minor fraction of the DOC pool in boreal lakes (Bade et al., 2007). The composition and quantity of DOC may differ between sites depending on climate and catchment properties such as vegetation, hydrology and soil properties (e.g. Clark et al., 2010). Lake-water DOC concentration and colour often show strong correlations (Pace and Cole, 2002; von Einem and Granéli, 2010) and their mutual increases over recent decades have been referred to as brownification (Granéli, 2012). Surface waters are variably coloured by humic substances, which are formed by terrestrial humification during degradation of soil organic matter and may comprise 50–75 % of the DOC pool (McDonald et al., 2004). Humic substances absorb solar ra-

diation, especially UV and short-wavelength visible radiation, and hence affect water temperature and aquatic primary productivity, with consequences for lake stratification and ecosystem functioning (Snucins and Gunn, 2000; Diehl et al., 2002; von Einem and Granéli, 2010). However, some studies have reported clear discrepancies between DOC concentration and colour in lake water (Erlandsson et al., 2008; Kritzberg and Ekström, 2012), indicating that the composition of DOC at the molecular level may be equally important for changes in water colour.

Here we present a detailed multi-proxy study based on well-dated sediment successions from two small nearby lakes in southern Sweden spanning the last approximately 800 years. One of them (Åbodasjön) is oligotrophic meso-humic with a mosaic landscape in its catchment area and with a long history of anthropogenic disturbance. The other lake (Lindhultsgöl) is oligotrophic polyhumic with a catchment area dominated by forest and wetlands, and is historically less influenced by anthropogenic disturbance (Bragée et al., 2013; Fredh et al., 2013). We applied a combination of palaeolimnological methods to the sediment sequences, including reconstruction of lake-water TOC concentration based on VNIRS (Rosén, 2005), diatom analysis to determine water pH, and pollen analysis and the Landscape Reconstruction Algorithm approach for reconstruction of catchment land-cover change (Sugita, 2007a, b). The aim of this study is to identify the major forcing mechanisms behind observed increases in TOC concentration in lakes of the upland area of southern Sweden during recent decades by comparing the impacts of changes in land use, sulfur and nitrogen deposition, and climate to long-term trends in lake-water TOC concentration since AD 1200. Particular focus is placed on the effects of differences in catchment characteristics and the degree of land-use intensity between the two study lakes. Ultimately, our findings may contribute to an enhanced understanding of lake-water TOC dynamics generally, on timescales beyond monitoring series, and to prediction of the future development of lake-water quality in boreal environments.

2 Study area and site descriptions

The two study lakes, Åbodasjön and Lindhultsgöl, are situated 6 km apart, about 30 km north-west of Växjö in the province of Småland, southern Sweden (Fig. 1). The crystalline bedrock is dominated by granite and gneiss (Wikman, 2000) and covered by sandy till of various thicknesses and scattered peat deposits (Daniel, 2009). The area is part of the boreo-nemoral zone characterized by mixed coniferous and deciduous forest (Sjörs, 1963; Gustafsson, 1996). The climate is generally maritime with a mean annual temperature of 6.4 °C (January 2.7 °C, July 15.9 °C) and an annual precipitation of 651 mm (January 52 mm, July 75 mm), based on reference normals from Växjö for 1961–

Table 1. Morphometric and hydrological characteristics of the two study lakes, sampled in July 2007 (von Einem and Granéli, 2010).

	Åbodasjön	Lindhultsgöl
Altitude (m)	221	212
Lake surface area (km^2)	0.5	0.07
Maximum depth (m)	9	5
Catchment area (km^2)	9.5	0.6
Residence time (yr)	0.5	–
pH	7.0	6.4
Alkalinity ($mEq\,L^{-1}$)	0.56	0.83
Chlorophyll a conc. ($\mu g\,L^{-1}$)	7.7	14.9
DOC conc. ($mg\,L^{-1}$)	11.0	23.8
Water colour ($mg\,Pt\,L^{-1}$)	40	960
Liming started	1984	1993

1990 (Alexandersson et al., 1991). The lakes are situated within the area of Sweden most significantly affected by increasing DOC concentrations since the 1990s (Löfgren et al., 2003). Lake size was also taken into account at the selection of study sites to enable reconstructions of local-scale land use based on fossil pollen records. The lakes are situated in the parish of Slätthög, established around AD 1000, and the first local population data are available from AD 1571, revealing 301 inhabitants (Andersson Palm, 2000). During the 1700s the population started to increase rapidly and a population peak was reached in the end of the 1800s, followed by a decrease in rural population due to industrialization.

Åbodasjön (Table 1, Fig. 1) is an oligotrophic mesohumic lake fed by two inlet streams, situated in the south and northeast, and with an outlet in the south-west. The village of Åboda (40 residents in 2004) is situated west of the lake, and the area around the lake margin is semi-open with mainly deciduous trees, grassland and cropland. The vegetation cover within the catchment area is dominated by managed coniferous woodland, wetlands, and patches of grassland and cropland.

Lindhultsgöl (Table 1, Fig. 1) is an oligotrophic polyhumic lake with no visible inlet streams. At least two artificial ditches drain into the lake from nearby wetlands and woodland, and there is an outlet consisting of an artificial ditch in the south. The catchment area is covered by managed coniferous forest and wetlands with shrubs and scattered pine trees.

3 Methods

3.1 Fieldwork, subsampling and dating

In early spring 2008 sequences of surface sediments were obtained from Åbodasjön and Lindhultsgöl at water depths of 8.6 and 5.2 m, respectively, using a gravity corer and a 1 m long Russian peat corer. Correlations between core seg-

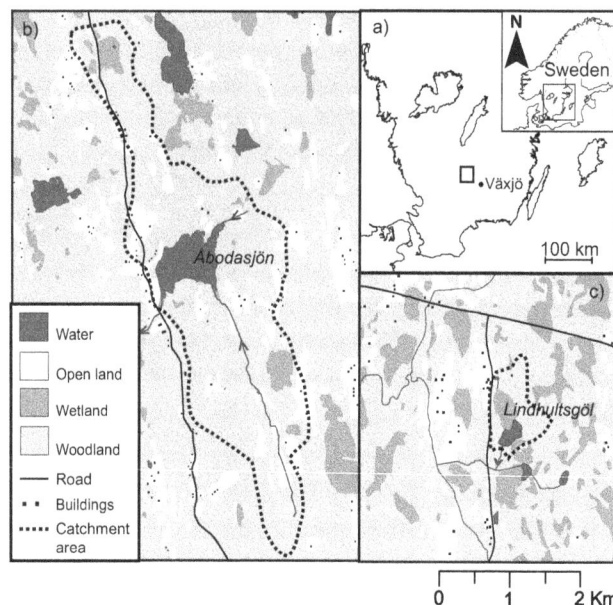

Figure 1. Location of study sites. (**a**) Map of Scandinavia and southern Sweden. The study area is marked by a square, and the closest city is Växjö. (**b**), (**c**) Maps of the studied lakes and the present-day land cover in their surroundings.

ments and surface sediments were based on mineral magnetic properties and X-ray fluorescence (XRF) measurements of element compositions. The uppermost 1 m parts of the sequences were subsampled into 0.5 cm contiguous sections for stratigraphic analyses. Age–depth models were based on ^{210}Pb dating along with ^{137}Cs, supplemented by radiocarbon dating of terrestrial plant remains and lead (Pb) pollution concentration variations (Bragée et al., 2013).

3.2 Visible–near-infrared spectroscopy (VNIRS)

Past changes of TOC concentration in the lake waters were reconstructed using a calibration model based on visible–near-infrared spectroscopy (VNIRS) of surface sediments from 140 Swedish lakes covering a TOC gradient from 0.7 to $24\,mg\,L^{-1}$ (Cunningham et al., 2011). The inferred TOC concentrations from Lindhultsgöl exceeded the range within the calibration set, and an additional set of 160 Canadian lakes with a DOC range of 0.6 to $39.6\,mg\,L^{-1}$ was also used (Rouillard et al., 2011). The model performance of the combined Swedish and Canadian calibration set is similar to the Swedish calibration set with an R^2 value of 0.6 between measured and predicted TOC concentration and a root-mean-squared error of prediction (RMSEP) of $4.1\,mg\,L^{-1}$ (10.5 % of the gradient).

3.3　Diatom analysis

Past changes in lake-water pH were reconstructed based on diatom assemblages in the sediment records. Diatom samples were prepared following standard methods (Battarbee et al., 2001). Following oxidization of freeze-dried sediment samples (0.01 g) with 15 % H_2O_2 solution for 24 h, 30 % H_2O_2 was added to digest organic matter using the water-bath technique described by Renberg (1990). For some samples HNO_3 was added to digest the remaining organic matter. To estimate diatom concentrations, known quantities of DVB (divinylbenzene) microspheres were added to the digested and cleaned samples (Battarbee and Kneen, 1982; Wolfe, 1997). Samples of 0.2 mL of the mixtures were evaporated onto cover slips and mounted onto microscope slides using the Zrax mounting medium (refractive index $= \sim 1.7+$). At least 400 diatom valves per sample were counted under a light microscope at $1000 \times$ magnification, using phase-contrast optics and identification keys (Krammer and Lange-Bertalot, 1986, 1988, 1991a, 1991b; Lange-Bertalot and Krammer 1989; Krammer , 1992). The diatom counts were expressed as relative abundances of each taxon. Diatoms were grouped into planktonic and benthic taxa for calculation of planktonic / benthic (P / B) ratios, indicative of light availability, as decreased light penetration reduces benthic growth.

Changes in pH were inferred from sedimentary assemblages (Di-pH) using a transfer function set, the online combined pH training set in the European Diatom Database (http://craticula.ncl.ac.uk/Eddi/jsp/). The calibration set for the model consists of 627 lakes with a pH range of 4.3–8.4. The diatom-inferred pH was based on locally weighted averaging and inverse deshrinking (Juggins and Birks, 2012). The model performance of the transfer function applied to Åbodasjön and Lindhultsgöl was assessed by leave-one-out cross-validation, which showed an R^2 value of 0.8 and an RMSEP of 0.4 pH units.

3.4　Carbon and nitrogen elemental analyses

The carbon–nitrogen (C / N) ratio of lake sediments gives an indication of the source (terrestrial and aquatic) of organic matter (Meyers and Lallier-Vergès, 1999). Acid-treated and freeze-dried sediment samples were analysed for sedimentary total organic carbon ($TOC_{sediment}$) and total nitrogen (TN) contents by combustion using a Costech ECS 4010 elemental analyser. The samples were pre-treated with 10 % HCl at 90 °C for 5–7 min for removal of potential trace amounts of $CaCO_3$. Elemental C / N ratios were converted to atomic ratios by multiplication by 1.167.

3.5　Trace element concentrations and X-ray fluorescence analysis (XRF)

Enhanced catchment erosion may be reflected by elevated concentrations of lithogenic elements in the sediment profile (Engstrom and Wright, 1984). Concentrations of phosphorus (P), zirconium (Zr) and titanium (Ti) in the sediments were measured by X-ray fluorescence (XRF) analysis (Boyle, 2000) followed by calculation of elemental Zr / Ti ratios for estimation of mineral grain-size variations within the lake sediments, as Zr is commonly associated with silt particles and Ti often occurs in the fine silt and clay fractions (Koinig et al., 2003; Taboada et al., 2005). Freeze-dried samples at 2–5 cm intervals of the sediment sequences were measured, using an S2 Ranger XRF spectrometer for total concentrations of 35 different major and trace elements. The spectrometer was calibrated using certified reference materials. Mass attenuation correction was based on theoretical alpha coefficients, with calculations taking organic matter concentrations into account.

3.6　Pollen analysis and Landscape Reconstruction Algorithm (LRA)

Changes in land use were quantified using LRA (Sugita, 2007a; b) based on pollen counts of dominant taxa in the sediment records from the two study sites and an additional lake (needed for the LRA calculation). A minimum of 1000 pollen grains of modelled arboreal and non-arboreal taxa were counted for contiguous 0.5 cm samples (1–10 samples) covering 20-year time spans.

The LRA allows the estimation of changes in the spatial coverage of 26 target taxa at regional and local scales. The pollen data, the LRA approach with its associated parameters, and the reconstructions of land use were described in detail by Fredh et al. (2013, 2014) and Mazier et al. (2014). In this paper, we focus on local land-use dynamics at 20-year intervals since AD 1200 at a spatial scale (modelled area) identified by Mazier et al. (2014) as a radius of 1740 m around Åbodasjön and 1440 m around Lindhultsgöl. The inferred covers of individual taxa are grouped into five different categories of land use according to Mazier et al. (2014): coniferous woodland, deciduous woodland, grassland, cropland and wetland. Although the LRA approach provides no information on the spatial distribution of the types of land use within the modelled areas – larger than the actual catchment areas – we assume that the changes in land use within the modelled areas broadly reflect catchment-scale vegetation changes.

3.7　Multivariate analyses

To explore the impact of various potential driving forces on the lake environment as reflected in the sediment record, we carried out canonical ordinations. The palaeolimnological parameters (VNIRS-inferred TOC concentration, sediment TOC_{sed} and TN, C / N ratio, sediment P content (ppm), Zr / Ti ratio and diatom inferred pH) were used as response variables. As potential forcing variables we used the pollen-inferred land-use categories (coniferous woodland, deciduous woodland, grassland, cropland and wetland) and in addition the cover of individual tree species (spruce and pine), total woodland cover (coniferous and deciduous), and the sum of cropland and grassland.

For the entire period after AD 1200, 20-year time slices were used for the analysis, using land cover as forcing. A mean value for each sedimentary variable was calculated over each 20-year time slice. A few time slices lacked measurements of P content and Zr / Ti ratio and were therefore left out of the analysis. For Di-pH, the analytical resolution was lower than 20 years (see Fig. 2), so linear interpolations between the available estimates were used to calculate average values for each 20-year time slice. A separate analysis was carried out for the period after AD 1880 including data on atmospheric deposition of sulfur (S), ammonium (NH_4) and nitrogen oxides (NO_x), and monitoring records of temperature and precipitation as potential forcing factors in addition to land cover. Annual climate data are available for the region from AD 1860, and deposition data are available at 5-year intervals from AD 1880. The temporal resolution of this ordination analysis was determined by the resolution of the inferred VNIRS-TOC reconstruction (see Fig. 3). Land cover was considered constant for each 20-year interval. For the other sediment parameters and for atmospheric deposition, linear interpolation was used to derive an estimated value for the year corresponding to each VNIRS-TOC sample. For the climatic variables (annual mean temperature and total annual precipitation), the value measured in the sample year and the mean value for the 10 years up to and including the sample year were both included in the analysis.

Ordinations were carried out using CANOCO v4.51. For all analyses, preliminary detrended canonical correspondence analysis showed the response data set had a gradient length < 1 standard deviation units, implying that linear based ordination techniques such as redundance analysis (RDA) were most suitable for these data sets (ter Braak and Smilauer, 1998).

Land-cover percentages were square-root-transformed, while the limnological parameters (which are measured in different units) were centred and standardized. Time was used as a co-variable to remove co-varying effects between, for example, changes in land use and atmospheric deposition. Manual forward selection was used to explore the explanatory power of the different forcing variables, and Monte Carlo tests with 999 unrestricted permutations were run to check their statistical significance in order to select the best explanatory variables for further analysis. The selected variables were checked for collinearity by inspecting their variance inflation factors, which were in all cases < 10, which

indicates that the selected parameters are not too closely correlated (Oksanen, 2011).

4 Results

Åbodasjön (Fig. 2): the inferred TOC reconstruction shows maximum inferred values of $14 \, mg \, L^{-1}$ around AD 1250, followed by a decrease to rather stable values at $9-10 \, mg \, L^{-1}$ after AD 1450. Around AD 1800 an increase was recorded, reaching peak values at ca. $12 \, mg \, L^{-1}$ between AD 1860 and 1910, followed by a sudden decrease, reaching a sequence minimum of ca. $7 \, mg \, L^{-1}$ in the 1980s. After AD 1990 an increase to $9-10 \, mg \, L^{-1}$ was recorded.

The diatom-inferred pH varies between 6.2 and 6.7, with a sample-specific standard error between 0.32 and 0.45. (Fig. 2 and Supplement). Periods of slightly elevated pH were recorded at AD 1350–1500 and AD 1700–1780, while lower values were recorded at AD 1520–1670 and after AD 1970. The diatom concentration increases to a peak around AD 1400, followed by a decrease to relatively stable values and a second decrease after AD 1850. The planktonic diatom taxa vary between 40 and 70 % of the diatom assemblage, and slightly elevated P / B ratios were recorded at AD 1250–1500 and in the top sample.

Sediment total organic carbon content (TOC_{sed}) and TN show slightly elevated values at AD 1250–1350, followed by a slight transient decrease and a gradual increase after AD 1450. In the 1800s TOC_{sed} content stabilizes at maximum values. The C / N ratio increases in AD 1200 to ca. 1300, followed by a slight decrease and a continuous increase from around AD 1450 to a sequence maximum at AD 1850–1900. There is a shift towards substantially lower TOC_{sed} and TN content, and C / N ratios at ca. AD 1850 (TN) and 1900 (TOC_{sed} and C / N). Thereafter, increasing trends in both TOC_{sed} and TN content is recorded from ca AD 1970 to the present, and C / N ratios after AD 1990.

P concentration decreases gradually from the beginning of the sequence interrupted by a shift to higher values at ca. AD 1440 and thereafter followed by continuously decreasing concentrations. The onset of the 1900s is characterized by an increase in P concentration peaking shortly after AD 1950. The Zr / Ti ratio record shows a period of elevated values at AD 1320–1450, followed by a temporary decease and continuously elevated values at AD 1600–1900. After around AD 1950 a slight decrease was recorded.

The LRA-inferred woodland (coniferous and deciduous) cover around Åbodasjön varies between 33 and 80 % since AD 1200. The cover of grassland and cropland together is 40–50 % at AD 1240–1400, followed by a decrease to a minimum of 15 % at AD 1520–1540, when deciduous and coniferous woodland reaches a peak in cover. After around AD 1540 grassland and cropland cover increases and reaches maxima of ca. 60 and 12 %, respectively, between AD 1820 and 1900. During the 1900s coniferous woodland, dominated

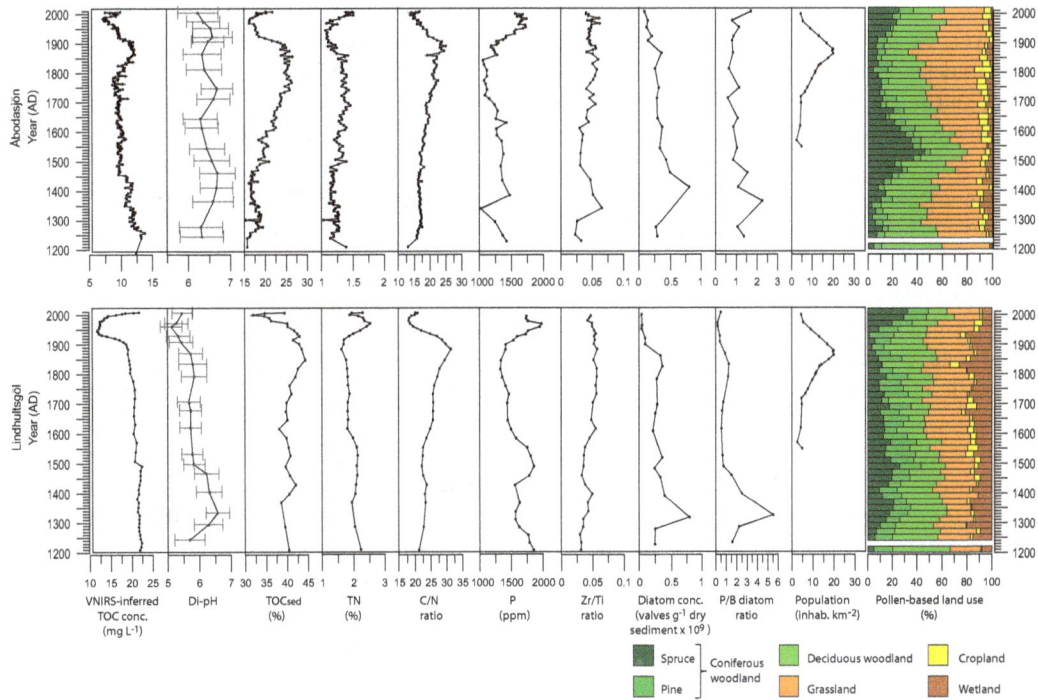

Figure 2. Records of VNIRS-inferred lake-water total organic carbon (TOC) concentration, diatom-inferred pH (Di-pH) (horizontal error bars represent ±1 SD), sediment total organic carbon (TOC$_{sed}$) and total nitrogen (TN) content, atomic carbon : nitrogen (C / N) ratio, elemental phosphorus (P) content, elemental zirconium : titanium (Zr / Ti) ratio, diatom valve concentration, diatom planctonic : benthic (P / B) ratio, documented population density, and pollen-based land use plotted against age from Åbodasjön (upper panel) and Lindhultsgöl (lower panel).

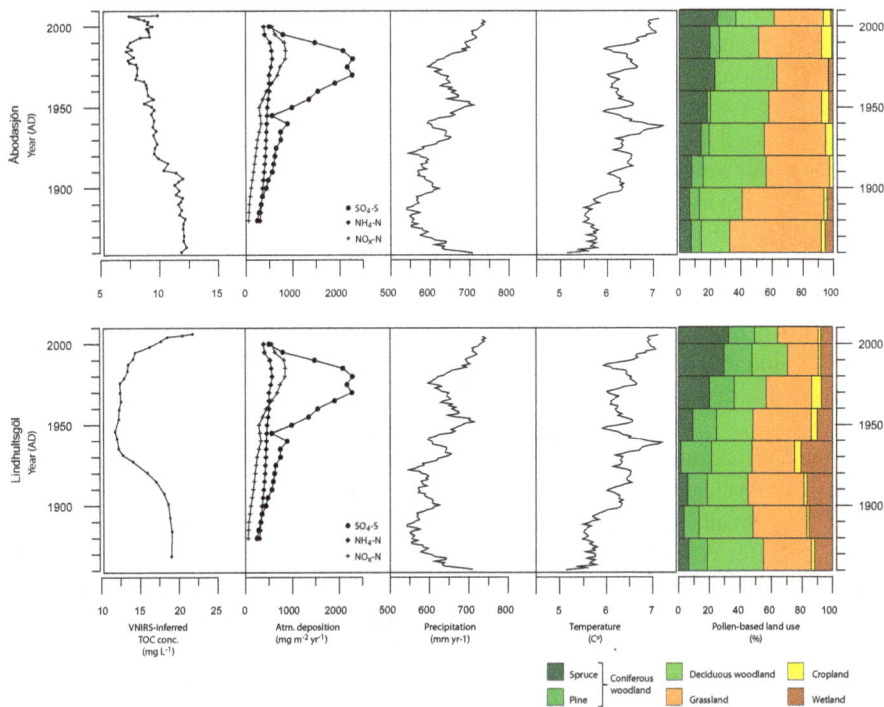

Figure 3. Records of VNIRS-inferred lake-water total organic carbon (TOC) concentration and pollen-based land use since AD 1900 from Åbodasjön and Lindhultsgöl plotted together with atmospheric sulfur (sulfate SO$_4$) and nitrogen (ammonium (NH$_4$) and nitrogen oxides (NOx)) deposition (from the Swedish Environmental Research Institute MAGIC model) and climate data from Växjö (annual precipitation and temperature expressed as 10-year running averages).

by spruce, increases from 10 to 30 %. Coniferous and deciduous woodland covers ca. 60 % of the lake catchment today.

RDA was used to describe the major gradients in the limnological data set and relate these patterns to the land-use variables during the last 800 years. Total woodland cover was identified as the most significant land-cover factor, explaining a statistically significant 13 %. Other land-cover variables that were significant when analysed on their own were spruce and cropland cover, each explaining 12 %; wetland and coniferous woodland, 10 %; and deciduous woodland, 6 %. When total woodland cover was included in the RDA analysis, deciduous woodland cover could still explain an additional 6 % of the variation, while no other land-cover parameters were statistically significant at the $P < 0.05$ level, as most of the variation they could explain was captured by the relationship to total woodland cover.

The ordination results are presented as a so-called triplot (Fig. 4a) showing the RDA scores for both the palaeolimnological response variables and the selected forcing variables, as well as the trajectory of down-core sample scores over time, along the first and second RDA axes. The figure indicates that the VNIRS-inferred TOC concentration along with sediment TOC_{sed}, C / N and Zr / Ti are all negatively related with woodland cover, which is correlated with the first RDA axis ($r = 0.71$). Deciduous tree cover is negatively correlated with the second axis ($r = -0.55$), along with diatom-inferred pH. Both seem to be negatively correlated with sediment TN.

For the period after AD 1880, five significant drivers were retained on the basis of forward selection. NH_4 deposition was identified as the main driver and explained 21 % of the variance, NO_x deposition (additional 12 %), 10-year mean annual precipitation (additional 3 %), sulfur deposition (additional 2 %) and 10-year mean annual temperature (additional 2 %). While some of the land-cover categories had significant effects if analysed individually, such as deciduous tree cover (18 %) and grassland cover (14 %), they were less important than the depositional and climatic parameters, and did not add significantly to the combined analysis. The RDA plot (Fig. 4b) indicates that Di-pH is negatively correlated with sediment TOC_{sed} and VNIRS-inferred TOC, which are both negatively related to NH_4- and S deposition and positively related to precipitation.

Lindhultsgöl (Fig. 2): the VNIRS-inferred TOC concentration exhibits high and stable values (21–22 mg L^{-1}) at AD 1200–1500, followed by a small but sudden decrease to values around 20 mg L^{-1}. After AD 1780 a gradual decrease was recorded, followed by a substantial decrease in AD 1900 to minimum values (12 mg L^{-1}) around AD 1930. An increase was recorded at AD 1980, which was accentuated after AD 1990, and reached pre-1900 values in the surface sediments.

Diatom-inferred pH varies between 5.0 and 6.8, with sample-specific standard errors between 0.31 and 0.47. The highest value was recorded following an increase around AD 1250 to above 6 between AD 1300 and 1450. The pe-

riod between AD 1500 and 1800 shows rather stable values around 5.8. In the 1900s, pH decreases to a minimum of 5.0 around AD 1960, followed by a slight increase until AD 2008. The pH reconstruction for Lindhultsgöl was influenced by a few dominating diatom taxa. The high values inferred in the lower parts were associated with the high abundance of the alkaliphilous (pH > 7) diatom taxon *Aulacoseira ambigua* (< 60 %), and the low pH in the 1900s was affected by the dominant acidophilous (pH < 7) taxon *Frustulia rhomboides* (< 60 %). Inference models are always associated with uncertainties and diatoms may respond to other variables than pH (Juggins, 2013). Therefore caution is necessary when interpreting the reconstructed pH data. The diatom concentration is high in the 1300s, followed by stable values until around AD 1850, when concentrations decrease. The planktonic diatom taxa vary between 15 and 85 % and the maximum in P / B ratio recorded in the 1300s was followed by rather stable ratios with a slight increase in the 1800s. Lowered P / B ratios were recorded after around AD 1900.

Relatively stable values were recorded for TOC_{sed} content at ca. AD 1200–1700 and for C / N ratios at ca. AD 1200–1550, followed by increasing values, peaking in the late 1800s. The total nitrogen (TN) content showed slightly decreasing values until AD 1900. At ca. AD 1900 significant decreases in both TOC_{sed} content and C / N ratio to minima in the 1980s to 1990s together with a subsequent increase in TN content to a coherent maxima in the 1980s were recorded. This was followed by reversed trends and a coherent increase after ca. AD 1990–2000 towards the top.

P concentration decreases gradually from the beginning of the sequence interrupted by a shift to higher values in the 1400s and thereafter followed by continuous decreasing concentrations. The onset of AD 1900 is characterized by an increase in P concentration peaking shortly after AD 1950. The Zr / Ti ratio shows a peak around AD 1350, following a gradual increase from ca. AD 1250. After a subsequent decrease, the Zr / Ti ratio increases around AD 1500 to rather stable values in the AD 1800s, followed by a decrease after AD 1930.

The woodland (coniferous and deciduous) cover around Lindhulsgöl varies between 44 and 70 % during the last 800 years. In contrast to Åbodasjön, wetlands cover more than 20 % during most of the period and decreases to less than 10 % after AD 1960. Grassland and cropland varies between 20 and 30 % at AD 1200–1580, followed by an increase to ca. 40 %. During the 1900s, coniferous woodland increases, and this land-use category covers ca. 50 % of the lake catchment today.

In the RDA analysis for the last 800 years, the forward selection for this site showed that spruce cover, the main explanatory variable, explains 20 % of the variance. After its inclusion in the RDA model, two other variables were found significant – cropland and wetland covers, explaining 7 % respectively 4 % of additional variance.

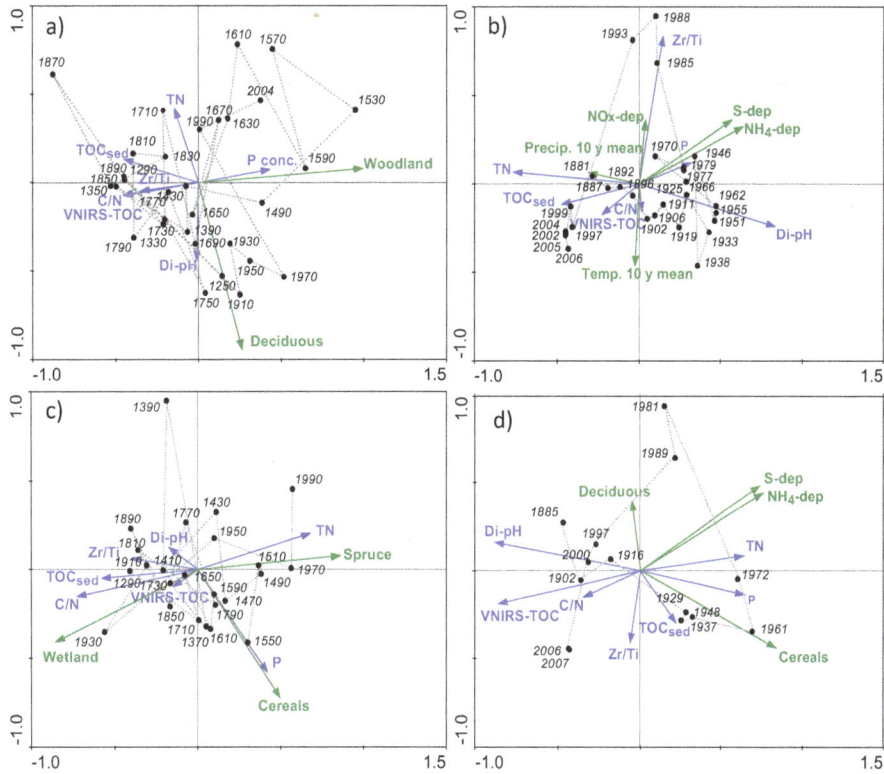

Figure 4. Scores for samples (black circles), palaeolimnological parameters (blue arrows) and driving forces (green arrows) on the first and second axes of the redundancy analyses for (**a**) Åbodasjön AD 1200–present, (**b**) Åbodasjön AD 1880–present, (**c**) Lindhultsgöl AD 1200–present and (**d**) Lindhultsgöl AD 1880–present. Sample ages represent the midpoint of each 20-year time slice.

A triplot showing the first and second RDA axes (Fig. 4c) indicates that TOC_{sed}, C / N and to a lesser extent VNIRS-inferred TOC concentration seem to be positively related to wetland cover, and negatively with spruce cover along the first axis. Sediment TN is positively correlated with spruce cover, while along the second axis sediment P content is correlated with cereal cover.

For the period after AD 1880, the RDA analysis indicates that cereal cover was the most important driver, alone explaining 28 % of the variation in the palaeolimnological data. The stepwise forward selection showed that three further variables could contribute significantly to the explanatory power of the model, i.e. S deposition (which could explain an additional 10 % of the variation if included together with cereal cover), NH_4 deposition (5 %) and deciduous tree cover (3 %). Ten-year mean precipitation has a small significant effect on its own, explaining 14 %, but is not significant once crop cover was included. No other climate parameter seems to have an effect at Lindhultsgöl, being overshadowed by the stronger effect of land use change at this site. The RDA plot (Fig. 4d) indicates that, like at Åbodasjön, VNIRS-inferred TOC concentration is negatively related to NH_4- and S deposition. But at this site, TOC and Di-pH seem to be positively related. The sediment TN and P content at this site

are positively related to cereal cover and the atmospheric N deposition.

5 Discussion

5.1 Impacts of land-use changes prior to AD 1900

In Åbodasjön, the highest lake-water TOC concentration was inferred at the beginning of the record, around AD 1200, and decreased during the following century, while human impact increased (Fig. 2). From AD 1260 the pollen record indicates an agricultural expansion with increased extent of croplands, meadows and pastures in the catchment (Fig. 2; Fredh et al., 2014). This expansion probably resulted in increased erosion and input of coarse lithogenic material, as indicated by elevated Zr / Ti ratios in the sediments. Also, elevated pH and diatom concentration suggest that more base cations and nutrients were released from the catchment, thereby reflecting cultural alkalinization (Renberg, 1990; Rosén et al., 2011). However, despite increased erosion, TOC concentration decreased in the lake water during the 1200s, which may be explained by decreased woodland cover, which lowered the terrestrial biomass production, from where a large portion of the TOC in lake water originates (Rosén et al., 2011).

At Lindhultsgöl, increasing anthropogenic impact was recorded during the 1200s from enhanced Zr/Ti ratios (Fig. 2) and increased charcoal concentrations (Fredh et al., 2014), reflecting increased erosion and land clearance by fire, respectively. During this time, the pH increased from 5.6 to 6.6, which was most likely caused by release of bases and nutrients from burning and grazing in the landscape (Renberg et al., 1993; Boyle, 2007). Moreover, the diatom concentration peaked, indicating temporarily enhanced aquatic productivity (cf. Rosén et al., 2011). However, the sediment record shows only a slight increase of open land, mainly grassland and cropland, around the lake. The persistently high TOC concentrations in the lake were probably related to the large proportion of wetlands within the catchment, as also indicated by the RDA results. High proportions of wetland are often associated with substantial supplies of DOC to nearby lakes (Rasmussen et al., 1989; Kortelainen, 1993; Xenopoulos et al., 2003; Mattsson et al., 2007). The lack of response in lake-water TOC concentration to catchment disturbance at Lindhultsgöl during the period of increased anthropogenic impact and potential alkalinization may be related to the unchanged proportion of open land, indicative of stable biomass production with a continuously high supply of DOC to the lake.

From ca. AD 1350 there was a reduction of human-induced catchment disturbance at Åbodasjön, as indicated by a decline in cropland and grassland cover (Fig. 2), and coniferous woodland, in particular spruce, increased substantially around AD 1400. This agricultural regression was followed by decreasing catchment erosion and stabilization of TOC concentrations in the lake water, an event that may be related to the Black Death pandemic, which struck Sweden in AD 1350. This was followed by several outbreaks throughout the 1400s, and as much as 60–70 % of the farms in the region were abandoned (Lagerås, 2007; Myrdal, 2012). At ca. AD 1450 there was a shift to lower lake-water TOC concentrations, accompanied by decreasing Zr/Ti ratio and diatom concentration. At Lindhultsgöl the regression led to decreases in catchment erosion, inferred pH and diatom concentration from ca. AD 1400 in response to increased cover of coniferous woodland.

From ca. AD 1450 to 1800 TOC concentrations in Åbodasjön were relatively stable, with only minor variations, despite major changes in land use. Following the increase at ca. AD 1350, coniferous woodland reached maximum cover of ca. 50 % around AD 1550, followed by a decrease related to the onset of a second agricultural expansion in the region (Lagerås, 2007). The pollen records from both lakes showed a gradual increase in cropland, meadows and pasture, more pronounced at Åbodasjön, together with enhanced erosion as reflected by increasing C/N and Zr/Ti ratios.

A substantial increase in lake-water TOC concentration was inferred at Åbodasjön from ca. AD 1800, peaking at AD 1860–1900, simultaneously with a substantial increase in population density (Fig. 2). The increase in rural population led to increased demands for land for crop cultivation, meadows and grazing, and areas previously regarded as less suitable for agriculture were cleared and drained (Myrdal, 1997). The pollen record shows a dominance of open-land taxa, and the open-land cover, predominantly grassland, reached a maximum of ca. 60 %. The RDA plot (Fig. 4a) also reflects that both total and deciduous woodland cover reached minimum values around this time. These changes were accompanied by maximum C/N ratios and TOC_{sed} values, reflecting an elevated input of terrestrial organic matter to the lake. From ca. AD 1700 improvements of the agrarian management in Sweden enhanced food productivity through a number of reforms, such as land divisions, crop rotation, irrigation, marling and better management of manure and urine (Emanuelsson, 2009). The introduction of agriculture in lake catchments, even at low proportions, is commonly associated with elevated DOC export and lake-water DOC concentrations (Correll et al., 2001; McTiernan et al., 2001; Mattsson et al., 2005). In contrast to the decrease in the inferred TOC concentrations in response to the early agricultural expansion, the new agricultural management in the 1800s improved organic productivity through the application of manure and fertilizers, leading to increased leaching of DOC to the lake water (cf. McTiernan et al., 2001). These agricultural reforms, in combination with the general increase in land-use pressure, may hence explain the substantial increase in the inferred TOC concentration at Åbodasjön.

At Lindhultsgöl broadly similar trends in C/N ratio and TOC_{sed} from the late 1700s to ca. AD 1900 as compared to Åbodasjön suggest increased land-use pressure and disturbance within the lake catchment. Coniferous woodland decreased, especially after AD 1800 and was partly replaced by deciduous woodland, indicating increased logging and expanding semi-open grazing areas. However, the high proportion of wetlands made the catchment less suitable for crop cultivation and resulted in a strikingly different development as compared to Åbodasjön. In the more marginal, forest-dominated area around Lindhultsgöl the increase in anthropogenic impact resulted in an increase in pH and a corresponding decrease in TOC concentration in the lake around AD 1800, probably reflecting decreased catchment biomass in a gradually more open woodland.

5.2 Forcing mechanisms during the last century

Around AD 1900 pronounced decreases in TOC concentrations were recorded in both of the study lakes (Figs. 2, 3). At Åbodasjön the decrease was slightly more gradual, reaching minimum values in the 1980s, while the inferred values at Lindhultsgöl declined rapidly to a sequence minimum around AD 1940 (Fig. 3). At AD 1980–1990 increasing trends were initiated at both lakes. These inferred variations in TOC concentration during the last century are in general inversely correlated with historically documented trends in sulfur deposition regionally in southern Sweden (Fig. 3), and

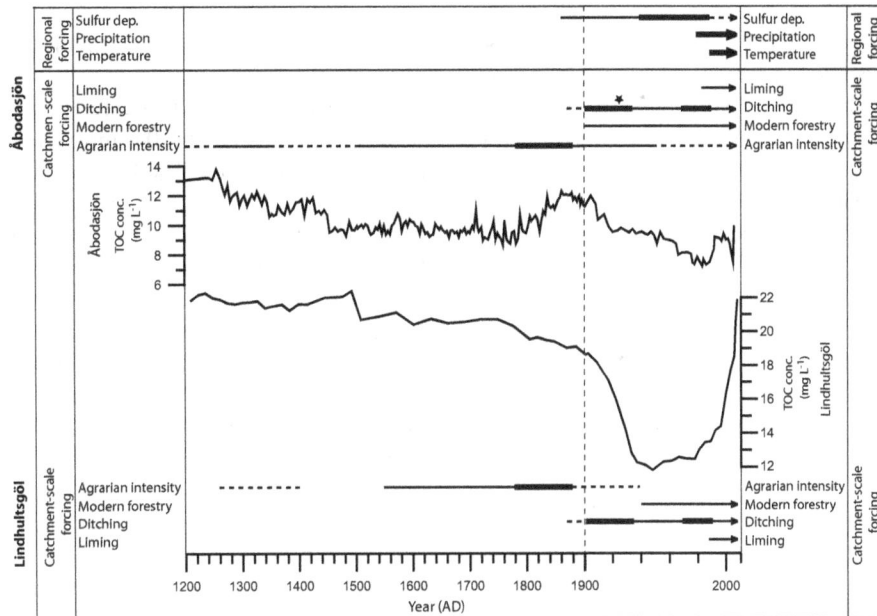

Figure 5. Records of VNIRS-inferred lake-water total organic carbon (TOC) concentration from Åbodasjön (upper graph) and Lindhultsgöl (lower graph) in the perspective of possible regional and catchment-scale forcings of TOC changes. Regional forcings include sulfur deposition, precipitation and temperature (Fig. 3). Local forcings include site-specific liming history, regional trends in ditching (Hånell, 2009) and changes in land use inferred from pollen data (Fig. 2) and historical accounts (agrarian intensity and modern forestry). Horizontal lines represent periods of activity, thick lines represent periods of increase or high intensity, and dashed lines represent periods of decrease or low intensity. Arrows indicate ongoing processes. The star marks a major drainage effort undertaken at the inlet of Åbodasjön in AD 1922. The vertical dashed lines represent AD 1900. Note the different scale for the period AD 1900–2010.

the RDA data also indicate that sulfur deposition is among the significant drivers of limnological changes in both lakes since AD 1880. Sulphur deposition started to increase at the onset of industrialization at the end of the 1800s, which led to acidification of soils and surface waters across large parts of Europe (Rohde et al., 1995). Thereafter, sulfur deposition increased significantly in the 1940s, peaking at AD 1980–1995 (Schöpp et al., 2003), followed during recent decades by progressively decreasing deposition and widespread recovery from acidification through decreasing sulfate concentrations in lakes and streams throughout Europe and North America (Evans et al., 2001, Skjelkvåle et al., 2003). The timing of this recovery is largely consistent with the increasing TOC concentrations in our two study lakes (Fig. 3) as well as with a study of TOC trends in Swedish rivers (Erlandsson et al., 2010). The deposition of nitrogen oxides, which also contribute to acidification, also showed a dramatic increase during the 1900s, with deposition peaking slightly later than for sulfur deposition, around AD 1990 (Fig. 3). In addition to contributing to acidification, deposition of nitrogen, both in the form of nitrogen oxides and ammonia, may contribute to eutrophication, and therefore can have an impact on limnic ecosystems. It has also been suggested that the response of soil microbial activity to nitrogen deposition may affect the export of humic matter to freshwaters (Findlay, 2005). Our analysis indicates that nitrogen deposition was among the

most significant drivers of change in the palaeolimnological record over the last century together with sulfur deposition.

Increases in lake-water DOC concentration have been linked to increased solubility of soil organic matter in response to declining acid deposition (Evans et al., 2006; Monteith et al., 2007), and, conversely, elevated sulfur deposition usually results in reduced transport of soil organic matter. In our lakes, declining VNIRS-inferred TOC concentrations were accompanied by decreasing C / N ratios, which suggests a reduction of terrestrial organic matter deposition with increased acid deposition. Decreasing values of inferred pH from the late 1800s to minimum values around AD 1960 at Lindhultsgöl also provide evidence of the acidification history. Although the diatom-based pH reconstruction indicates continued acidification until the 1960s, the minimum in VNIRS-inferred TOC concentration was reached already around AD 1940, a few decades before sulfur deposition peaked. This may be explained by the high proportion of wetlands in the catchment of Lindhultsgöl. Evans et al. (2012) showed that already acidic soils may exhibit limited responses to enhanced acid deposition as DOC leaching stabilizes at a certain pH, below which no further decrease in DOC concentration occurs.

Despite the general negative correlation between sulfur deposition and inferred TOC concentration at our study sites, major changes in land use during the last century may also have had important effects on DOC export to the lakes. The onset of industrialization in the late 1800s led to urbanization and the documented decrease in rural population. Traditional types of land use were abandoned, in particular meadows and pastures, which were typically converted into spruce plantations and cultivated fields (Antonsson and Jansson, 2011). This development is clearly reflected in our pollen records as concomitant decreases in grassland and increases in coniferous woodland cover in the 1900s (Fig. 3). This land-use change is most pronounced at Lindhultsgöl, where grassland, cropland and wetland cover are reduced at the expense of woodland, and the RDA indicates a significant effect of especially the cereal cover reduction at this site. A significant reduction of the supply of terrestrial organic matter, as indicated by decreasing C / N ratios at both lakes, may be partly explained by the increase in sulfur deposition, which suppressed leaching of soil organic matter. However, reduced catchment erosion may also have been a direct effect of stabilization of previously disturbed soils following the rural population decrease and woodland succession. At the transition to commercial forestry and crop cultivation around AD 1900, new management practices with possible effects on lake-water TOC concentration, such as ditching, drainage and clear-cut harvesting, were introduced. However, ditching and drainage of forests and crop cultivations may involve complex responses of surface-water DOC concentrations, as some studies report increases (Ecke, 2008), while others provide evidence of decreases (Åström et al., 2001; McTiernan et al., 2001). In Sweden, ditching and drainage operations started in the late 1800s, with substantial increases from AD 1900 to the 1930s (Hånell et al., 2009). A major artificial deepening of the inlet stream at Åbodasjön in the early 1920s resulted in enhanced export of lithogenic material from the adjacent croplands for a few decades (Bragée et al., 2013). The interruption of decreasing TOC concentrations in the lake at ca. AD 1920–1960 (Fig. 3) was possibly reinforced by increased supply of soil-derived DOC through enhanced export from the inlet surroundings. Crop cultivation along the inlet was abandoned in the 1950s, which led to decreased supply of lithogenic material (Bragée et al., 2013) and an accelerated decrease in TOC concentration in the lake. Previous studies have attributed variations in the release of DOC to surface waters to changing forestry practices, and clear cutting can significantly affect stream-water DOC levels in boreal forests (Lepistö et al., 2008; Laudon et al., 2009). Considering the increased areal distribution of woodland and forestry activities within the catchments during the last century, this may constitute a potential source of increased DOC supply. Given the increase in the extent of clear cuts between AD 1946 and 2005, from 1 to 20 % of the modelled land-use area at Åbodasjön and from 0 to 13 % at Lindhultsgöl (Mazier et al., 2014), this process may have contributed to the elevated TOC concentrations in the lakes in the 1990s. However, ditching and clear-cutting probably result in only temporary increases in the supply of DOC, af-

fecting at least the following growth season (Laudon et al., 2009) and may therefore be difficult to distinguish in palaeolimnological records.

The increase in VNIRS-inferred TOC concentration at both lakes around AD 1990 is most likely linked to the recovery from acidification. The low sample resolution in the uppermost parts of the diatom records precludes detailed evaluation of recent changes in pH in response to decreased sulfur deposition, although the slight increase in the uppermost part of the record from Lindhultsgöl indicates a recent recovery. However, pH is not a straightforward measure of recovery from acidification (Skjelkvåle et al., 2003; SanClements et al., 2012), and the inconsistent responses in our records may be explained by the contemporary increase in lake-water TOC concentration as organic acids usually have an acidifying effect (Evans et al., 2001). Soil conditions are important for the solubility of organic matter, and the high proportion of coniferous woodland at both lakes and wetlands at Lindhultsgöl, typically associated with organic-rich soils, may have induced increased leaching of DOC in response to decreasing sulfur deposition during recent decades (Evans et al., 2012). Site-specific catchment soil properties may therefore be important for explaining the observed increases in TOC concentration in our study lakes after AD 1990 compared to other lakes in the region that show unchanged or even decreasing trends in DOC concentration (von Einem and Granéli, 2010). In addition, wetland areas in the catchments of both lakes have been treated by liming on a yearly basis to mitigate acidification, starting in AD 1984 at Åbodasjön and in AD 1993 at Lindhultsgöl, which may have contributed to the effects of declining sulfur deposition by accelerated leaching of DOC to the lakes (cf. Hindar et al., 1996).

In addition to changes in sulfur deposition and land management practices, climate may affect DOC concentration of lake waters through a variety of processes, including temperature-driven soil organic productivity and decomposition as well as precipitation-driven water table fluctuations and transport of organic carbon from terrestrial soils (e.g. Sobek et al., 2007). Increases in precipitation and temperature have been brought forward as potential causes of observed increases in DOC concentration in lake waters during the last three decades in several studies (Freeman et al., 2001; Hongve et al., 2004; Sarkkola et al., 2009). Future climate predictions for northern Europe include higher seasonal amounts and intensity of precipitation, as well as increasing mean annual air temperatures (Alcamo et al., 2002), which may result in continued increases in DOC export to lake waters (Larsen et al., 2010). Available meteorological data from Växjö (Fig. 1), reaching back to AD 1860, show an increase in annual precipitation from ca. AD 1980 and an increase in mean annual temperature from ca. AD 1990 (Fig. 3). Hence, climate change may have contributed to the observed and reconstructed increases in lake-water TOC concentration over recent decades, and the RDA data indicate that, at least at Åbodasjön, both precipitation and temperature have had an impact on the lake over the last century, while these effects seem to be less important at Lindhultsgöl. A possible explanation may be the larger catchment of Åbodasjön, making it more sensitive to changes in run-off, erosion and transport of terrestrial organic matter. The large proportion of wetland around Lindhultsgöl may also have a dampening effect on increased precipitation. At Lindhultsgöl, changes in land use have played a more important role at the centennial timescale. Changes in sulfur deposition during the last century have been a main driver for limnological change at both sites, despite their different land use and catchment characteristics (Figs. 3, 4), which supports the interpretation that this is a key factor behind the regional changes observed in lake-water TOC concentrations. This demonstrates the importance of applying a long-term perspective on lake-water DOC dynamics in order to differentiate between causal relationships.

5.3   Recent brownification and future implications

Our reconstructions indicate that TOC concentrations in the lakes were generally high during the past eight centuries, reaching similar or higher concentrations than those observed during recent decades. Commonly, there is a correlation between water colour (usually measured as absorbance at ca. 420–436 nm or using the platinum scale) and DOC concentration in lake waters. However, colour is strongly influenced by the composition of DOC, and a recent study has demonstrated that declining acidification in southern Sweden has led to increased leaching from soils of mobile, hydrophobic and aromatic DOC that contains relatively large and strongly coloured molecular compounds (Ekström et al., 2011). Moreover, iron has a strong influence on water colour, and elevated iron concentrations have been observed with the recent brownification in the UK (Neal et al., 2008) as well as in Sweden (Huser et al., 2011; Kritzberg and Ekström, 2012). Therefore, the VNIRS-inferred changes in TOC concentration in our two study lakes may not necessarily reflect changes in colour, although monitoring data from Åbodasjön indicate that this was indeed the case during recent decades (County Administrative Board of Kronoberg, unpublished data), consistent with increases in water colour observed in several other lakes in the study region. This is supported by high abundances of the diatom *Aulacoseira tenella* (> 20 %) in surface sediments from Åbodasjön, a species often associated with high DOC concentrations and strongly coloured lake waters (Huttunen and Turkia, 1994).

In contrast, the elevated TOC concentrations recorded in Åbodasjön during the late 1800s were most likely not associated with a corresponding increase in water colour, as indicated by unchanged diatom planktonic : benthic (P / B) ratios. Benthic and planktonic diatom communities are likely to respond to changes in the input of terrestrial organic matter through associated effects on the transparency of the water column, as the benthic community is primarily limited

by light in nutrient-poor lakes (Rosén et al., 2009; Karlsson et al., 2009). At this site, the pronounced increase in agricultural intensity in the late 1800s probably resulted in enhanced export of DOC compounds with relatively low molecular weights, which are typically associated with agriculture (cf. Cronan et al., 1999; Dalzell et al., 2011). A dominance of this type of DOC would not result in any significant increase in water colour as DOC derived from agricultural areas is in general structurally less complex and less coloured than DOC from forest soils (Wilson and Xenopoulos, 2009).

The early agricultural expansion in the 1200s resulted in a change in the diatom community towards elevated P / B ratios and a dominance of planktonic taxa typically favoured by high pH (Fig. 2). Hence, this diatom response to increased nutrient transport to the lake was most likely associated with early land use and not with any major increase in water colour caused by increased input of terrestrial organic carbon.

At Lindhultsgöl, minimum P / B ratios were recorded during the period of maximum sulfur emissions at AD 1950–1990, which indicates a decrease in water colour associated with the corresponding minima in inferred lake-water TOC concentration and pH.

Based on our results we can conclude that the increases in TOC concentration and water colour in our study lakes during the past three decades have been driven mainly by declining atmospheric sulfur deposition. This suggests a recovery from the phase of maximum sulfur emissions, which resulted in exceptionally low TOC concentrations in the lakes at ca. AD 1930–90. The RDA data obtained from the palaeolimnological records over the period since AD 1880 also indicate a recovery. At both sites, the temporal development of the RDA scores during this period (Fig. 4b and d) show that the youngest samples fall near the oldest, indicating a return to pre-industrial conditions, following a time of highly anomalous conditions. At Åbodasjön, there was first a period of low inferred lake-water TOC concentration and high pH in the 1930s–1960s, followed by decreasing pH but high Zr / Ti ratios (perhaps indicating enhanced erosion) in the 1980s and early 1990s, and then increasing lake-water TOC concentration and low Zr / Ti ratios towards the present. At Lindhultsgöl, the temporal development of the RDA scores also illustrates a partial recovery with similar scores, especially on the first axis, for young and old samples, while the period 1930–1980 was characterized by high first axis scores, associated with sedimentary indicators of low pH and VNIRS-inferred TOC concentration, while the driving forces were characterized by high sulfur and nitrogen deposition values and relatively high cereal cover in the catchment.

Our long-term records demonstrate that the TOC concentrations of the study lakes were strongly influenced by changes in agricultural practices, general land-use pressure, and associated variations in forest cover during the last 800 years (Fig. 5). The historical differences in the extent of agricultural activity at the sites establish that site-specific catchment characteristics and land-use dynamics are of great importance for lake-water DOC variations. The recently initiated increase in TOC concentration in the lakes may continue in the near future depending on the quantity of organic carbon stored in catchment soils due to suppression of DOC leaching during the acidification episode. However, the recovery of lake-water TOC concentrations has now reached levels that are comparable to the situation before the onset of 20th century acidification, which may lead to a levelling-off of the increasing trend. Given the reduction of atmospheric sulfur emissions during recent decades, it is likely that previously suppressed or masked effects of changes in land management and climate during the last century will become progressively more important drivers of lake-water DOC concentrations in the future.

Acknowledgements. This work was funded by the Swedish Research Council Formas (grant to W. Granéli). The authors are grateful to Shinya Sugita for input on the quantitative vegetation reconstructions, and Sofia Holmgren, Linda Randsalu-Wendrup and Christian Bigler for helpful support with diatom preparation and categorization. We are very grateful to all members of the NordForsk network LANDCLIM (coordinated by M. J. Gaillard, Linnaeus University, Sweden) for useful and inspiring discussions during the numerous workshops (2009–2011). We acknowledge the Swedish Meteorological Institute (SMHI) for precipitation and temperature data, and the Swedish Environmental Research Institute and MAGIC for sulfur and nitrogen deposition data. Constructive comments by the reviewers improved the final presentation.

Edited by: C. P. Slomp

References

Alcamo, J., Mayerhofer, P., Guardans, R., van Harmelen, T., van Minnen, J., Onigkeit, J., Posch, M., and de Vries, B.: An integrated assessment of regional air pollution and climate change in Europe: findings of the AIR-CLIM Project, Environ. Sci. Policy, 5, 257–272, 2002.

Alexandersson, H., Karlström, C., and Larsson-McCann, S.: Temperature and precipitation in Sweden, 1961–90, Reference normals, SMHI report 81, SMHI, Norrköping, 1991.

Andersson Palm, L.: Folkmängden i Sveriges socknar och kommuner 1571–1997, Books-on-Demand, Göteborg, 385 pp., 2000.

Antonsson, H. and Jansson, U. (Eds.): Agriculture and forestry in Sweden since 1900, The Royal Swedish Academy of Agriculture and Forestry, Stockholm, pp. 512, 2011.

Armstrong, A., Holden, J., Kay, P., Francis, B., Foulger, M., Gledhill, S., McDonald, A. T., and Walker, A.: The impact of peatland drain-blocking on dissolved organic carbon loss and discolouration of water; results from a national survey, J. Hydrol., 381, 112–120, 2010.

Arvola, L., Rask, M., Ruuhijärvi, J., Tulonen, T., Vuorenmaa, J., Ruoho-Airola, T., and Tulonen, J.: Long-term patterns in pH and colour in small acidic boreal lakes of varying hydrological and landscape settings, Biogeochemistry, 101, 269–279, 2010.

Åström, M., Aaltonen, E.-K., and Koivusaari, J.: Effect of ditching operations on stream-water chemistry in a boreal forested catchment, Sci. Total Environ., 279, 117–129, 2001.

Bade, D. L., Carpenter, S. R., Cole, J. J., Pace, M. L., Kritzberg, E., Van de Bogert, M. C., Cory, R. M., and McKnight, D. M.: Sources and fates of dissolved organic carbon in lakes as determined by whole-lake carbon isotope additions, Biogeochemistry, 84, 115,-129, 2007.

Battarbee, R. W. and Kneen, M. J.: The use of electronically counted microspheres in absolute diatom analysis, Limnol. Oceanogr. 27, 184–188, 1982.

Battarbee, R. W., Jones, V. J., Flower, R. J., Cameron, N. G., Bennion, H., Carvalho, L., and Juggins, S.: Diatoms, in: Tracking environmental change using lake sediments, Volume 3: terrestrial, algal and siliceous indicators, edited by: Smol, J. P., Birks, H. J. B., and Last, W. M., Kluwer Academic, Dortrecht, 155–202, 2001.

Boyle, J. F.: Rapid elemental analysis of sediment samples by isotope source XRF, J. Paleolimnol., 23, 213–221, 2000.

Boyle, J. F.: Loss of apatite caused irreversible early-Holocene lake acidification, The Holocene, 17, 543–547, 2007.

Bragée, P., Choudhary, P., Routh, J. Boyle, J. F., and Hammarlund, D.: Lake ecosystem responses to catchment disturbance and airborne pollution: an 800-year perspective in southern Sweden, J. Paleolimnol., 50, 545–560, 2013.

Clark, J. M., Bottrell, S. H., Evans, C. D., Monteith, D. T., Bartlett, R., Rose, R., Newton, R. J., and Chapman, P. J.: The importance of the relationship between scale and process in understanding long-term DOC dynamics, Sci. Total Environ., 408, 2768–2775, 2010.

Cole, J. J., Prairie, Y. T., Caraco, N. F., McDowell, W. H., Tranvik, L. J., Striegl, R. G., Duarte, C. M., Kortelainen, P., Downing J. A., Middelburg, J. J., and Melack, J.: Plumbing the global carbon cycle: integrating inland waters into the terrestrial carbon budget, Ecosystems, 10, 172–185, 2007.

Correll, D. L., Jordan, T. E., and Weller, D. E.: Effects of precipitation, air temperature, and land use on organic carbon discharges from Rhode River watersheds, Water Air Soil Poll., 128, 139–159, 2001.

Cronan, C. S., Piampiano, J. T., and Patterson, H. H.: Influence of Land Use and Hydrology on Exports of Carbon and Nitrogen in a Maine River Basin, J. Environ. Qual., 28, 953–961, 1999.

Cunningham, L., Bishop, K., Mettävainio, E., and Rosén, P.: Paleoecological evidence of major declines in total organic carbon concentrations since the nineteenth century in four nemoboreal lakes, J. Paleolimnol., 45, 507–518, 2011.

Dalzell, B. J., King, J. Y., Mulla, D. J., Finlay, J. C., and Sands, G. R.: Influence of subsurface drainage on quantity and composition of dissolved organic matter export from agricultural landscapes, J. Geophys. Res., 116, G02023, doi:10.1029/2010JG001540, 2011.

Daniel, E.: Beskrivning till jordartskartan 5E Växjö NV, K 168 Sveriges Geologiska Undersökning (SGU), 77 pp., 2009.

Diehl, S., Berger, S., Ptacnik, R., and Wild, A.: Phytoplankton, light, and nutrients in a gradient of mixing depths: field experiments, Ecology, 83, 399–411, 2002.

Ecke, F.: Drainage ditching at the catchment scale affects water quality and macrophyte occurrence in Swedish lakes, Freshwater Biol., 54, 119–126, 2008.

Ekström, S. M., Kritzberg, E. S., Kleja, D. B., Larsson, N., Nilsson, P. A., Graneli, W., and Bergkvist, B.: Effect of acid deposition on quantity and quality of dissolved organic matter in soil–water, Environ. Sci. Technol., 45, 4733–4739, 2011.

Emanuelsson, U. (Ed.): The rural landscapes of Europe: how man has shaped European nature, Swedish Research Council Formas, Stockholm, Sweden, pp. 384, 2009.

Engstrom D. R. and Wright H. E. J.: Chemical stratigraphy of lake sediments as a record of environmental change, in: Lake Sediments and Environmental History, edited by: Haworth E. Y. and Lund J. W. G., Leicester University Press, Bath, 11–68, 1984.

Erlandsson, M., Buffam, I., Fölster, J., Laudon, H., Temnerud, J., Weyhenmeyer, G. A., and Bishop, K.: Thirty-five years of synchrony in the organic matter concentrations of Swedish rivers explained by variation in flow and sulphate, Global Change Biol., 14, 1191–1198, 2008.

Erlandsson, M., Cory, N., Köhler, S., and Bishop, K.: Direct and indirect effects of increasing dissolved organic carbon levels on pH in lakes recovering from acidification, J. Geophys. Res., 115, G03004, doi:10.1029/2009JG001082, 2010.

Evans, C. D., Cullen, J. M., Alewell, C., Kopácek, J., Marchetto, A., Moldan, F., Prechtel, A., Rogora, M., Vesely, J., and Wright, R.: Recovery from acidification in European surface waters, Hydrol. Earth Syst. Sc., 5, 283–298, 2001.

Evans, C. D., Monteith, D. T., and Cooper, D. M.: Long-term increases in surface water dissolved organic carbon: Observations, possible causes and environmental impacts, Environ. Pollut., 137, 55–71, 2005.

Evans, C. D., Chapman, P. J., Clark, J. M., Monteith, D. T., and Cresser, M. S.: Alternative explanations for rising dissolved organic carbon export from organic soils, Glob. Change Biol., 12, 2044–2053, 2006.

Evans, C. D., Jones, T. G., Burden, A., Ostle, N., Zieliński, P., Cooper, M. D., Peacock, M., Clark, J. M., Oulehle, F., Cooper, D., and Freeman, C.: Acidity controls on dissolved organic carbon mobility in organic soils, Glob. Change Biol., 18, 3317–3331, 2012.

Findlay, S. E.: Increased carbon transport in the Hudson River: unexpected consequence of nitrogen deposition?, Front. Ecol. Environ., 3, 133–137, 2005.

Fredh, D., Broström, A., Rundgren, M., Lagerås, P., Mazier, F., and Zillén, L.: The impact of land-use change on floristic diversity at regional scale in southern Sweden 600 BC–AD 2008, Biogeosciences, 10, 3159–3173, doi:10.5194/bg-10-3159-2013, 2013.

Fredh, D., Mazier, F., Bragée, P., Lagerås, P., Rundgren M., Hammarlund, D., and Broström, A.: The effect of local land-use on floristic diversity during the past 1000 years in southern Sweden, The Holocene, submitted, 2014.

Freeman, C., Evans, C. D., Monteith, D. T., Reynolds, B., and Fenner, N.: Export of organic carbon from peat soils, Nature, 412, 785, 2001.

Granéli, W.: Brownification of Lakes, in:. Encyclopedia of Lakes and Reservoirs, edited by: Bengtsson, L., Herschy, R. W., and Fairbridge, R. W., Springer Science, New York, 117–119, 2012.

Gustafsson, L.: Geographical classifications of plants and animals, in:, National Atlas of Sweden, Geography of plants and animals, edited by: Gustafsson, L. and Ahlén, I., SNA publishing, Stockholm, 25–28, 1996.

Haaland, S., Hongve, D., Laudon, H., Riise, G., and Vogt, R. D.: Quantifying the drivers of the increasing colored organic matter in boreal surface waters, Environ. Sci. Technol., 44, 2975–2980, 2010.

Hånell, B.: Möjlighet till höjning av skogsproduktionen i Sverige genom dikesrensning, dikning och gödsling av torvmarker, in: Skogsskötsel för ökad tillväxt, edited by: Fahlvik, N., Johansson, U., and Nilsson, U., Faktaunderlag till MINT-utredningen, SLU, Rapport, Bilaga 4, 1–28, 2009.

Hindar, A., Kroglund, F., Lydersen, E., Skiple, A., and Høgberget, R.: Liming of wetlands in the acidified Lake Røynelandsvatn catchment in southern Norway: effects on stream water chemistry, Can. J. Fish. Aquat. Sci., 53, 985–993, 1996.

Hongve, D., Riise, G., and Kristiansen, J.: Increased colour and organic acid concentrations in Norwegian forest lakes and drinking water – a result of increased precipitation?, Aquat. Sci. 66, 231–238, 2004.

Huser, B. J., Köhler, S. J., Wilander, A., Johansson, K., and Fölster, J.: Temporal and spatial trends for trace metals in streams and rivers across Sweden (1996–2009), Biogeosciences, 8, 1813–1823, doi:10.5194/bg-8-1813-2011, 2011.

Huttunen, P. and Turkia J.: Diatoms as indicators of alkalinity and TOC in lakes: Estimation of optima and tolerances by weighted averaging, in: Proceedings of the 11th International Diatom Symposium, San Francisco, U.S.A., 12–17 August 1990, Memoirs of the California Academy of Sciences, No 17, edited by: Kociolek J. P., 649–658, 1994.

Juggins, S.: Quantitative reconstructions in palaeolimnology: new paradigm or sick science?, Quaternary Sci. Rev., 64, 20–32, 2013.

Juggins, S. and Birks, H.J.B.: Quantitative environmental reconstructions from biological data, in: Tracking environmental change using lake sediments, Volume 5: data handling and numerical techniques, edited by: Birks, H. J. B,, Lotter, A. F., Juggins, S., and Smol, J. P., Springer, Dordrecht, 431–494, 2012.

Karlsson, J., Bystrom, P., Ask, J., Ask, P., Persson, L., and Jansson, M.: Light limitation of nutrient-poor lake ecosystems, Nature, 460, 506–509, 2009.

Koinig, K. A., Shotyk, W., Lotter, A. F., Ohlendorf, C., and Sturm, M.: 9000 years of geochemical evolution of lithogenic major and trace elements in the sediment of an alpine lake–the role of climate, vegetation, and land-use history, J. Paleolimnol., 30, 307–320, 2003.

Kortelainen, P.: Content of total organic carbon in Finnish lakes and its relationship to catchment characteristics, Can. J. Fish. Aquat. Sci., 50, 1477–1483, 1993.

Krammer, K.: Pinnularia: Eine Monographie der Europäischen Taxa, Bibliotheca Diatomologica, Vol. 26, Cramer J. Gebrüder Borntraeger Verlag, Berlin/Stuttgart, pp. 353, 1992.

Krammer, K. and Lange-Bertalot, H.: Bacillariophyceae, 1. Naviculaceae, in: Süsswasserflora von Mitteleuropa, Vol. 2, edited by:

Ettl, H., Gärtner, G., Gerloff, J., Heynig, H., and Mollenhauer D., Gustav Fischer Verlag, Stuttgart, pp. 876, 1986.

Krammer, K. and Lange-Bertalot, H.: Bacillariophyceae, 2. Bacillariaceae, in: Süsswasserflora von Mitteleuropa, Vol. 2, edited by: Ettl, H., Gärtner, G., Gerloff, J., Heynig, H., and Mollenhauer D., Gustav Fischer Verlag, Stuttgart, pp. 596, 1988.

Krammer, K. and Lange-Bertalot, H.: Bacillariophyceae, 3. Centrales, Fragilariaceae, Eunotiaceae, in: Süsswasserflora von Mitteleuropa, Vol. 2, edited by: Ettl, H., Gärtner, G., Gerloff, J., Heynig, H., and Mollenhauer D. (Eds.), Gustav Fischer Verlag, Stuttgart, pp. 576, 1991a.

Krammer, K. and Lange-Bertalot, H.: Bacillariophyceae, 4. Achnanthaceae, Kritische Ergänzungen zu Navicula (Lineolatae) und Gomphonema, in: Süsswasserflora von Mitteleuropa, Vol. 2, edited by: Ettl, H., Gärtner, G., Gerloff, J., Heynig, H., and Mollenhauer D., Gustav Fischer Verlag, Stuttgart, pp. 437, 1991b.

Kritzberg, E. S. and Ekström, S. M.: Increasing iron concentrations in surface waters – a factor behind brownification?, Biogeosciences, 9, 1465–1478, doi:10.5194/bg-9-1465-2012, 2012.

Lagerås, P.: The ecology of expansion and abandonment-Medieval and post-medieval land-use and settlement dynamics in a landscape perspective, National Heritage Board, Stockholm, pp. 256, 2007.

Lange-Bertalot, H. and Krammer, K.: Achnanthes, eine Monographie der Gattung: mit Definition der Gattung Cocconeis und Nachträgen zu den Naviculaceae, Bibliotheca Diatomologica, Vol. 18, Cramer J. Gebrüder Borntraeger Verlag, Berlin/Stuttgart, pp. 393, 1989.

Larsen, S., Andersen, T. O. M., and Hessen, D. O.: Climate change predicted to cause severe increase of organic carbon in lakes, Global Change Biol., 17, 1186–1192, 2010.

Laudon, H., Hedtjärn, J., Schelker, J., Bishop, K., Sørensen, R., and Ågren, A.: Response of Dissolved Organic Carbon following Forest Harvesting in a Boreal Forest, AMBIO, 38, 381–386, 2009.

Ledesma, J. L. J., Köhler, S. J., and Futter, M. N.: Long-term dynamics of dissolved organic carbon: Implications for drinking water supply, Sci. Total Environ., 432, 1–11, 2012.

Lepistö, A., Kortelainen, P., and Mattsson, T.: Increased organic C and N leaching in a northern boreal river basin in Finland, Global Biogeochem. Cy., 22, GB3029, doi:10.1029/2007GB003175, 2008.

Löfgren, S., Forsius, M., and Andersen, T.: The color of water – climate induced water color increase in Nordic lakes and streams due to humus, Nordic council of Ministers brochure, Copenhagen, pp. 12, 2003.

Mattsson, T., Kortelainen, P., and Räike, A.: Export of DOM from boreal catchments: impacts of land use cover and climate, Biogeochemistry, 76, 373–394, 2005.

Mattsson, T., Kortelainen, P., Lepistö, A., and Räike, A.. Organic and minerogenic acidity in Finnish rivers in relation to land use and deposition, Sci. Total Environ., 383, 183–192, 2007.

Mazier, F., Broström, P., Bragée, P., Fredh, D., Stenberg, L., Thiere, G., Sugita, S., and Hammarlund, D.: Two hundred years of land-use change in South Swedish Uplands: comparison of historical map-based estimates with pollen-based reconstruction using the Landscape Reconstruction Algorithm, Rev. Palaeobot. Palyno., in press, 2014.

McDonald, S., Bishop, A. G., Prenzler, P. D., and Robards, K.: Analytical chemistry of freshwater humic substances, Anal. Chim. Ac., 527, 105–124, 2004.

McTiernan, K. B., Jarvis, S. C., Scholefield, D., and Hayes, M. H. B.: Dissolved organic carbon losses from grazed grasslands under different management regimes, Water Res., 35, 2565–2569, 2001.

Meyers, P. A. and Lallier-Verges, E.: Lacustrine sedimentary organic matter records of Late Quaternary paleoclimates, J. Paleolimnol., 21, 345–372, 1999.

Monteith, D. T., Stoddard, J. L., Evans, C. D., De Wit, H. A., Forsius, M., Hogasen, T., Wilander, A., Skjelkvale, B. L., Jeffries, D. S., Vuorenmaa, J., Keller, B., Kopacek, J., and Vesely, J.: Dissolved organic carbon trends resulting from changes in atmospheric deposition chemistry, Nature, 450, 537–540, 2007.

Myrdal, J.: En agrarhistorisk syntes, in: Agrarhistoria, edited by: Larsson, B. M. P., Morell, M., and Myrdal, J., LTs Förlag, Stockholm, 302–322, 1997.

Myrdal, J.: Scandinavia, in: Agrarian change and crisis in Europe, 1200–1500, edited by: Kitsikopoulos, H., Routledge, New York, 204–249, 2012.

Neal, C., Lofts, S., Evans, C. D., Reynolds, B., Tipping, E., and Neal, M.: Increasing iron concentrations in UK upland waters, Aquat. Geochem., 14, 263–288, 2008.

Oksanen, J.: Multivariate analysis of ecological communities in R: vegan tutorial, R package version 1.7., 2011.

Pace, M. L. and Cole, J. J.: Synchronous variation of dissolved organic carbon and color in lakes, Limnol. Oceanogr., 47, 333–342, 2002.

Rasmussen, J. B., Godbout, L., and Schallenberg, M: The humic content of lake water and its relationship to watershed and lake morphometry, Limnol. Oceanogr., 34, 1336–1343, 1989.

Renberg, I.: A procedure for preparing large sets of diatom slides from sediment cores., J. Paleolimnol., 4, 87–90, 1990.

Renberg, I., Korsman, T., and Birks, H. J. B.: Prehistoric increases in the pH of acid-sensitive Swedish lakes caused by land-use changes, Nature, 362, 824–827, 1993.

Rohde, H., Grennfelt, P., Wisniewski, J., Ågren, C., Bengtsson, G., Johansson, K., Kauppi, P., Kucera, V., Rasmussen, l., Rosseland, B., Schotte, l., and Selldén, G.: Acid Reign 95 – Conference Summary Statement, Water Air Soil Pollut., 85, 1–14, 1995.

Rosén, P.: Total organic carbon (TOC) of lake water during the Holocene inferred from lake sediments and near-infrared spectroscopy (NIRS) in eight lakes from northern Sweden, Biogeochemistry, 76, 503–516, 2005.

Rosén, P., Cunningham, L., Vonk, J., and Karlsson, J.: Effects of climate on organic carbon and the ratio of planktonic to benthic primary producers in a subarctic lake during the past 45 years, Limnol. Oceanogr., 54, 1723–1732, 2009.

Rosén, P., Bindler, R., Korsman, T., Mighall, T., and Bishop, K.: The complementary power of pH and lake-water organic carbon reconstructions for discerning the influences on surface waters across decadal to millennial time scales, Biogeosciences, 8, 2717–2727, doi:10.5194/bg-8-2717-2011, 2011.

Rouillard, A., Rosén, P., Douglas, M. S., Pienitz, R., and Smol, J. P.: A model for inferring dissolved organic carbon (DOC) in lakewater from visible-near-infrared spectroscopy (VNIRS) measures in lake sediment, J. Paleolimnol., 46, 187–202, 2011.

SanClements, M. D., Oelsner, G. P., McKnight, D. M., Stoddard, J. L., and Nelson, S. J.: New insights into the source of decadal increases of dissolved organic matter in acid-sensitive lakes of the Northeastern United States, Environ. Sci. Technol., 46, 3212–3219, 2012.

Sarkkola, S., Koivusalo, H., Laurén, A., Kortelainen, P., Mattsson, T., Palviainen, M., Piirainen, S., Starr, M., and Finér, L: Trends in hydrometeorological conditions and stream water organic carbon in boreal forested catchments, Sci. Environ., 408, 92–101, 2009.

Schöpp, W., Posch, M., Mylona, S., and Johansson, M.: Long-term development of acid deposition (1880–2030) in sensitive freshwater regions in Europe, Hydrol. Earth Syst. Sci., 7, 436–446, 2003,
http://www.hydrol-earth-syst-sci.net/7/436/2003/.

Sjörs, H.: Amphi-Atlantic zonation, Nemoral to Arctic, in: North Atlantic biota and their history, edited by: Löve, A. and Löve, D., Pergamon Press, Oxford, 109–125, 1963.

Skjelkvåle, B. L., Evans, C., Larssen, T., Hindar, A., and Raddum, G. G.: Recovery from acidification in European surface waters: A view to the future, AMBIO, 32, 170–175, 2003.

Snucins, E. and Gunn, J.: Interannual variation in the thermal structure of clear and colored lakes, Limnol. Oceanogr., 45, 1639–1646, 2000.

Sobek, S., Tranvik, L. J., Prairie, Y. T., Kortelainen, P., and Cole, J. J.: Patterns and regulation of dissolved organic carbon: An analysis of 7500 widely distributed lakes, Limnol. Oceanogr., 52, 1208–1219, 2007.

Stoddard, J. L., Kahl, J. S., Deviney, F. A., DeWalle, D. R., Driscoll, C. T., Herlihy, A. T., Kellogg, J. H., Murdoch, P. S., Webb, J. R., and Webster, K. E.: Response of surface water chemistry to the Clean Air Act Amendments of 1990, Report EPA 620/R-03/001, US Environmental Protection Agency, North Carolina, 2003.

Sugita, S.: Theory of quantitative reconstruction of vegetation I: pollen from large sites REVEALS regional vegetation composition, The Holocene, 17, 229–241, 2007a.

Sugita, S.: Theory of quantitative reconstruction of vegetation II: all you need is LOVE, The Holocene, 17, 243–257, 2007b.

Taboada, T., Cortizas, A. M., García, C., and García-Rodeja, E.: Particle-size fractionation of titanium and zirconium during weathering and pedogenesis of granitic rocks in NW Spain, Geoderma, 131, 218–236, 2005.

Ter Braak, C. J. F. and Smilauer, P.: CANOCO reference manual and User's guide to CANOCO for Windows – Softward for Canonica Community Ordination (version 4), Centra for Biometry, Wageningen, pp. 351, 1998.

von Einem, J. and Granéli, W.: Effects of fetch and dissolved organic carbon on epilimnion depth and light climate in small forest lakes in southern Sweden, Limnol. Oceanogr., 55, 920–930, 2010.

Vuorenmaa, J., Martin, F., and Jaakko, M.: Increasing trends of total organic carbon concentrations in small forest lakes in Finland from 1987 to 2003, Sci. Total Environ., 365, 47–65, 2006.

Wetzel, R.: Limnology, 3 edition: Lake and River Ecosystems, Academic Press, San Diego, pp. 1006, 2001.

Wikman, H.: Beskrivning till berggrundskartorna 5E Växjö NO och NV, Af 201 and 216, Sveriges Geologiska Undersökning (SGU), 108 pp., 2000.

Wilson, H. F. and Xenopoulos, M. A.: Effects of agricultural land use on the composition of fluvial dissolved organic matter, Nat. Geosci., 2, 37–41, 2009.

Wolfe, A. P.: On diatom concentrations in lake sediments: results from an inter-laboratory comparison and other tests performed on a uniform sample, J. Paleolimnol., 18, 261–268, 1997.

Worrall, F. and Burt, T. P.: Trends in DOC concentration in Great Britain, J. Hydrol., 346, 81–92, 2007.

Xenopoulos, M. A., Lodge, D. M., Frentress, J., Kreps, T. A., Bridgham, S. D., Grossman, E., and Jackson, C. J.: Regional comparisons of watershed determinants of dissolved organic carbon in temperate lakes from the Upper Great Lakes region and selected regions globally, Limnol. Oceanogr., 48, 2321–2334, 2003.

Yallop, A. R., Clutterbuck, B., and Thacker, J.: Increases in humic dissolved organic carbon export from upland peat catchments: the role of temperature, declining sulphur deposition and changes in land management, Clim. Res., 45, 43–56, 2011.

8

Seasonal variations in concentration and lability of dissolved organic carbon in Tokyo Bay

A. Kubo[1], M. Yamamoto-Kawai[2], and J. Kanda[1]

[1]Department of Ocean Sciences, Tokyo University of Marine Science and Technology, 4-5-7 Konan, Minato-ku, Tokyo, 108-8477, Japan
[2]Center for Advanced Science and Technology, Tokyo University of Marine Science and Technology, 4-5-7 Konan, Minato-ku, Tokyo, 108-8477, Japan

Correspondence to: A. Kubo (kuboatsushi0412@gmail.com)

Abstract. Concentrations of recalcitrant and bioavailable dissolved organic carbon (DOC) and their seasonal variations were investigated at three stations in Tokyo Bay, Japan, and in two freshwater sources flowing into the bay. On average, recalcitrant DOC (RDOC), as a remnant of DOC after 150 days of bottle incubation, accounted for 78 % of the total DOC in Shibaura sewage treatment plant (STP) effluent, 67 % in the upper Arakawa River water, 66 % in the lower Arakawa River water, and 78 % in surface bay water. Bioavailable DOC (BDOC) concentrations, defined as DOC minus RDOC, were lower than RDOC at all stations. In freshwater environments, RDOC concentrations were almost constant throughout the year. In the bay, RDOC was higher during spring and summer than in autumn and winter because of freshwater input and biological production. The relative concentration of RDOC in the bay derived from phytoplankton, terrestrial, and open-oceanic waters was estimated to be 8–10, 21–32, and 59–69 %, respectively, based on multiple regression analysis of RDOC, salinity, and chl a. In addition, comparison with previous data from 1972 revealed that concentrations of RDOC and BDOC have decreased by 33 and 74 % at freshwater sites and 39 and 76 % in Tokyo Bay, while the ratio of RDOC to DOC has increased. The change in DOC concentration and composition was probably due to increased amounts of STP effluent entering the system. Tokyo Bay exported mostly RDOC to the open ocean because of the remineralization of BDOC.

1 Introduction

The dissolved organic carbon (DOC) pool is the largest organic carbon reservoir in the ocean and contains 662 Pg of carbon, which is roughly equivalent to that stored in the atmosphere in the form of carbon dioxide (Hansell et al., 2009). In open oceans, DOC production is ultimately constrained by primary production (e.g., Carlson, 2002). In coastal waters, DOC consists of diverse mixtures of carbon with varying timescales of lability formed by primary production and materials of terrestrial origin. Riverine DOC export to the open ocean has been estimated to range from 0.21 to 0.25 PgC yr^{-1} (Meybeck, 1993; Ludwig et al., 1996; Hedges et al., 1997; Cauwet, 2002), without considering loss or gain of DOC in coastal waters. Coastal waters are typically considered passive conduits in regional and global carbon budgets (Cole et al., 2007; Aufdenkampe et al., 2011; Regnier et al., 2013). However, degradation of terrestrial DOC and biological production of DOC in coastal regions can significantly modify the flux of DOC to the open ocean. Dai et al. (2012) recently reported that riverine DOC export to the open ocean would be reduced to 0.17 PgC yr^{-1} if 10% was degraded in coastal waters. However, their assumption of 10 % was based on the results of only a few bottle incubation experiments (Amon and Benner, 1996; Raymond and Bauer, 2000; Moran et al., 1999). Therefore, to better understand DOC export to the open ocean, experimental data describing DOC lability, preferably from different environmental locations and different seasons, are needed.

In this study, we measured seasonal variations in the concentration and lability of DOC in Tokyo Bay, Japan, to evaluate the significance of DOC degradation to the carbon budget in coastal waters and carbon export to the open ocean. The bay is semi-enclosed, with an area of about $922 \, \mathrm{km^2}$ and a mean water depth of 19 m. The residence time of water in the bay is estimated to be about 50 days (Takada et al., 1992). The bay is located in central Japan and surrounded by metropolitan areas, with a total population of about 29 million. Tokyo Bay represents typical highly urbanized coastal waters, which are rapidly expanding worldwide (Nellemann et al., 2008). We also compared our results with those obtained by Ogura (1975), who carried out an investigation of Tokyo Bay in the 1970s and found that DOC in coastal waters could be divided into bioavailable DOC (BDOC) and recalcitrant DOC (RDOC). Owing to his investigation, BDOC and RDOC data from 1972 are available for Tokyo Bay.

2 Materials and methods

Freshwater samples were collected two and eight times from the upper and lower Arakawa River, respectively, and five times from effluent of the Shibaura sewage treatment plant (STP; Fig. 1) between December 2011 and October 2013. Freshwater samples were collected using a bucket, transferred into HCl acid-washed 1 L polyethylene bottles, and kept in the dark until being processed in the laboratory. The bucket and sample bottles were rinsed three times with sample water before being filled. Within 2 h of sample collection, the freshwater samples were carried back to the laboratory. DOC and the degradation experiment samples were filtered immediately after arrival in the laboratory through GF/F filters (nominal pore size $0.7 \, \mu\mathrm{m}$) that had been precombusted at $450 \, °\mathrm{C}$ for 3 h. Surface seawater of Tokyo Bay was collected monthly in 8 L Niskin bottles mounted on a conductivity–temperature–depth (CTD) rosette on the R/V *Seiyo-maru* of Tokyo University of Marine Science and Technology at three stations from January 2012 to December 2012 (Fig. 1). Within 1 h of sample collection, DOC and the degradation experiment samples were filtered through precombusted GF/F filters on board. Then, samples were kept in the dark and carried back to the laboratory within 4 h. We assumed that GF/F filters allow the passage of a significant fraction of free-living bacteria into DOC samples (e.g., Bauer and Bianchi, 2011). In addition, Tranvik and Höfle (1987) investigated the interactions between bacterial assemblages and DOC consumption using batch cultures and found that the DOC bioavailability was independent of the inoculum. Tanaka et al. (2011) also showed that the mineralization rate of the BDOC fraction in a coral reef was not different from natural waters and waters filtrated by GF/F; nevertheless, the initial bacterial abundance in the incubated waters filtrated by GF/F was about 30–50 % of bacteria abundance

in natural waters. Therefore, we did not add the microbial community. We also did not add nutrients for the degradation experiment because we assumed nutrients were not limiting microbial growth (see Sect. 3.1). Degradation experiment samples were then transferred to 600 mL amber glass bottles and stored at room temperature ($20 \, °\mathrm{C}$) in total darkness until analysis. The 100 mL headspace in each glass bottle contains about $900 \, \mu\mathrm{mol}$ oxygen. The highest initial DOC concentration in this study was $430 \, \mu\mathrm{mol \, L^{-1}}$ (Table 1). If we assume that 1 mol of oxygen is consumed when 1 mol of organic carbon is mineralized into CO_2, the oxygen in the headspace should have provided sufficient oxygen supply for heterotrophic decomposition by bacteria. The degradation experiments were conducted based on a total of seven incubations (0, 5, 10, 20, 50, 100, and 150 days) per field sampling event. After incubation, samples were dispensed into glass vials that had been prewashed with HCl and pure water (Milli-Q water, Millipore Corp., Bedford, MA, USA) and then precombusted. Freshwater samples were preserved with $6 \, \mathrm{mol \, L^{-1}}$ HCl at a concentration corresponding to 1 % of the sample volume, then stored in a refrigerator ($5 \, °\mathrm{C}$). Tokyo Bay samples were frozen ($-25 \, °\mathrm{C}$) without adding HCl. DOC samples were measured at least in triplicate with a total organic carbon (TOC) analyzer (TOC-$\mathrm{V_{CSH}}$, Shimadzu, Kyoto, Japan). Potassium hydrogen phthalate (Wako Pure Industries, Osaka, Japan) was used as a standard for the measurement of DOC. DOC blank, including pure water, instrument blank, and any carbon derived from the vials, was about $3 \, \mu\mathrm{mol \, L^{-1}}$ in total.

RDOC was here defined as the concentration of DOC remaining at 150 days, and BDOC was obtained by subtracting RDOC from the initial DOC (Lønborg et al., 2009). The degradation rate of DOC was described by a first-order exponential decay model with a constant RDOC pool:

$$\mathrm{DOC}(t) = \mathrm{BDOC} \cdot \exp(-k \cdot t) + \mathrm{RDOC}, \tag{1}$$

where $\mathrm{DOC}(t)$ is the amount of DOC remaining at time t (day), k is the degradation rate constant ($\mathrm{day^{-1}}$), and RDOC is the remaining DOC pool after 150 days of incubation. BDOC is the bioavailable DOC ($\mu\mathrm{mol \, L^{-1}}$) at the beginning of incubation and practically equals the subtraction of RDOC from initial DOC. Using BDOC and RDOC concentrations, k can be estimated by fitting the observed $\mathrm{DOC}(t)$ values to Eq. (1) using Matlab 2012a. For comparison with the results reported by Lønborg and Álvarez-Salgado (2012), we used the following equation to normalize the degradation rate to the rate at $15 \, °\mathrm{C}$:

$$k\left(15 \, °\mathrm{C}\right) = k(T) \cdot (Q_{10})^{\frac{T-15}{10}}, \tag{2}$$

where $k(15 \, °\mathrm{C})$ and $k(T)$ are the degradation rate constants at $15 \, °\mathrm{C}$ and $T \, °\mathrm{C}$ ($20 \, °\mathrm{C}$ for our experiment). Q_{10} is the temperature coefficient. In this study, we used a value of 2.2, based on Lønborg and Álvarez-Salgado (2012).

Temperature and salinity were measured in the field using a YSI EC 300 (YSI/Nanotech Inc., Yellow Springs,

Table 1. Temperature ($^\circ$C), salinity, chl a concentrations (μg L^{-1}), DOC concentrations (μmol L^{-1}) $\pm$ standard deviation, and POC concentrations (μmol L^{-1}) at the upper Arakawa River (upper AR), the lower Arakawa River (lower AR), and Shibaura STP stations.

Station	Date	Temp.	Sal.	chl a	DOC	POC
Upper AR	Apr 2013	10.9	0.0	0.2	33 ± 0	13
Upper AR	Oct 2013	17.4	0.0	0.2	42 ± 1	7
Lower AR	Dec 2011	12.1	0.6	2.0	247 ± 4	178
Lower AR	Jan 2012	7.0	0.2	7.6	290 ± 5	145
Lower AR	Feb 2012	7.2	0.2	49.3	355 ± 3	313
Lower AR	May 2012	23.6	0.2	33.9	205 ± 1	168
Lower AR	Jul 2012	24.2	0.2	1.5	213 ± 2	84
Lower AR	Aug 2012	23.9	0.0	1.2	185 ± 2	59
Lower AR	Nov 2012	17.4	0.2	2.0	236 ± 2	63
Lower AR	Dec 2012	11.8	0.2	10.7	155 ± 1	76
Shibaura STP*	Jan 2012	14.9	0.4	0.9	387 ± 2	191
Shibaura STP*	Feb 2012	17.2	1.2	0.1	305 ± 3	79
Shibaura STP	May 2012	27.6	2.9	3.7	430 ± 4	71
Shibaura STP	Jul 2012	27.9	1.9	0.5	366 ± 3	38
Shibaura STP	Aug 2012	27.7	1.9	2.2	292 ± 2	48
Shibaura STP	Nov 2012	20.5	4.0	0.3	341 ± 3	76
Shibaura STP	Dec 2012	17.2	0.4	1.2	366 ± 3	83

* Sampling for water properties only (no degradation experiment).

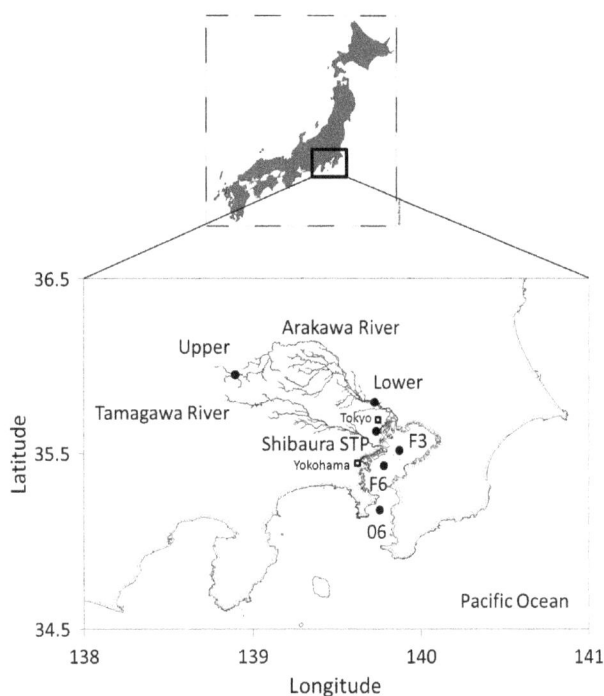

Figure 1. Map of Tokyo Bay. Locations of sampling sites are indicated by black circles.

OH, USA) at freshwater sites and a CTD (Falmouth Scientific Inc., Bourne, MA, USA) for sites in the bay. Water samples for chlorophyll a (chl a) measurement were filtered through precombusted (450 $^\circ$C, 3 h) GF/F filters. Af-

ter filtration, chlorophyllous pigments were extracted using N, N-dimethylformamide, and the concentrations of chl a were determined by the fluorometric method (Suzuki and Ishimaru, 1990; fluorometer used: TD-700, Turner Designs, Sunnyvale, CA, USA). Samples for particulate organic carbon (POC) were filtered through precombusted (450 $^\circ$C, 3 h) GF/F filters, after which the filters were stored at -80 $^\circ$C until analysis. The samples for POC analyses were dried at 60 $^\circ$C and acidified with vapor at 12 mol L^{-1} HCl to remove carbonate before analysis. POCs were measured using an isotope ratio mass spectrometer (Hydra 20-20, SerCon Ltd., Crewe, UK) coupled to an elemental analyzer (ANCA-GSL, SerCon Ltd., Crewe, UK).

3 Results and discussion

3.1 Nutrient conditions in Tokyo Bay

Nutrient concentrations in freshwater and Tokyo Bay sites were high throughout the year (Tables S1 and S2 in the Supplement). During summer, the phosphorus concentration generally decreased and the nitrogen / phosphorus ratio was higher than the Redfield ratio of 16 (Redfield et al., 1963), suggesting that phosphorus acts as a limiting factor of primary production in the bay. A degradation experiment with phosphate (KH$_2$PO$_4$, 2 μmol L^{-1}) was conducted in July 2012 to ensure that phosphorus was not a limiting factor; at this time, the concentration of phosphate was the lowest in the year (0.1 μmol L^{-1}; Tables S1 and S2). The results of the degradation experiment with added phosphorus were not significantly different from those of the degradation experiment without added phosphorus ($y = 1.1x - 8.2$, $R^2 = 0.97$, $p < 0.05$). We did not add nutrients for the degradation experiment because we assumed nutrients were not limiting microbial growth.

3.2 Lability and sources of freshwater DOC flowing into Tokyo Bay

The lowest chl a, DOC, and POC concentrations were observed at the upper Arakawa River station, which is considered to be pristine (Table 1). The average concentration of DOC was $38 \mu mol L^{-1}$ at the upper Arakawa River station. Headstream water sources in Japan are mostly surface runoff from neighboring watersheds and groundwater input which runs through the mineral soil horizon before entering surface water (Nakamura et al., 2011). Precipitation is characterized by very low DOC concentrations (Avery Jr. et al., 2003). Groundwater inputs through the mineral soil horizon typically have low DOC concentrations because mineral soils have the ability to adsorb a significant amount of DOC (Aitkenhead et al., 2003). Such low concentrations of DOC in headstream waters have commonly been reported in Japan (e.g., Maki et al., 2010), as well as in other countries (e.g., Yamashita et al., 2011). The results of the DOC degradation experiments at the upper Arakawa River station are shown in Fig. 2a. Rapid degradation of the labile pool was observed within the first 20 days of incubation. Additionally, the average concentration of RDOC was $25 \mu mol L^{-1}$, which was the lowest value in freshwater and Tokyo Bay sites, and its contribution to the total DOC was 67 %.

Relatively high temperatures and DOC values were observed at Shibaura STP, while seasonal variations in chl a and POC were relatively small (Table 1). The average concentration of DOC was $355 \mu mol L^{-1}$, which was about 9 times higher than the value at the upper Arakawa River station. The annual mean concentration of RDOC was $278 \mu mol L^{-1}$, while the mean contribution of RDOC to the total DOC was 78 % (Fig. 2b). The RDOC concentrations did not vary greatly between observation months, and a significant linear relationship was observed between BDOC and DOC ($R^2 = 0.976$, $p < 0.001$, slope $= 1.16$), indicating that the seasonal variations in DOC were mostly due to variations in the bioavailable fraction. Typically, STP effluents have high organic carbon concentrations and a large bioavailable fraction (Servais et al., 1995, 1999; Kaushal and Belt, 2012). In contrast, effluent of Shibaura STP showed a high proportion of RDOC (67–93 %). These findings suggest that most of the BDOC were degraded before being discharged. This likely occurred because STPs in Japan conduct secondary treatment, which consists of the removal of wastewater suspended solids by sedimentation and degradation of dissolved organic matter by activated sludge treatment (Kadlec and Wallace, 2008).

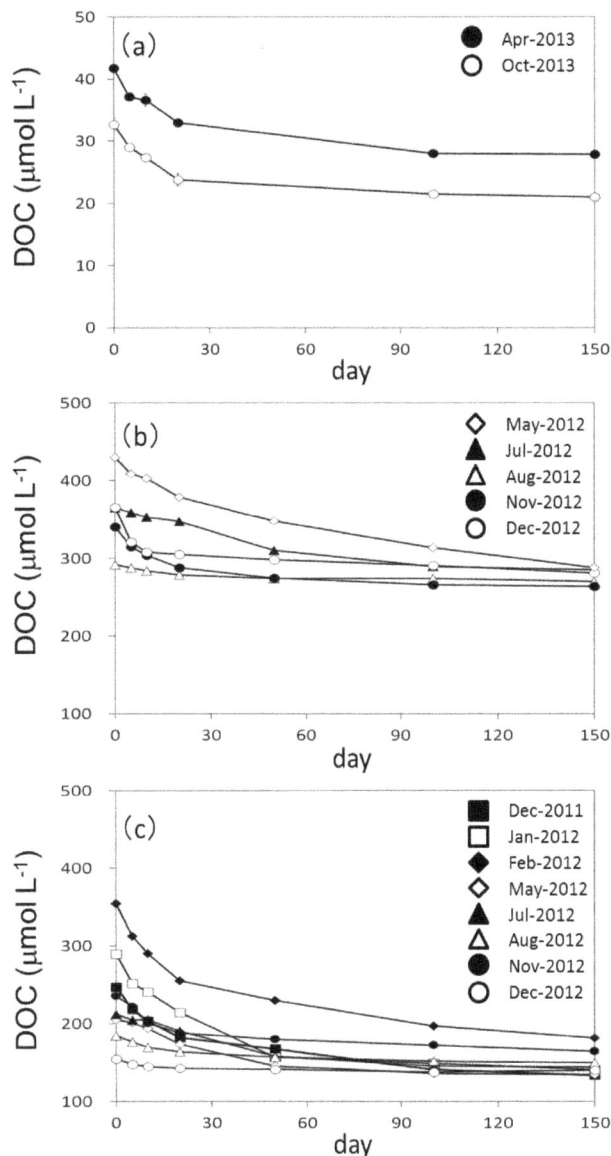

Figure 2. Changes in dissolved organic carbon ($\mu mol L^{-1}$) in surface water of (**a**) the upper Arakawa River station, (**b**) Shibaura STP, and (**c**) the lower Arakawa River station. Error bars represent the standard deviations.

Relatively high chl a and POC concentrations were observed at the lower Arakawa River station (Table 1). The maximum concentrations of chl a, DOC, and POC were observed in spring. The average concentration of DOC was $235 \mu mol L^{-1}$, while the annual mean concentration of RDOC was $149 \mu mol L^{-1}$ and the mean contribution of RDOC to the total DOC was 66 % (Fig. 2c). The concentrations of DOC were more than 6 times higher than those at the upper Arakawa River station. High concentrations of nutrients were also observed at the lower Arakawa River station (see Tables S1 and S2), which was likely a result of inputs of DOC and nutrients from STPs between observation sites.

The RDOC concentrations did not show large differences between observation months, and a significant linear relationship between BDOC and DOC was observed ($R^2 = 0.942$, $p < 0.001$, slope $= 1.12$), indicating that the seasonal variations of DOC at the lower Arakawa River station were due to variations in the bioavailable fraction.

Freshwater flowing into Tokyo Bay primarily consists of a mixture of river water and STP effluent. The total discharge ratio of river water to STP effluent in the bay is about 1 : 1 (Japan Sewage Works Association, 2010; Bureau of Sewerage, 2013). Assuming that the ratio of river water to STP effluent is 1 : 1 and that data collected at the upper Arakawa River station and Shibaura STP represent these two sources, the average concentrations of RDOC and BDOC in freshwater would be 152 and 47 μmol L^{-1}, respectively. These values are comparable with those observed at the lower Arakawa River station (149 and 86 μmol L^{-1}, respectively). Arakawa River, which is the largest river flowing into the bay, accounts for about 30 % of the freshwater discharge (Nihei et al., 2007a). Most rivers flowing into the bay have similar water quality because of similar land use within the drainage basin (Nihei et al., 2007b); accordingly, we can reasonably assume that observed RDOC and BDOC concentrations at the lower Arakawa River station represent concentrations of total river water flowing into Tokyo Bay.

Table 2 summarizes the first-order decay constants obtained by fitting the exponential degradation of DOC with time. The annual average degradation rate constant normalized to 15 °C at the lower Arakawa River station was 0.031 ± 0.005 day^{-1}, which was similar to other coastal waters (0.066 ± 0.065 day^{-1}; Lønborg and Álvarez-Salgado, 2012).

3.3 Tokyo Bay

Seasonal variations in temperature, salinity, chl a, POC, and DOC at the three stations in Tokyo Bay are presented in Fig. 3. High values of temperature, chl a, POC, and DOC were observed during spring and summer, while low values were observed during autumn and winter. Salinity was higher during autumn and winter than in spring and summer. DOC concentrations ranged from 81 to 182, 76 to 153, and 60 to 108 μmol L^{-1} at stations F3, F6, and 06, respectively (Fig. 3). The concentrations of DOC were generally lower than these at the lower Arakawa River station.

3.3.1 Lability of DOC

Rapid degradation of the labile pool occurred within the first 20 days of incubation, indicating that BDOCs were remineralized during the residence time of the bay water (Fig. 4). The seasonal variations in DOC, RDOC, and BDOC concentrations at the three stations in Tokyo Bay are shown in Fig. 5. RDOC ranged from 70 to 120 μmol L^{-1} at F3, 58 to 130 μmol L^{-1} at F6, and 48 to 80 μmol L^{-1} at 06. The

Table 2. Degradation constants for DOC (k_{20}) and normalized degradation constants at 15 °C (k_{15}) $\pm$ standard deviation at the upper Arakawa River (upper AR), the lower Arakawa River (lower AR), and Shibaura STP stations. R^2 indicates coefficient of determination.

Station	Date	k_{20} (day^{-1})	R^2	k_{15} (day^{-1})
Upper AR	Apr 2013	0.072 ± 0.006	0.99	0.049 ± 0.004
Upper AR	Oct 2013	0.053 ± 0.007	0.98	0.036 ± 0.005
Lower AR	Dec 2011	0.038 ± 0.004	0.97	0.025 ± 0.003
Lower AR	Jan 2012	0.040 ± 0.004	0.99	0.027 ± 0.003
Lower AR	Feb 2012	0.038 ± 0.003	0.96	0.026 ± 0.002
Lower AR	May 2012	0.028 ± 0.004	0.99	0.019 ± 0.003
Lower AR	Jul 2012	0.025 ± 0.005	0.99	0.017 ± 0.004
Lower AR	Aug 2012	0.045 ± 0.010	0.99	0.031 ± 0.007
Lower AR	Nov 2012	0.052 ± 0.005	0.97	0.035 ± 0.004
Lower AR	Dec 2012	0.110 ± 0.014	0.97	0.071 ± 0.010
Shibaura STP	May 2012	0.019 ± 0.005	0.99	0.013 ± 0.004
Shibaura STP	Jul 2012	0.021 ± 0.006	0.99	0.014 ± 0.004
Shibaura STP	Aug 2012	0.040 ± 0.021	0.97	0.027 ± 0.015
Shibaura STP	Nov 2012	0.062 ± 0.006	0.99	0.042 ± 0.004
Shibaura STP	Dec 2012	0.110 ± 0.005	0.92	0.072 ± 0.004

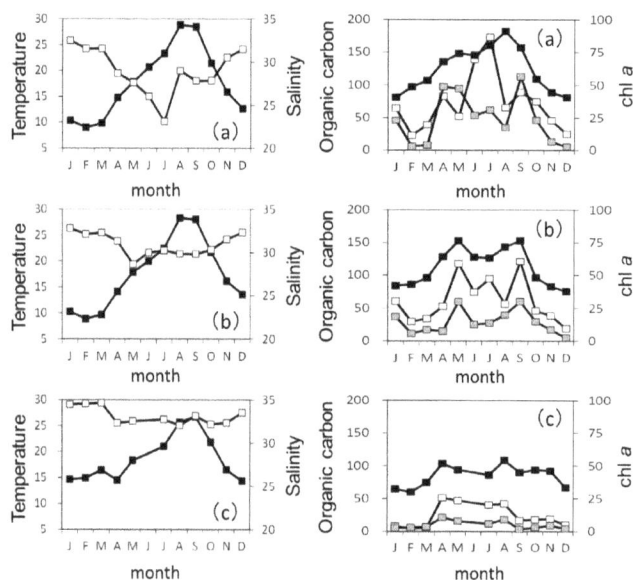

Figure 3. Seasonal variations in salinity (white square), temperature (°C; black square), dissolved organic carbon (μmol L^{-1}; black square), particulate organic carbon (μmol L^{-1}; white square), and chlorophyll a (μg L^{-1}; gray square) at stations (**a**) F3, (**b**) F6, and (**c**) 06.

mean contributions of RDOC to the total DOC were 81 % at F3, 77 % at F6, and 72 % at 06. Both RDOC and BDOC showed similar seasonal variations as DOC, with high variations being observed in spring and summer and low variations in autumn and winter. The contribution of RDOC to the total DOC was higher than that of BDOC at all stations for the entire observation period. The RDOC concentrations of the surface water were significantly higher than

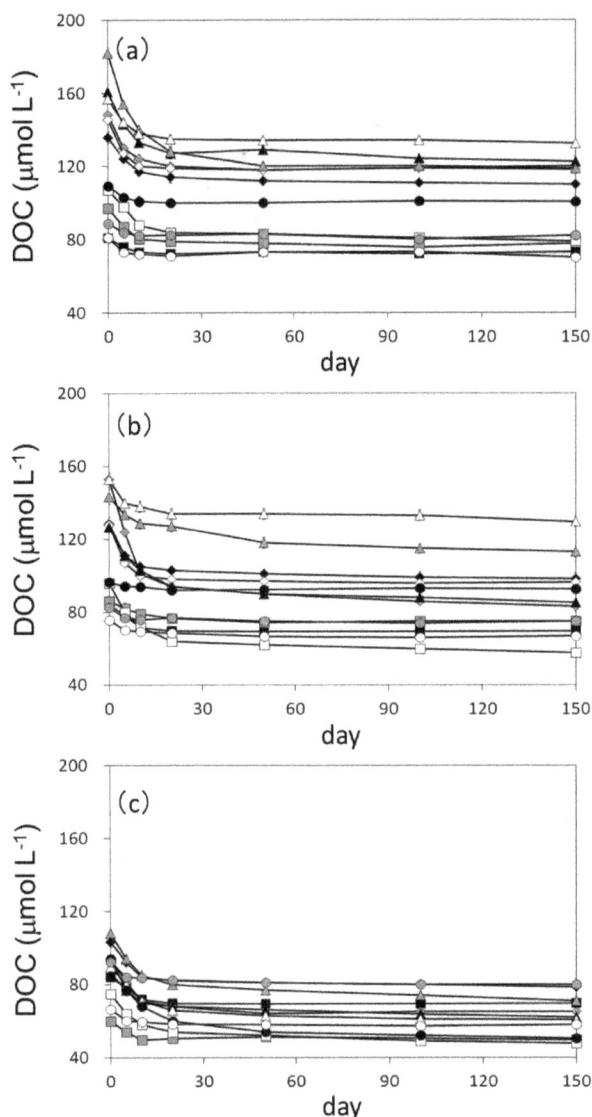

Figure 4. Changes in dissolved organic carbon (μmol L^{-1}) in the surface water of stations **(a)** F3, **(b)** F6, and **(c)** 06. Black square: January 2011; gray square: February 2011; white square: March 2011; black diamond: April 2011; gray diamond: May 2011; white diamond: June 2011; black triangle: July 2011; gray triangle: August 2011; white triangle: September 2011; black circle: October 2011; gray circle: November 2011; white circle: December 2011. Error bars represent the standard deviations.

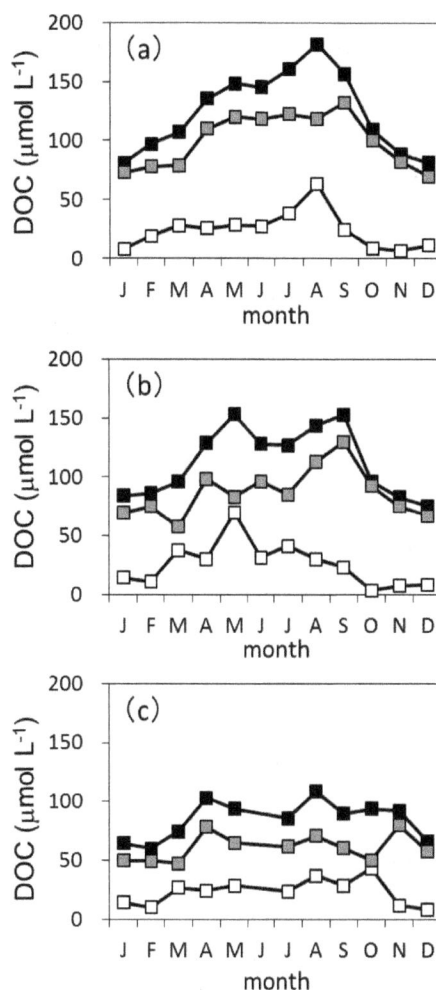

Figure 5. Seasonal variations in DOC (black square), bioavailable DOC (BDOC; white square), and recalcitrant DOC (RDOC; gray square) at stations **(a)** F3, **(b)** F6, and **(c)** 06. Error bars represent the standard deviations.

those of the bottom water at 06 (see Table S3 in the Supplement). Thus, our RDOC results likely include a fraction of semi-labile DOC. Degradation of this semi-labile DOC fraction would occur by bacterial mineralization with a longer time span, photodegradation (Moran and Zepp, 1997; Opsahl and Benner, 1997; Obernosterer and Benner, 2004), aggregation (Sholkovitz, 1976; Mulholland, 1981), and/or sorption to particles (Chin et al., 1998; Kerner et al., 2003). However, the results of this study did not change significantly

when DOCs were divided into BDOC, semi-labile DOC, and RDOC. The lifetime of semi-labile DOC is about 1.5 years (Hansell, 2013), which is considerably longer than the residence time of Tokyo Bay (Takada et al., 1992). Therefore, in our analysis, there was no problem with the inclusion of semi-labile DOC in RDOC. In addition, Ogura (1975) only divided DOC into BDOC and RDOC; therefore, we divided DOC in the same way to enable comparison with that study.

Table 3 summarizes the degradation constants of DOC for the bay surface waters. The annual average degradation rate constants normalized to 15 °C at F3, F6, and 06 were 0.128 ± 0.014, 0.094 ± 0.016, and 0.083 ± 0.010 day^{-1}, respectively. Most degradation rate constants for the bay water were higher than those of freshwater (Table 2). The half-lives of BDOC were calculated from the degradation rate constant. The annual average half-lives of BDOC at F3, F6, and 06 were 5.4, 7.4, and 8.4 days, respectively. BDOC produced

Table 3. Degradation constants for DOC (k_{20}) and normalized degradation constants at 15 °C (k_{15}) ± standard deviation in Tokyo Bay (stations F3, F6, and 06). R^2 indicates coefficient of determination.

Station	Date	k_{20} (day^{-1})	R^2	k_{15} (day^{-1})
F3	Jan 2012	0.236 ± 0.032	0.98	0.159 ± 0.022
F3	Feb 2012	0.162 ± 0.012	0.99	0.110 ± 0.008
F3	Mar 2012	0.093 ± 0.007	0.97	0.063 ± 0.005
F3	Apr 2012	0.120 ± 0.012	0.99	0.081 ± 0.008
F3	May 2012	0.203 ± 0.009	0.99	0.137 ± 0.006
F3	Jun 2012	0.286 ± 0.007	0.99	0.193 ± 0.005
F3	Jul 2012	0.127 ± 0.010	0.97	0.086 ± 0.007
F3	Aug 2012	0.109 ± 0.005	0.99	0.074 ± 0.004
F3	Sep 2012	0.153 ± 0.010	0.99	0.103 ± 0.007
F3	Oct 2012	0.301 ± 0.025	0.98	0.203 ± 0.017
F3	Nov 2012	0.269 ± 0.024	0.87	0.182 ± 0.017
F3	Dec 2012	0.221 ± 0.019	0.97	0.149 ± 0.014
F6	Jan 2012	0.150 ± 0.016	0.98	0.101 ± 0.011
F6	Feb 2012	0.095 ± 0.020	0.99	0.064 ± 0.014
F6	Mar 2012	0.100 ± 0.003	0.99	0.067 ± 0.002
F6	Apr 2012	0.151 ± 0.011	0.98	0.102 ± 0.008
F6	May 2012	0.115 ± 0.005	0.98	0.077 ± 0.004
F6	Jun 2012	0.209 ± 0.007	0.99	0.141 ± 0.005
F6	Jul 2012	0.083 ± 0.006	0.99	0.056 ± 0.004
F6	Aug 2012	0.050 ± 0.012	0.97	0.033 ± 0.008
F6	Sep 2012	0.120 ± 0.016	0.95	0.081 ± 0.011
F6	Oct 2012	0.188 ± 0.054	0.84	0.127 ± 0.040
F6	Nov 2012	0.243 ± 0.025	0.91	0.164 ± 0.020
F6	Dec 2012	0.170 ± 0.021	0.92	0.115 ± 0.015
06	Jan 2012	0.043 ± 0.007	0.94	0.029 ± 0.005
06	Feb 2012	0.223 ± 0.012	0.97	0.150 ± 0.008
06	Mar 2012	0.091 ± 0.007	0.98	0.061 ± 0.005
06	Apr 2012	0.133 ± 0.010	0.99	0.089 ± 0.007
06	May 2012	0.134 ± 0.007	0.99	0.090 ± 0.005
06	Jul 2012	0.089 ± 0.010	0.97	0.060 ± 0.007
06	Aug 2012	0.085 ± 0.006	0.97	0.057 ± 0.004
06	Sep 2012	0.094 ± 0.008	0.99	0.063 ± 0.006
06	Oct 2012	0.085 ± 0.005	0.99	0.057 ± 0.004
06	Nov 2012	0.146 ± 0.021	0.94	0.098 ± 0.014
06	Dec 2012	0.240 ± 0.013	0.87	0.162 ± 0.009

by phytoplankton in the bay water might have led to faster degradation rates because the half-lives of BDOC were about 5 times faster than the residence time of the bay water.

RDOC concentrations in Tokyo Bay were negatively correlated with salinity and positively correlated with chl a (Table 4). In the bay, salinity was lower in spring and summer than in autumn and winter (Fig. 3) because of high freshwater input during spring and summer. The freshwater RDOC concentration was higher than that of Tokyo Bay water; therefore, a negative relationship between RDOC and salinity was observed. RDOC is also produced directly by phytoplankton (Kragh and Søndergaard, 2009). Hence, the positive relationship between RDOC and chl a observed in this study likely reflected RDOC produced by phytoplankton.

Table 4. Correlation coefficients (R^2) of the significant ($p < 0.05$) linear regressions between DOC and hydrological data in Tokyo Bay (stations F3, F6, 06, and total data). X indicates dependent variable. N.s. indicates not significant.

Station	X	Salinity	chl a
F3	DOC	0.54	0.36
F3	RDOC	0.68	0.56
F3	BDOC	n.s.	n.s.
F6	DOC	0.74	0.64
F6	RDOC	0.64	0.59
F6	BDOC	0.37	0.31
06	DOC	0.81	0.51
06	RDOC	0.54	0.47
06	BDOC	0.29	n.s.
Total data	DOC	0.68	0.52
Total data	RDOC	0.73	0.62
Total data	BDOC	0.11	n.s.

3.3.2 RDOC sources

To estimate the sources of RDOC in Tokyo Bay, multiple linear regression analysis with salinity and chl a as the independent variables was applied to all RDOC data observed at three stations in Tokyo Bay. BDOC in Tokyo Bay was not well correlated with salinity and chl a (Table 4), so multiple linear regression analysis was not applied to the BDOC data. We obtained the following multiple linear regression equation (Model I):

$$[\text{RDOC}] = (259 \pm 38) - (5.96 \pm 1.20) \cdot [\text{Sal}]$$
$$+ (0.597 \pm 0.20) \cdot [\text{Chl } a] \qquad (3)$$
$$\left(r^2 = 0.79, \ P < 0.001, \ n = 35 \right),$$

where [RDOC] is the RDOC concentration (μmol L^{-1}), [Sal] is salinity, and [Chl a] is the chlorophyll a concentration (μg L^{-1}) of each sample. The end-member of terrestrial RDOC ([RDOC$_{\text{terr-end}}$]) was as follows when the salinity was 0:

$$[\text{RDOC}_{\text{terr-end}}] = (259 \pm 38) + (0.597 \pm 0.20) \cdot [\text{Chl } a_{\text{river}}], \quad (4)$$

where [Chl a_{river}] is the chl a concentration (μg L^{-1}) at the freshwater site. The end-member of terrestrial RDOC was higher than the average RDOC concentration at the lower Arakawa River station (149μmol L^{-1}) and was similar to that of Shibaura STP (278μmol L^{-1}). The ratio of river water to STP effluent was 1 : 1 (Japan Sewage Works Association, 2010; Bureau of Sewerage, 2013), and data collected at the upper Arakawa River station and Shibaura STP represent these two sources (see Sect. 3.2). It is possible that freshwater inputs into Tokyo Bay were more strongly influenced by STPs than headstream waters. Alternatively, if we assume that the RDOC concentration at salinity = 0 and chl $a = 0$ was close to the average RDOC concentration actually observed at the lower Arakawa River station (149μmol L^{-1}), we obtain the following multiple regression equation (Model II):

$$[\text{RDOC}] = 149 - (2.65 \pm 0.26) \cdot [\text{Sal}]$$
$$+ (1.03 \pm 0.40) \cdot [\text{Chl } a] \qquad (5)$$
$$\left(r^2 = 0.71, \; P < 0.001, \; n = 35\right).$$

The end-member of terrestrial RDOC ($[\text{RDOC}_{\text{terr-end}}]$) is as follows when salinity is 0:

$$[\text{RDOC}_{\text{terr-end}}] = 149 + (1.03 \pm 0.40) \cdot [\text{Chl } a_{\text{river}}]. \qquad (6)$$

In this study, we assumed that [Chl a_{river}] was $6.0 \, \mu\text{g L}^{-1}$ (Ministry of the Environment: http://www.env.go.jp), which was the average value of surface waters in Arakawa River. Although [Chl a_{river}] is usually lower than $10 \, \mu\text{g L}^{-1}$ throughout the year, phytoplankton blooms occasionally persist (Ministry of the Environment: http://www.env.go.jp). The calculation of the RDOC sources using the minimum and maximum chl a concentration at the lower Arakawa River station (Table 1) resulted in estimated RDOC sources that did not differ significantly from the minimum and maximum concentrations.

The concentrations of RDOC in the open ocean ($[\text{RDOC}_{\text{ocean-end}}]$) can be estimated by assuming that salinity and chl a in the open ocean were 34.5 (Okada et al., 2007) and $1.0 \, \mu\text{g L}^{-1}$, respectively (Japan Meteorological Agency: http://www.jma.go.jp/jma/index.html), which were the average values of surface waters offshore from Tokyo Bay. The $[\text{RDOC}_{\text{ocean-end}}]$ values were 54.0 and $58.6 \, \mu\text{mol L}^{-1}$ for Models I and II, respectively, which were comparable to the annual average RDOC concentration of the bottom water at 06 (see Table S3). Following the method described Ogawa and Ogura (1990), we estimated the contributions of RDOC from different sources (RDOC from the open ocean [$\text{RDOC}_{\text{ocean origin}}$], terrestrial RDOC [$\text{RDOC}_{\text{terr}}$], and RDOC from phytoplankton [$\text{RDOC}_{\text{phyto}}$]), using two models of the multiple linear regression analysis. The RDOC concentrations can be expressed as follows:

$$[\text{RDOC}] = \left[\text{RDOC}_{\text{phyto}}\right] + \left[\text{RDOC}_{\text{ocean origin}}\right] + [\text{RDOC}_{\text{terr}}]. \qquad (7)$$

The equation describing RDOC derived from the open ocean ($[\text{RDOC}_{\text{ocean origin}}]$) is as follows:

$$\left[\text{RDOC}_{\text{ocean origin}}\right] = [\text{RDOC}_{\text{ocean-end}}] \cdot [\text{Sal}]/34.5. \qquad (8)$$

The terrestrial RDOC ($[\text{RDOC}_{\text{terr}}]$) is as follows:

$$[\text{RDOC}_{\text{terr}}] = [\text{RDOC}_{\text{terr-end}}] \cdot (34.5 - [\text{Sal}])/34.5. \qquad (9)$$

The RDOC derived from phytoplankton ($[\text{RDOC}_{\text{phyto}}]$) can be estimated from Eq. (7):

$$[\text{RDOC}_{\text{phyto}}] = [\text{RDOC}] - \left[\text{RDOC}_{\text{ocean origin}}\right] - [\text{RDOC}_{\text{terr}}]. \qquad (10)$$

For each multiple linear regression equation (Eqs. 3 and 5), the two-sided 95 % confidence bounds of each coefficient and intercept were estimated. For the concentrations of RDOC originating from phytoplankton, terrestrial, and open-oceanic waters, we estimated the upper and lower bounds by changing an equation within its error range. The relative concentrations of RDOC (%) with error in the bay originating from phytoplankton, terrestrial, and open-oceanic waters at the three stations are presented in Table 5. The results show that the open ocean is the major source of RDOC in Tokyo Bay. At station F3, which is located close to land, terrestrial RDOC was comparable to that from the open ocean. The concentration of terrestrial RDOC was significantly higher than that of RDOC from phytoplankton at all stations, even at the bay mouth.

The influx of terrestrial TOC (POC + DOC) from the rivers to Tokyo Bay was estimated using a mass balance model ($8.1 \times 10^{10} \, \text{gC yr}^{-1}$; Yanagi et al., 1993), and the DOC / TOC ratio at the freshwater site was 0.62 (Kubo, unpublished data). Hence, the influx of terrestrial DOC was estimated to be $5.0 \times 10^{10} \, \text{gC yr}^{-1}$, and RDOC accounted for 66 % of terrestrial DOC (see Sect. 3.2; $3.3 \times 10^{10} \, \text{gC yr}^{-1}$). The efflux of TOC from the surface bay to the open ocean was estimated using a mass balance model ($9.4 \times 10^{10} \, \text{gC yr}^{-1}$; Yanagi et al., 1993), and the DOC / TOC ratio in the surface bay mouth was 0.69 (Kubo, unpublished data). Hence, the efflux of DOC was estimated to be $6.5 \times 10^{10} \, \text{gC yr}^{-1}$, and RDOC accounted for 73 % in the surface bay mouth (see Sect. 3.3; $4.7 \times 10^{10} \, \text{gC yr}^{-1}$). Assuming that terrestrial and phytoplankton RDOC were exported outside of the bay in the same ratio as at the bay mouth (Table 5), Tokyo Bay exported mostly terrestrial RDOC to the open ocean, owing to the high concentration of terrestrial RDOC and remineralization of BDOC. Moreover, the ratio of terrestrial RDOC input into the bay ($3.3 \times 10^{10} \, \text{gC yr}^{-1}$) and terrestrial RDOC efflux to the open ocean (0.9×10^{10} and $0.6 \times 10^{10} \, \text{gC yr}^{-1}$, respectively, for Models I and II) was 28 and 17 %, respectively. Residual terrestrial RDOC in the bay may be removed from the water column by photodegradation (Moran and Zepp, 1997; Opsahl and Benner, 1997; Obernosterer and Benner; 2004), aggregation (Sholkovitz, 1976; Mulholland, 1981), and/or sorption to particles (Chin et al., 1998; Kerner et al., 2003).

The fate of terrestrial DOC in the coastal ocean and the open ocean has long been the subject of debate (Hedges et al., 1997). For example, biomarkers (e.g., lignin phenols) and the stable carbon isotopic composition of DOC are commonly used to estimate the contribution of terrestrial DOC to the open ocean (Druffel et al., 1992; Hedges et al., 1997; Raymond and Bauer, 2001; Bauer and Bianchi, 2011). Lignin phenol analysis indicated that terrestrial DOC comprises only a small fraction (4–10 %) of the total DOC in the open ocean (Meyers-Schulte and Hedges, 1986; Opsahl and Benner, 1997; Hernes and Benner, 2006). In addition, the stable carbon isotopic composition of DOC also indicated

Table 5. Relative concentration of RDOC (%) ± error* in Tokyo Bay derived from phytoplankton, terrestrial, and open-oceanic waters, estimated from two multiple linear regressions. *See Sect. 3.3.2.

Station	Model I			Model II		
	Phyto.	Terr.	Ocean	Phyto.	Terr.	Ocean
F3	12 ± 7	42 ± 3	46 ± 5	14 ± 6	29 ± 3	57 ± 4
F6	8 ± 6	35 ± 3	57 ± 5	10 ± 6	23 ± 2	67 ± 3
06	4 ± 4	20 ± 4	76 ± 4	5 ± 4	12 ± 2	83 ± 4
Total data	8 ± 6	32 ± 3	59 ± 5	10 ± 5	21 ± 3	69 ± 4

that terrestrial DOC represents less than 10 % of the total DOC (Bauer et al., 2002). As a result, most terrestrial DOC is remineralized in coastal waters, and only a small fraction is exported to the open ocean. In this study, terrestrial RDOC in the surface bay mouth accounted for less than 20 % of the total RDOC (Table 5). Although these levels were slightly higher than those reported in previous studies using lignin phenols and stable carbon isotopic compositions of DOC, they are probably reasonable given that exported terrestrial RDOC were further diluted with open-oceanic water once outside the bay. Nevertheless, more complete information regarding the sources and lability of DOC are important to enable a better understanding of the fate of DOC in the coastal ocean and open ocean.

3.4   Change of DOC over 4 decades

Ogura (1975) investigated the concentrations of RDOC and BDOC in Tokyo Bay and freshwater sources flowing into the bay in the 1970s using GF/C filters (nominal pore size 1.2 μm) to collect filtrate of degradation samples and found that the contribution of the DOC fraction from 0.45 μm (Millipore HA filter, Millipore Corp., Bedford, MA) to 1.2 μm was about 10 % of the total DOC in Tokyo Bay. Ogawa and Ogura (1992) also showed that the low-molecular-weight DOC (< 10 000 Dalton; < 0.2 μm) in the bay comprised a major portion of the total DOC filtered by 1.2 μm (78–97 %). Hence, the DOC fraction from 0.7 to 1.2 μm comprised a minor proportion of the DOC in Tokyo Bay. Ogura (1975) used a wet chemical oxidation method to measure the samples, while Ogawa and Ogura (1992) showed that both a wet chemical oxidation method and high-temperature catalytic oxidation method for measuring DOC concentrations of Tokyo Bay waters generated similar results. Ogura (1975) conducted degradation experiments with a short incubation time (about 50 days) compared to our experiment (150 days). However, BDOC can be consumed by bacteria over short periods (days to weeks; Lønborg and Álvarez-Salgado, 2012), and, therefore, the remaining DOC pool after 150 days of incubation, used in Eq. (1), is not significantly different from RDOC concentration at 50 days. Actually, degradation rate constants calculated using data from the first 50 days (k_{50})

and those using all data from 150 days of experiment (k_{150}) are not significantly different ($k_{50} = 0.90 \cdot k_{150} + 0.016$, $R^2 = 0.86$, $p < 0.01$). Therefore, we assume that our degradation rate constants are comparable to those reported by Ogura (1975).

In 1972, the average concentrations of RDOC and BDOC were 224 and 337 μmol L^{-1} (40 and 60 % of the total DOC, respectively) in the freshwater environment of the lower Tamagawa River, which flows into Tokyo Bay (Ogura, 1975). The present RDOC and BDOC concentrations at the lower Arakawa River station (149 and 86 μmol L^{-1}) are lower than those reported by Ogura (1975). If we assumed that the amount of freshwater discharge into the bay has increased by 24 % (Okada et al., 2007), the amount of RDOC and BDOC flowing into the bay would have decreased by 17 and 68 %, respectively. Ogura (1975) also estimated a degradation rate constant (k_{15}) of 0.087 day^{-1}, which is much higher than that observed in the present study (Table 2). These changes are consistent with the fact that the proportion of treated wastewater to the total freshwater inflow to the bay increased from 11 to 28 % from 1970 to 2000 (National Institute for Land and Infrastructure Management, 2004). The degradation of DOC at STPs before being discharged should lower the BDOC fraction more than RDOC. Overall, our results indicate that the quantity of DOC flowing into the bay has decreased, and the quality of DOC becomes more recalcitrant.

In Tokyo Bay, the concentrations of DOC at station F3 decreased from 287 μmol L^{-1} in 1972 (Ogura, 1975) to 124 μmol L^{-1} in 2012, most likely because of a decrease in DOC discharge from rivers and a decrease in primary production (Yamaguchi and Shibata, 1979; Yamaguchi et al., 1991; Bouman et al., 2010). The concentrations of RDOC and BDOC observed in this study (100 and 24 μmol L^{-1}, respectively) were lower than those estimated by Ogura (1975) in 1972 (165 and 100 μmol L^{-1}, respectively). Conversely, the contribution of RDOC to the total DOC in this study (80.6 %) is higher than the value observed in 1972 (57.5 %; Ogura, 1975). The concentrations of RDOC and BDOC in Tokyo Bay have decreased because of a decrease in DOC load from the land, especially for BDOC. As a result, DOC becomes more recalcitrant. In addition, decreasing nutrient loads in the bay have caused decreasing primary production (Yamaguchi and Shibata, 1979; Yamaguchi et al., 1991; Bouman et al., 2010). Therefore, DOC produced by phytoplankton should also have decreased.

4   Summary

Rapid degradation of the labile pool was observed at freshwater sites and Tokyo Bay within the first 20 days of incubation. BDOCs are remineralized during the residence time of the bay water. The contribution of RDOC to the total DOC was higher than that of BDOC at all stations for the entire observation period and accounted for 77 % of the total.

Accordingly, Tokyo Bay exported mostly terrestrial RDOC to the open ocean owing to the high concentration of terrestrial RDOC and faster half-lives of BDOC relative to the residence time of the bay water. The concentrations of RDOC and BDOC have decreased in the last 40 years at freshwater sites and Tokyo Bay, during which time DOC has become more recalcitrant because of improved sewage treatment. Since organic carbon degradation occurs at STPs before being discharged, DOC flowing into the bay has decreased, especially the BDOC fraction.

Acknowledgements. We thank Chinatsu Oouchida, as well as other scientists, officers, and crew members on board the R/V *Seiyo-maru* for their help in sampling. This work was supported by a Grant-in-Aid for Scientific Research (C) (24510009) from the Ministry of Education, Culture, Sports, Science and Technology, Japan, and by a Canon Foundation grant. The authors are grateful to the anonymous reviewers and the associate editor Silvio Pantoja, who provided valuable comments on the manuscript.

Edited by: S. Pantoja

References

Aitkenhead, J. A., McDowell, W. H., and Neff, J. C.: Sources, production, and regulation of allochthonous dissolved organic matter inputs to surface waters, in: Aquatic Ecosystems: Interactivity of Dissolved Organic Matter, edited by: Findlay, S. E. G. and Sinsabaugh, R. L., Academic Press, San Diego, 25–70, 2003.

Amon, R. M. W. and Benner, R.: Photochemical and microbial consumption of dissolved organic carbon and dissolved oxygen in the Amazon River system, Geochim. Cosmochim. Ac., 60, 1783–1792, 1996.

Aufdenkampe, A. K., Mayorga, E., Raymond, P. A., Melack, J. M., Doney, S. C., Alin, S. R., Aalto, R. E., and Yoo, K.: Riverine coupling of biogeochemical cycles between land, oceans, and atmosphere, Front. Ecol. Environ., 9, 53–60, 2011.

Avery Jr., G. B., Willey, J. D., Kieber, R. J., Shank, G. C., and Whitehead, R. F.: Flux and bioavailability of Cape Fear River and rainwater dissolved organic carbon to Long Bay, southeastern United States, Global Biogeochem. Cy., 17, 1042, doi:10.1029/2002GB001964, 2003.

Bauer, J. E. and Bianchi, T. S.: Dissolved organic carbon cycling and transformation, in: Treatise on estuarine and coastal science, edited by: Wolanski, E. and McLusky, D. S., Academic Press, Sam Diego, 5, 7–67, 2011.

Bauer, J. E., Druffel, E. R. M., Wolgast, D. M., and Griffin, S.: Temporal and regional variability in sources and cycling of DOC and POC in the northwest Atlantic continental shelf and slope, Deep-Sea Res. Pt. II, 49, 4387–4419, 2002.

Bouman, H. A., Nakane, T., Oka, K., Nakata, K., Kurita, K., Sathyendranath, S., and Platt, T.: Environmental controls on phytoplankton production in coastal ecosystems: A case study from Tokyo Bay, Estuar. Coast. Shelf S., 87, 63–72, 2010.

Bureau of Sewerage: Management plan 2013, Bureau of Sewerage, Tokyo Metropolitan Government, 117 pp., 2013 (in Japanese).

Carlson, C. A.: Production and removal processes, in: Biogeochemistry of Marine Dissolved Organic Matter, edited by: Hansell, D. A. and Carlson, C. A., Academic Press, San Diego, 91–151, 2002.

Cauwet, G.: DOM in the coastal zone, in: Biogeochemistry of Marine Dissolved Organic Matter, edited by: Hansell, D. A. and Carlson, C. A., Academic Press, San Diego, 579–609, 2002.

Chin, W. C., Orellana, M. V., and Verdugo, P.: Spontaneous assembly of marine dissolved organic matter into polymer gels, Nature, 391, 568–572, 1998.

Cole, J. J., Prairie, Y. T., Caraco, N. F., McDowell, W. H., Tranvik, L. J., Striegl, R. G., Duarte, C. M., Kortelainen, P., Downing, J. A., Middelburg, J. J., and Melack, J.: Plumbing the global carbon cycle: integrating inland waters into the terrestrial carbon budget, Ecosystems, 10, 171–184, 2007.

Dai, M., Yin, Z., Meng, F., Liu, Q., and Cai, W. J.: Spatial distribution of riverine DOC inputs to the ocean: an updated global synthesis, Current Opinion in Environmental Sustainability, 4, 170–178, 2012.

Druffel, E. R. M., Williams, P. M., Bauer, J. E., and Ertel, J. R.: Cycling of dissolved and particulate organic matter in the open ocean, J. Geophys. Res., 97, 15639–15659, 1992.

Hansell, D. A.: Recalcitrant dissolved organic carbon fractions, Annu. Rev. Mar. Sci., 5, 3.1–3.25, 2013.

Hansell, D. A., Carlson, C. A., Repeta, D. J., and Schlitzer, R.: Dissolved organic matter in the ocean: A controversy stimulates new insights, Oceanography, 22, 202–211, 2009.

Hedges, J., Keil, R. G., and Benner, R.: What happens to terrestrial organic matter in the ocean?, Org. Geochem., 27, 195–212, 1997.

Hernes, P. J. and Benner, R.: Terrigenous organic matter sources and reactivity in the North Atlantic Ocean and a comparison to the Arctic and Pacific oceans, Mar. Chem., 100, 66–79, 2006.

Japan Sewage Works Association: Sewage statistics, Japan Sewage Works Association, Tokyo, 2010 (in Japanese).

Kadlec, R. H. and Wallace, S. D.: Treatment wetlands second edition, CRC Press, Boca Raton, Florida, 1000 pp., 2008.

Kaushal, S. S. and Belt, K. T.: The urban watershed continuum: evolving spatial and temporal dimensions, Urban Ecosystems, 15, 409–435, 2012.

Kerner, M., Hohenberg, H., Ertl, S., Reckermann, M., and Spitzy, A.: Self-organization of dissolved organic matter to micelle-like microparticles in river water, Nature, 422, 150–154, 2003.

Kragh, T. and Søndergaard, M.: Production and decomposition of new DOC by marine plankton communities: carbohydrates, refractory components and nutrient limitation, Biogeochemistry, 96, 177–187, 2009.

Lønborg, C. and Álvarez-Salgado, X. A.: Recycling versus export of bioavailable dissolved organic matter in the coastal ocean and efficiency of the continental shelf pump, Global Biogeochem. Cy., 26, GB3018, doi:10.1029/2012GB004353, 2012.

Lønborg, C., Davidson, K., Álvarez-Salgado, X. A., and Miller, A. E. J.: Bioavailability and bacterial degradation rates of dissolved organic matter in a temperate coastal area during an annual cycle, Mar. Chem., 113, 219–226, 2009.

Ludwig, W., Probst, J. L., and Kempe, S.: Predicting the oceanic input of organic carbon by continental erosion, Global Biogeochem. Cy., 10, 23–41, 1996.

Maki, K., Kim, C., Yoshimizu, C., Tayasu, I., Miyajima, T., and Nagata, T.: Autochthonous origin of semi-labile dissolved organic carbon in a large monomictic lake (Lake Biwa): carbon stable isotopic evidence, Limnology, 11, 143–153, 2010.

Meybeck, M.: C, N, P and S in rivers: From sources to global inputs, in: Interactions of C, N, P and S Biogeochemical Cycles and Global Change, edited by: Wollast, R., Mackenzie, F. T., and Chou, L., Springer-Verlag, Berlin, 163–193, 1993.

Meyers-Schulte, K. J. and Hedges, J. I.: Molecular evidence for a terrestrial component of organic matter dissolved in ocean water, Nature, 321, 61–63, 1986.

Moran, M. A. and Zepp, R. G.: Role of photoreactions in the formation of biologically labile compounds from dissolved organic matter, Limnol. Oceanogr., 42, 1307–1316, 1997.

Moran, M. A., Sheldon Jr., W. M., and Sheldon, J. E.: Biodegradation of riverine dissolved organic carbon in five estuaries of the southeastern United States, Estuaries, 22, 55–64, 1999.

Mulholland, P. J.: Formation of particulate organic carbon in water from a southeastern swamp-stream, Limnol. Oceanogr., 26, 790–795, 1981.

Nakamura, T., Osaka, K., Hiraga, Y., and Kazama, F.: Nitrogen and oxygen isotope composition of nitrate in stream water of Fuji River basin, J. Jpn. Assoc. Hydrol. Sci., 41, 79–89, 2011 (in Japanese with English abstract).

National Institute for Land and Infrastructure Management: Information of port environment, 2004, available at: http://www.nilim.go.jp/, last access: 25 June 2013.

Nellemann, C., Hain, S., and Alder, J.: Rapid response assessment in dead water: Merging of climate change with pollution, overharvest, and infestations in the world's fishing grounds, United Nations Environment Programme, 61 pp., 2008.

Nihei, Y., Takamura, T., and Watanabe, N.: Issues on discharge monitoring in main influence rivers into Tokyo Bay, Journal of JSCE, 54, 1226–1230, 2007a (in Japanese with English abstract).

Nihei, Y., Ehara, K., Usuda, M., Sakai, A., and Shigeta, K.: Water quality and pollutant load in the Edo River, Ara River and Tama River, Journal of JSCE, 54, 1221–1225, 2007b (in Japanese with English abstract).

Obernosterer, I. and Benner, R.: Competition between biological and photochemical processes in the mineralization of dissolved organic carbon, Limnol. Oceanogr., 49, 117–124, 2004.

Ogawa, H. and Ogura, N.: Source and behavior of organic carbon of seawater in Tokyo Bay, Chikyukagaku, 24, 27–41, 1990 (in Japanese with English abstract).

Ogawa, H. and Ogura, N.: Comparison of two methods for measuring dissolved organic carbon in sea water, Nature, 356, 696–698, 1992.

Ogura, N.: Further studies on decomposition of dissolved organic matter in coastal seawater, Mar. Biol., 31, 101–111, 1975.

Okada, T., Takao, T., Nakayama, K., and Furukawa, K.: Change in freshwater discharge and residence time of seawater in Tokyo Bay, Journal of JSCE, 63, 67–72, 2007 (in Japanese with English abstract).

Opsahl, S. and Benner, R.: Distribution and cycling of terrigenous dissolved organic matter in the ocean, Nature, 386, 480–482, 1997.

Raymond, P. A. and Bauer, J. E.: Bacterial consumption of DOC during transport through a temperate estuary, Aquat. Microb. Ecol., 22, 1–12, 2000.

Raymond, P. A. and Bauer, J. E.: Riverine export of aged terrestrial organic matter to the North Atlantic Ocean, Nature, 409, 497–500, 2001.

Redfield, A. C., Ketchum, B. H., and Richards, F. A.: The influence of organisms on the composition of sea-water, in: The Sea, edited by: Hill, M. N., Interscience, New York, 26–77, 1963.

Regnier, P., Friedlingstein, P., Ciais, P., Mackenzie, F. T., Gruber, N., Janssens, I. A., Laruelle, G. G., Lauerwald, R., Luyssaert, S., Andersson, A. J., Arndt, S., Arnosti, C., Borges, A. V., Dale, A. W., Gallego-Sala, A., Goddéris, Y., Goossens, N., Hartmann, J., Heinze, C., Ilyina, T., Joos, F., LaRowe, D. E., Leifeld, J., Meysman, F. J. R., Munhoven, G., Raymond, P. A., Spahni, R., Suntharalingam, P., and Thullner, M.: Anthropogenic perturbation of the carbon fluxes from land to ocean, Nat. Geosci., 6, 597–607, 2013.

Servais, P., Barillier, A., and Garnier, J.: Determination of the biodegradable fraction of dissolved and particulate organic carbon in waters, Ann. Limnol.-Int. J. Lim., 31, 75–80, 1995.

Servais, P., Garnier, J., Demarteau, N., Brion, N., and Billen, G.: Supply of organic matter and bacteria to aquatic ecosystems through waste water effluents, Water Res., 33, 3521–3531, 1999.

Sholkovitz, E. R.: Flocculation of dissolved organic and inorganic matter during the mixing of river water and seawater, Geochim. Cosmochim. Ac., 40, 831–845, 1976.

Suzuki, R. and Ishimaru, T.: An improved method for the determination of phytoplankton chlorophyll using N, N-Dimethylformamide, Journal of the Oceanographical Society of Japan, 46, 190–194, 1990.

Takada, H., Ishiwatari, R., and Ogura, N.: Distribution of linear alkylbenzenes (LABs) and linear alkylbenzene sulphonates (LAS) in Tokyo Bay sediments, Estuar. Coast. Shelf S., 35, 141–156, 1992.

Tanaka, Y., Ogawa, H., and Miyajima, T.: Production and bacterial decomposition of dissolved organic matter in a fringing coral reef, J. Oceanogr., 67, 427–437, 2011.

Tranvik, L. and Höfle, M. G.: Bacterial growth in mixed cultures on dissolved organic carbon from humic and clear waters, Appl. Environ. Microb., 53, 482–488, 1987.

Yamaguchi, Y. and Shibata, Y.: Recent status of primary production in Tokyo Bay, Bull. Coast. Oceanogr., 16, 106–111, 1979 (in Japanese).

Yamaguchi, Y., Satoh, H., and Aruga, Y.: Seasonal changes of organic carbon and nitrogen production by phytoplankton in the estuary of river Tamagawa, Mar. Pollut. Bull., 23, 723–725, 1991.

Yamashita, Y., Kloeppel, B. D., Knoepp, J., Zausen, G. L., and Jaffé, R.: Effect of watershed history on dissolved organic matter characteristics in headwater streams, Ecosystems, 14, 1110–1122, 2011.

Yanagi, T., Saino, T., Ishimaru, T., and Uye, S.: A carbon budget in Tokyo Bay, J. Oceanogr., 49, 249–256, 1993.

9

Authigenic apatite and octacalcium phosphate formation due to adsorption–precipitation switching across estuarine salinity gradients

J. F. Oxmann[1] **and L. Schwendenmann**[2]

[1]GEOMAR Helmholtz Centre for Ocean Research Kiel, Marine Biogeochemistry, 24148 Kiel, Germany
[2]School of Environment, The University of Auckland, Auckland 1010, New Zealand

Correspondence to: J. F. Oxmann (joxmann@geomar.de)

Abstract. Mechanisms governing phosphorus (P) speciation in coastal sediments remain largely unknown due to the diversity of coastal environments and poor analytical specificity for P phases. We investigated P speciation across salinity gradients comprising diverse ecosystems in a P-enriched estuary. To determine P load effects on P speciation we compared the high P site with a low P site. Octacalcium phosphate (OCP), authigenic apatite (carbonate fluorapatite, CFAP) and detrital apatite (fluorapatite) were quantitated in addition to Al / Fe-bound P (Al / Fe-P) and Ca-bound P (Ca-P). Gradients in sediment pH strongly affected P fractions across ecosystems and independent of the site-specific total P status. We found a pronounced switch from adsorbed Al / Fe-P to mineral Ca-P with decreasing acidity from land to sea. This switch occurred at near-neutral sediment pH and has possibly been enhanced by redox-driven phosphate desorption from iron oxyhydroxides. The seaward decline in Al / Fe-P was counterbalanced by the precipitation of Ca-P. Correspondingly, two location-dependent accumulation mechanisms occurred at the high P site due to the switch, leading to elevated Al / Fe-P at pH < 6.6 (landward; adsorption) and elevated Ca-P at pH > 6.6 (seaward; precipitation). Enhanced Ca-P precipitation by increased P loads was also evident from disproportional accumulation of metastable Ca-P (Ca-P_{meta}) at the high P site. Here, sediments contained on average 6-fold higher Ca-P_{meta} levels compared with the low P site, although these sediments contained only 2-fold more total Ca-P than the low P sediments. Phosphorus species distributions indicated that these elevated Ca-P_{meta} levels resulted from transformation of fertilizer-derived Al / Fe-P to OCP and CFAP in nearshore areas. Formation of CFAP as well as its precursor, OCP, results in P retention in coastal zones and can thus lead to substantial inorganic P accumulation in response to anthropogenic P input.

1 Introduction

Desorption and precipitation of phosphate along salinity gradients are influenced by redox potential (Eh) and pH (van Beusekom and de Jonge, 1997). Typically, Eh decreases and pH increases from land to the sea (Clarke, 1985; Huang and Morris, 2005; Sharp et al., 1982). Seawater inundation induces the Eh gradient by limiting oxygen diffusion into the sediment, thereby initiating anaerobic respiration. Sediments regularly inundated by seawater tend to have higher pH values than terrestrial soils because soils naturally acidify due to vegetation-derived inputs (effects of enhanced carbonic acid production, root exudate release, litter decomposition, proton extrusion). Human activities such as N fertilization can also contribute to soil acidification (Fauzi et al., 2014; Hinsinger et al., 2009; Richardson et al., 2009). The acid generated is neutralized downstream by the high alkalinity of seawater.

Changes in pH and Eh facilitate phosphorus (P) desorption from particulate matter and generally account for the non-conservative behaviour of dissolved reactive P during admixing of water along salinity gradients. Particulate P includes a significant amount of inorganic P, which mainly consists of calcium-bound P (Ca-P; detrital and authigenic) and aluminium / iron-bound P (Al / Fe-P). The Al / Fe-P fraction contains adsorbed inorganic P (Al / Fe-(hydr)oxide-bound P), which can be partly released to solution (e.g. Slomp, 2011). The Al / Fe-P fraction of oxidized, acidic sediment usually comprises relatively large proportions of adsorbed P. Phosphate desorption induced by pH and Eh gradients generally results in progressively decreasing concentrations of the Al / Fe-P from upper to lower intertidal zones (Andrieux-Loyer et al., 2008; Coelho et al., 2004; Mortimer, 1971; Jordan et al., 2008; Paludan and Morris, 1999; Sutula et al., 2004).

Decreasing Eh in sediment from upper to lower zones involve critical levels for reduction of ferric iron compounds (e.g. Gotoh and Patrick, 1974). These critical levels facilitate desorption of Fe-(hydr)oxide-bound P due to less efficient sorption of P by iron in the Fe(II) state compared to the Fe(III) state (Hartzell and Jordan, 2012; Sundareshwar and Morris, 1999). Desorption from metal (hydr)oxides with increasing pH (Oh et al., 1999; Spiteri et al., 2008), on the other hand, is driven by the decreasing surface electrostatic potential with increasing pH (Barrow et al., 1980; Sundareshwar and Morris, 1999). This effect may be partly offset by the increasing proportion of the strongly sorbing divalent phosphate ion (HPO_4^{2-}) with increasing pH until pH 7 ($\sim pK_2$, which decreases with salinity increase; Atlas, 1975). In the alkaline pH range, however, this offset is less pronounced thus allowing stronger desorption (Bolan et al., 2003; Bowden et al., 1980; Haynes, 1982). Similarly, studies on soils attributed desorption with pH to increasing competition between hydroxyl and phosphate ions for sorption sites, or to less sorption sites due to Al hydroxide precipitation (Anjos and Roswell, 1987; Smyth and Sanchez, 1980).

The release of P adsorbed on Al / Fe-(hydr)oxide facilitates Ca-P formation at higher pH (e.g. during early diagenesis in marine sediment; Heggie et al., 1990; Ruttenberg, 2003; Ruttenberg and Berner, 1993; Slomp, 2011). Consequently, desorption at higher pH does not necessarily increase soluble P (van Cappelen and Berner, 1988; Reddy and Sacco, 1981). This agrees with a switch from low phosphate concentrations in equilibrium with adsorbed P at acidic pH to low phosphate concentrations in equilibrium with mineral Ca-P under alkaline conditions (e.g. Murrmann and Peech, 1969). Low equilibrium concentrations under alkaline conditions are a result of the decreasing solubility of Ca-P phases, such as carbonate fluorapatite (CFAP) and octacalcium phosphate (OCP), with increasing pH (Hinsinger, 2001; Murrmann and Peech, 1969). Precipitation of Ca-P may therefore mitigate a desorption-derived P release from sediment (e.g. van Beusekom and de Jonge, 1997). Similarly, Ca-P precipitation is likely to result in the occasionally observed and apparently conflicting decrease of available P by liming to neutral or alkaline pH (Bolan et al., 2003; Haynes, 1982; Naidu et al., 1990). Accordingly, concentrations of Ca-P usually increase seaward as a consequence of enhanced precipitation (Andrieux-Loyer et al., 2008; Coelho et al., 2004; Paludan and Morris, 1999; Sutula et al., 2004).

Authigenic Ca-P is widely dispersed in marine sediment, but its solubility in seawater remains difficult to predict. Because seawater has been proposed as being largely undersaturated or close to saturation with respect to CFAP, both a possible formation or dissolution of CFAP in seawater cannot be entirely excluded at present (Atlas and Pytkowicz, 1977; Baturin, 1981; Bentor, 1980; Faul et al., 2005; Lyons et al., 2011). In contrast, detrital fluorapatite (FAP) is unlikely to dissolve in seawater (Ruttenberg, 1990; Howarth et al., 1995). In addition to the dependence on species-specific saturation states (Atlas, 1975; Gunnars et al., 2004), the occurrence of Ca-P minerals depends on their formation kinetics (Atlas and Pytkowicz, 1977; Gulbrandsen et al., 1984; Gunnars et al., 2004; Jahnke et al., 1983; Schenau et al., 2000; Sheldon, 1981) and inhibitors such as Mg^{2+} ions (Golubev et al., 1999; Gunnars et al., 2004; Martens and Harriss, 1970). In general, the first solid to form is the one which is thermodynamically least favoured (Ostwald step rule; see Morse and Casey, 1988 and Nancollas et al., 1989).

Given slow or inhibited direct nucleation (Golubev et al., 1999; Gunnars et al., 2004; Martens and Harriss, 1970), species of the apatite group may form by transformation of metastable precursors that are less susceptible to inhibitory effects of Mg^{2+} such as OCP (Oxmann, 2014; Oxmann and Schwendenmann, 2014). Precursor phases form more readily (e.g. days to weeks for OCP; Bell and Black, 1970) and can promote successive crystallization until the thermodynamically favoured but kinetically slow apatite formation occurs (ten to some thousand years; Schenau et al., 2000; Jahnke et al., 1983; Gulbrandsen et al., 1984). Several studies presented field and experimental evidence for this mode of apatite formation in sediment systems (Gunnars et al., 2004; Jahnke et al., 1983; Krajewski et al., 1994; Oxmann and Schwendenmann, 2014; Schenau et al., 2000; van Cappellen and Berner, 1988). A systematic comparison of P K-edge XANES (X-ray absorption near-edge structure spectroscopy) fingerprints from reference materials and marine sediment particles also provided evidence for the occurrence of OCP in sediment (Oxmann, 2014). However, despite significant progress in the determination of different matrix-enclosed Ca-P phases, it is not yet clear whether specific conditions at certain locations facilitate successive or direct crystallization of apatite (Slomp, 2011).

Provided more soluble Ca-P minerals such as OCP or less stable CFAP form in coastal environments, these minerals might mirror short-term changes of human alterations to the P cycle. Conversely, sparingly soluble apatite minerals may reflect long-term changes due to slow precipitation. Hence, the proportion of more soluble Ca-P should increase relative to total Ca-P in response to increased P inputs. This human alteration to the solid-phase P speciation may have implications for P fluxes and burial. To better describe P transformations from terrestrial to marine systems and to track the fate of anthropogenic P inputs, we analysed effects on P fractions and species across different ecosystems of a high P

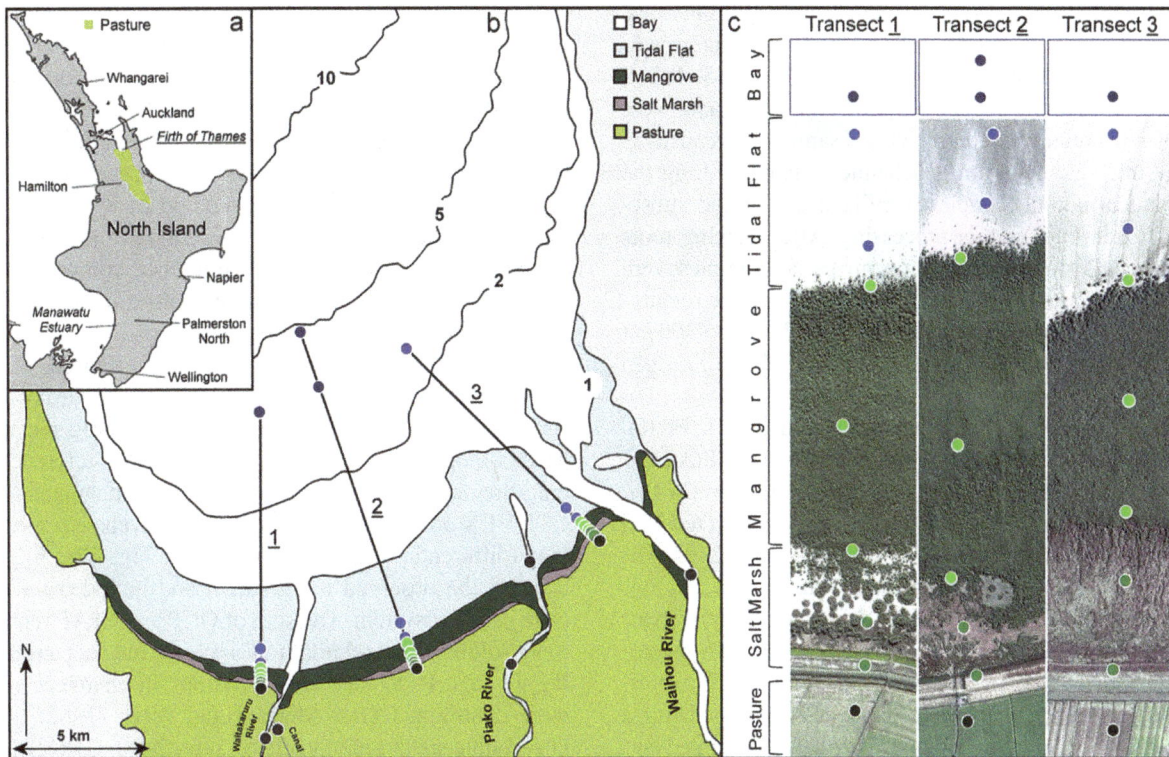

Figure 1. Study area. (**a**) Location of the Firth of Thames, North Island, New Zealand. The area of the catchment area, which is predominantly used for pastoral agriculture (1.3 million ha), is shown in green. (**b**) Firth of Thames transects across different ecosystems. (**c**) Plots ($n = 28$) along transects were located in the following ecosystems: bay (dark blue), tidal flat (blue), mangrove (light green), salt marsh (green) and pasture (dark green). Tidal flat plots close to mangrove forests included mangrove seedlings. Five additional plots were located at rivers (grey). Isolines indicate elevations in metres below mean sea level. Google Earth images for areas from pasture to tidal flat in (**c**).

site (Firth of Thames, New Zealand). We then compared the findings with results from a low P site (Saigon River delta, Vietnam) to distinguish speciation differences related to increased P loads. Octacalcium phosphate, authigenic apatite and detrital apatite were determined using a recently validated conversion–extraction method (CONVEX; Oxmann and Schwendenmann, 2014).

2 Materials and methods

2.1 Study area, Firth of Thames, New Zealand

The Firth of Thames, a meso-tidal, low-wave energy estuary of the Waikato region rivers Waihou and Piako, is located at the southern end of the Hauraki Gulf (37° S, 175.4° E; Fig. 1). It is the largest shallow marine embayment in New Zealand (800 km^2; < 35 m depth). The tides are semi-diurnal with a spring tide range of 2.8 m and a neap tide range of 2.0 m (Eisma, 1997). The southern shore of the bay (∼ 7800 ha) is listed as a wetland of international importance under the Ramsar Convention. The Firth of Thames encompasses large tidal flats (up to 4 km wide) and extensive areas of mangroves (*Avicennia marina* subsp. *australasica*) at

the southern end of the embayment (Brownell, 2004). Mangroves have been expanding seawards leading to a 10-fold increase in area since the mid-1900s (Swales et al., 2007). Mangrove expansion has been related to sediment accumulation and nutrient enrichment but may also coincide with climatic conditions (Lovelock et al., 2010; Swales et al., 2007). The upper coastal intertidal zone is covered by salt marshes. Behind the levee ca. 1.3 million ha is used for pastoral agriculture (∼ half of the total area of the Waikato region; Hill and Borman, 2011; Fig. 1a).

2.2 Sampling, field measurements and sample preparation

We established 28 plots along three transects (Fig. 1b, c). Transects extended across the entire tidal inundation gradient and across different ecosystems including bay ($n = 4$), tidal flat ($n = 6$), mangrove ($n = 9$), salt marsh ($n = 6$) and pasture ($n = 3$). Transects were at least 300 m from rivers to exclude areas affected by sediment aeration. Five additional plots were located along rivers. Sediment cores were taken during low tide using a polycarbonate corer (one core per plot; length: 40 cm; diameter: 9 cm). Immediately after core sampling, sediment pH, Eh and temperature were measured in situ at 0–5, 10–15, 30–35 and 35–40 cm depth intervals. Cores were divided into the following surface, intermediate and deeper sections: 0–5, 10–15, 30–35 and 35–40 cm. Longer core sections reduce vertical variability and were chosen for the relatively coarse vertical sampling because the focus of this study was on geochemical changes along the land–sea continuum. Samples were kept on ice and subsequently frozen until further processing. After thawing roots were removed from the sediment samples. Subsamples were then taken for particle size and salinity analysis. The remaining material was dried, ground and sieved (37 °C; < 300 μm mesh; PM 100; Retsch, Haan, Germany) for P and nitrogen analyses.

Temperature, pH and Eh were measured with a Pt-100 temperature sensor, sulfide resistant SensoLyt SEA/PtA electrodes and pH/Cond340i and pH 3310 mV meters (WTW, Weilheim, Germany). The mV meters were connected to a computer with optoisolators (USB-isolator; Serial: 289554B; Acromag Inc., Wixom, USA) for data visualization and logging (MultiLab pilot; WTW, Weilheim, Germany). Topographic elevation at the plots was measured with a total station (SET530R; Sokkia Co., Atsugi, Japan) relative to a reference point and converted to geo-referenced elevation using a global navigation satellite system (Trimble R8; Trimble Navigation Ltd., Sunnyvale, USA). Inundation duration was calculated from measured elevation above mean sea level and local tide tables (Waikato Regional Council, Hamilton East, New Zealand).

2.3 Sediment analyses

Phosphorus fractions and total P were analysed using three different methods. (i) The relative proportion of more soluble Ca-P was determined by preferential extraction of this fraction using the Morgan test method (Morgan, 1941); (ii) Al / Fe-P and Ca-P fractions were determined by sequential extraction of P after Kurmies (1972); and (iii) total P (TP) was analysed after Andersen (1976) as modified by Ostrofsky (2012). The Morgan test, commonly used to determine available P, preferentially extracts more soluble Ca-P phases using a pH 4.8 buffered acetic acid (see Sect. 4.4). Hence, the term metastable Ca-P (Ca-P_{meta}) is used for Morgan P in this paper. The method of Kurmies includes initial wash steps with KCl / EtOH to eliminate OCP precipitation prior to the alkaline extraction and Na_2SO_4 extractions to avoid readsorption. It therefore provides an accurate means of determining Al / Fe-P and Ca-P using NaOH and H_2SO_4, respectively (Supplement Fig. S2c; steps 2a–3c). Total inorganic P (TIP) was defined as the sum of inorganic P fractions (Ca-P, Al / Fe-P). Organic P was calculated by subtracting TIP from TP.

Octacalcium phosphate, CFAP (authigenic apatite) and FAP (detrital apatite) were quantitated using the CONVEX method (Oxmann and Schwendenmann, 2014). This method employs a conversion procedure by parallel incubation of sediment subsamples at different pH values (approximate pH range 3 to 8) in 0.01 M $CaCl_2$ for differential dissolution of OCP, CFAP and FAP (Fig. S2c). The concentration of OCP, CFAP and FAP is determined by the difference of Ca-P concentrations before and after differential dissolution of OCP, CFAP and FAP, respectively. These Ca-P concentrations are determined by the method of Kurmies. Differential dissolution was verified by standard addition experiments. For these experiments, reference compounds were added to the sediment subsamples before incubation using polyethylene caps loaded with 2 μmol P g^{-1} (ultra-micro balance XP6U; Mettler Toledo GmbH, Greifensee, Switzerland). Reference compounds included OCP, hydroxylapatite (HAP), various CFAP specimens, FAP and biogenic apatite. Methodology, instrumentation and the suite of reference minerals are described in Oxmann and Schwendenmann (2014). CONVEX analysis was conducted for seven sediment samples (differential dissolution shown in Fig. S2a, b), which covered the observed pH gradient and included sediments from each ecosystem. The sum of OCP and CFAP represents more soluble Ca-P (similar to Ca-P_{meta}) and was termed Ca-$P_{OCP+CFAP}$. Phosphate concentrations in chemical extracts were determined after Murphy and Riley (1962) using a UV–visible spectrophotometer (Cintra 2020; GBC Scientific Equipment, Dandenong, Australia).

Particle size was analysed using laser diffractometry (Mastersizer, 2000; Malvern Instruments Ltd., Malvern, UK; sediment dispersed in 10 % sodium hexametaphosphate solution). Salinity was determined by means of a TetraCon 325 electrode (WTW, Weilheim, Germany; wet sediment to deionized water ratio: 1 : 5). Nitrogen content was measured using a C / N elemental analyzer (TruSpec CNS; LECO soil 1016 for calibration; LECO Corp., St. Joseph, USA).

Concentrations of P fractions and proportions of more soluble Ca-P phases in sediments of the Firth of Thames site were compared with those of a contrasting low P site in the Saigon River delta (Oxmann et al., 2008, 2010). The site was located in the UNESCO Biosphere Reserve Can Gio close to the South China Sea and was not significantly influenced by anthropogenic P inputs. The region is not used for agriculture and the Saigon River downriver from Ho Chi Minh City (ca. 50 km from the study site) did not contain high levels of P (Schwendenmann et al., unpublished data). In contrast, the physical–chemical sediment characteristics measured at the two sites were comparable. For example, pH, Eh and salinity showed similar gradients along the land–sea transects of both sites and these parameters had similar ranges and mean values for mangrove sediments of both sites (Sect. 3.4). An area of acid sulfate sediments at the low P site was analysed separately and confirmed results of the site comparison despite its significantly lower pH values (Sect. 3.4).

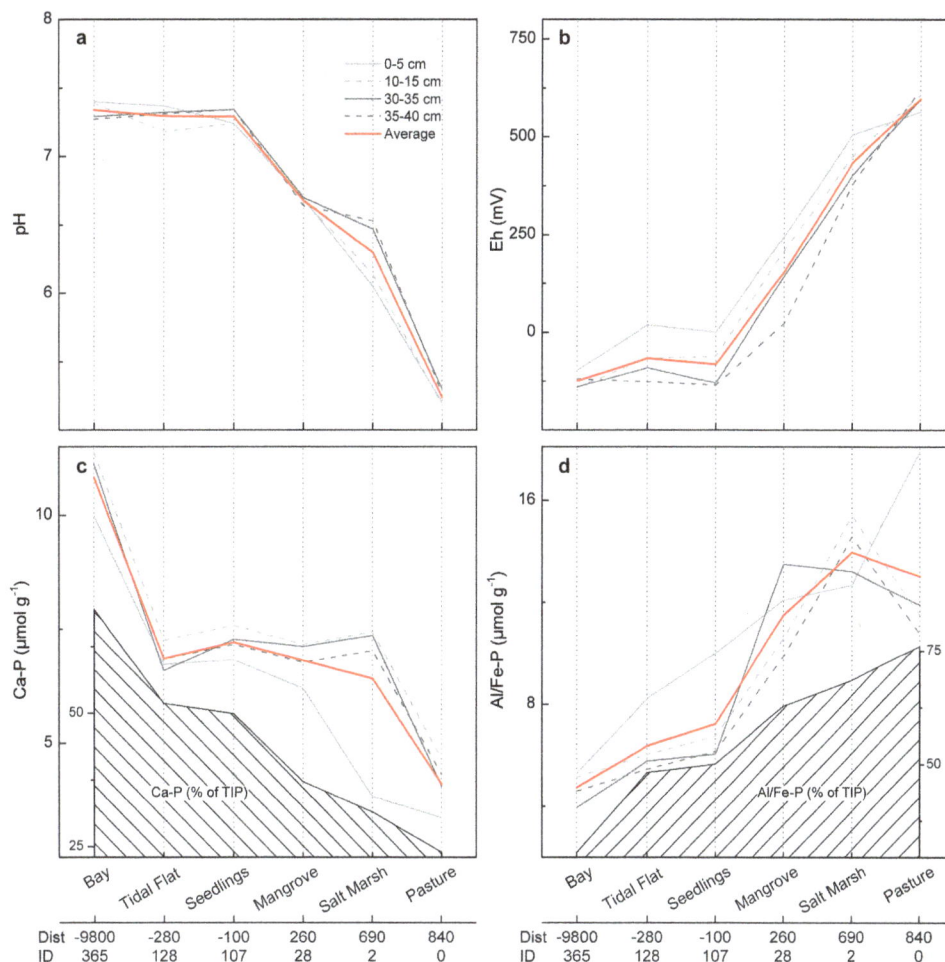

Figure 2. Physical–chemical sediment characteristics – (**a**): pH; (**b**): Eh – and sediment phosphorus fractions – (**c**): Ca-P; (**d**): Al / Fe-P – across ecosystems at the Firth of Thames. Each depth interval includes average values of several plots of each ecosystem across the entire site (all transects; see Table S1 for the number of averaged data and their values). Mean values of all depth intervals (112 samples) are also shown for each parameter. Mean fraction concentrations in % of total inorganic P (TIP) are given in (**c**) and (**d**); x axis labels include mean distance (Dist, metres, mangrove seaward margin is set as zero) and mean inundation durations (ID; days yr^{-1}); r and p values for correlations among these parameters are given in Table 1.

3   Results

3.1   Particle size, pH and Eh

Inundation duration ranged from 365 days yr^{-1} in the bay to 0 days yr^{-1} in the pastures (Supplement Table S1). Clay, silt and sand fractions varied between 0 and 20, 60 and 80 and 0 and 30 %, respectively (not shown). Particle size distribution differed only slightly among transects. In contrast, considerable differences in particle size distribution were found along transects with higher silt and lower sand contents in mangrove plots. Salinity decreased from the bay (32 ‰) to tidal flats (21 ‰). The highest values were measured in the mangroves (35 ‰), which declined to 25 ‰ in the salt marshes and 0 ‰ in the pastures (Table S1). Sediment water content ranged between 60 and 70 % in the bay, tidal flat and man-

grove fringe plots and decreased to 20–30 % in the pastures (not shown).

Sediment pH ranged from 5.18 in the pastures to 7.4 in the bay (Fig. 2a, b; Table S1). Redox potential varied between 621 mV in the pastures and −141 mV in the bay (Fig. 2a, b; Table S1). This pronounced and relatively constant pH increase and Eh decrease towards the bay was closely correlated with inundation duration (Fig. 2a, b; $p < 0.0001$). While systematic differences among depth intervals were less apparent for pH, Eh typically decreased with increasing sediment depth (Fig. 2b).

3.2   Phosphorus fractions: transformations

Ecosystem-averaged Ca-P concentrations varied considerably and ranged from $3.37\,\mu mol\,g^{-1}$ (pastures, 0–5 cm)

Table 1. Correlations between physical–chemical sediment characteristics and P fractions at P-enriched transects of the Firth of Thames, New Zealand.

Correlation			n	Subset[a]	r	p[b]
pH	vs.	Eh	112		−0.83561	2.2×10^{-30}
Ca-P	vs.	pH	112		0.60525	1.6×10^{-12}
Ca-P	vs.	Ca-P$_{meta}$	112		0.57074	5.0×10^{-11}
Al/Fe-P	vs.	Eh	112		0.54742	4.2×10^{-10}
Al/Fe-P	vs.	pH	28	0–5 cm	−0.81043	1.7×10^{-07}[c]
Ca-P$_{meta}$	vs.	pH	112		0.44885	6.9×10^{-07}
Al/Fe-P	vs.	pH	112		−0.43497	1.6×10^{-06}[c]
Al/Fe-P	vs.	Eh	77	pH > 6.6	0.50890	2.3×10^{-06}
Al/Fe-P	vs.	Eh	28	Surface	0.75979	2.7×10^{-06}
Al/Fe-P	vs.	pH	77	pH > 6.6	−0.48749	6.9×10^{-06}
Ca-P	vs.	Salinity	112		0.39824	1.4×10^{-05}
Ca-P$_{meta}$	vs.	Eh	112		−0.26717	4.4×10^{-03}
Ca-P	vs.	Al/Fe-P	112		−0.26236	5.2×10^{-03}[d]
Al/Fe-P	vs.	Eh	35	pH < 6.6	0.15496	NS
Al/Fe-P	vs.	Ca-P$_{meta}$	112		0.09103	NS
Al/Fe-P	vs.	Salinity	112		−0.13359	NS

Al/Fe-P: Al/Fe-bound P; Ca-P: calcium-bound P; Ca-P$_{meta}$: metastable Ca-P. [a] Blank rows indicate complete sample set analysed. [b] All correlations with organic P non-significant. [c] Surface layer showed a stronger correlation between Al/Fe-P and pH than the complete data set because other depth intervals showed a peak at pH 6.6 (see Fig. 3c). [d] Note that r and p values of Ca-P vs. Al/Fe-P strongly depend on selected pH intervals (cf. Fig. 3c). NS: non-significant

Table 2. Phosphorus fractions at different pH intervals in sediments of the high P site (Firth of Thames) and low P site (Saigon River delta).[a]

Site	n		pH		Ca-P μmol g^{-1}		Al/Fe-P μmol g^{-1}	
	< 6.6	> 6.6	< 6.6	> 6.6	< 6.6	> 6.6	< 6.6	> 6.6
High P	35	72	5.9	7.0	5.92	7.69	13.65	8.90
Low P	66	23	6.0	6.9	3.55	4.10	8.03	7.89
% Increase					+49	**+88**	**+70**	+13

Ca-P: calcium-bound P (mean); Al/Fe-P: Al/Fe-bound P (mean). [a] The analysis was restricted to sediments at overlapping pH intervals for both sites (pH < 6.6: 4.83–5.99; pH > 6.6: 6.01–7.47) to compare the increase for similar mean pH values at the lower (average pH ~ 6) and upper (average pH ~ 7) pH intervals. Note the pH-dependent accumulation of Al/Fe-P and Ca-P (bold; Sect. 4.4).

to 11.37μmol g^{-1} (bay, 10–15 cm) (Fig. 2c, d; Table S1). In contrast, Al/Fe-P was highest in the pastures (17.9μmol g^{-1}; 0–5 cm) and lowest in the bay (3.94μmol g^{-1}; 30–35 cm). On average, the lowest Ca-P and the highest Al/Fe-P concentrations were measured in 0–5 cm depth (Fig. 2c, d). Averaged percentages of Ca-P (% of TIP) steadily increased and averaged percentages Al/Fe-P (% of TIP) steadily decreased from pastures to bay (Fig. 2c, d). Along the marked downstream transition from Al/Fe-P (2.7-fold decrease) to Ca-P (2.6-fold increase), the average drop in Al/Fe-P from pastures to bay approximately matched the average Ca-P increase (−8.27 vs. +6.73 μmol g^{-1}; averages across all plots and depth intervals of each ecosystem; Table S1). Furthermore, mean Al/Fe-P concentrations in the different systems were negatively correlated with those of Ca-P ($r = -0.66$, $p < 0.001$;

Table S1). In addition, the decline in Al/Fe-P with depth was counterbalanced by the Ca-P increase with depth (−1.16 vs. +1.18 μmol g^{-1}; Table S1). Pastures were excluded from estimating these changes with depth because here the large loss of Al/Fe-P with depth was not counterbalanced by Ca-P (apparent surface runoff: −6.00 vs. +0.69 μmol g^{-1}; Table S1).

3.3 Phosphorus fractions: P load, pH and Eh effects

Mean sediment Ca-P concentration at the high P site was approximately twice the level of Ca-P at the low P site (Fig. 3a). Mean sediment Al/Fe-P concentration was approximately 30 % higher at the high P site compared to the low P site. However, at both sites Ca-P increased strongly with pH (Fig. 3a). Al/Fe-P showed a peak at ~ pH 6.6

Table 3. Phosphorus fractions in mangrove sediments of the high P site (Firth of Thames) and low P site (Saigon River delta).

Mangrove site	n	pH	Eh mV	Ca-P μmol g^{-1}	Al / Fe-P μmol g^{-1}	Ca-P$_{meta}$ μmol g^{-1}	Ca-P$_{meta}$ (% of Ca-P)
High P	48	6.8	95	6.91	10.44	2.35	34
Low P	64	6.4	66	3.85	8.36	0.40	10
Low P$^a_{acid}$	(32)	(5.0)	(240)	(2.78)	(6.26)	(0.27)	(10)
% Increase[b]				**+80** (+150)	**+25** (+65)	**+482** (+800)	**+240** (+240)

Ca-P: calcium-bound P (mean); Al / Fe-P: Al / Fe-bound P (mean); Ca-P$_{meta}$: metastable Ca-P (mean). [a] Area of acid sulfate sediments in mangroves of the low P site. [b] Percentages of P fraction increase at the high P site in comparison to the low P site (bold), which had similar pH and Eh values. Values for a comparison of the high P site with an area of acid sulfate sediments (Low P$_{acid}$) in parentheses.

Table 4. Correlation coefficients (for $p < 0.05$) between concentrations of P fractions, OCP, CFAP, FAP and pH in sediments analysed for particular Ca-P species.[a]

	Al / Fe-P	Res. P	pH	FAP	CFAP	OCP	Ca-P	OCP+CFAP	Ca-P$_{meta}$
Res. P	0.88****								
pH	[b]	–							
FAP	–	–	–						
CFAP	–	0.69**	–	–					
OCP	–	–	–	–	–				
Ca-P	[b]	–	0.68*	–	0.77**	0.81***			
OCP+CFAP	–	–	0.88****	–	0.65*	0.67*	0.79**		
Ca-P$_{meta}$	–	–	–	–	0.84**[c]	–	0.76*[c]	0.74*[c]	
TIP	0.95****	0.91****	–	–	0.66*	–	–	–	0.69*[c]

TIP: total inorganic P; Al / Fe-P: Al / Fe-bound P; Ca-P: calcium-bound P; Ca-P$_{meta}$: metastable Ca-P; OCP: octacalcium phosphate; CFAP: carbonate fluorapatite; FAP: fluorapatite; Res. P: residual P. NS is non-significant; * is 0.05 level; ** is 0.01 level; *** is 0.001 level; **** is 0.0001 level. [a] Species distributions shown in Fig. 5. [b] See Fig. 3 for correlations among Ca-P, Al / Fe-P and pH using a larger set of fraction data. [c] $n = 9$; for all other correlations $n = 13$.

(Fig. 3b). Thus, despite large differences in P fraction concentrations between the two sites pH dependencies of both fractions were similar, except for Al / Fe-P concentrations in 0–5 cm depth. Concentrations of Al / Fe-P in this depth range showed a continuous decrease with pH at the high P site due to high Al / Fe-P levels in acidic surface sediment of the pastures (linear regression in Fig. 3b; Table 1; $r = -0.81$, $p < 0.0001$).

In sediments with pH < 6.6 the average concentration of Al / Fe-P was 70 % higher at the high P site than at the low P site (Table 2; Fig. 3b). Despite these largely elevated levels of Al / Fe-P at topographically higher areas, Al / Fe-P was only slightly increased (13 %) in the lower intertidal zones and the bay (pH > 6.6; Table 2; Fig. 3b). Calcium phosphate in contrast showed the opposite pattern of enrichment at the high P site. In comparison to the low P site the average concentration of Ca-P was only 49 % higher in the upper intertidal zones and pastures (pH < 6.6) but increased by 88 % in the lower intertidal zones and the bay (pH > 6.6; Table 2; Fig. 3a).

Although Ca-P and Al / Fe-P clearly showed opposite trends along the three transects of the high P site (Fig. 2c, d), both fractions increased strongly with pH below pH 6.6 (Fig. 3a, b). Both fractions were positively correlated at

pH < 6.6 (shown in Fig. 3c for lower depth of both sites). At pH > 6.6, however, Ca-P increased further, whereas Al / Fe-P abruptly decreased (cf. Fig. 3a and b). Because this switch occurred in the landward to seaward direction, it is in agreement with the observed Ca-P increase and Al / Fe-P decrease towards the bay (Fig. 2c, d).

3.4 Metastable calcium phosphate

Metastable Ca-P (Ca-P$_{meta}$) increased strongly with pH (Fig. 4a; Firth of Thames cross-data-set correlations in Table 1), similar to Ca-P (Fig. 3a), and correlated with Ca-P at both sites (Fig. 4b; Table 1). Yet sediments of the high P site contained on average 6-fold higher concentrations of Ca-P$_{meta}$ compared with the low P site (Fig. 4c). In contrast, sediments of the high P site contained only 2-fold more total Ca-P than those of the low P site (Fig. 3a). On average, Ca-P$_{meta}$ comprised ca. 35 % of total Ca-P at the high P site and only 10 % at the low P site (Fig. 4d).

To verify that the higher Ca-P$_{meta}$ concentrations were not a consequence of site-specific differences in vegetation or physical–chemical sediment conditions we restricted the comparison to mangrove plots, which showed similar ranges and mean values of pH, Eh and salinity at both sites (Firth of

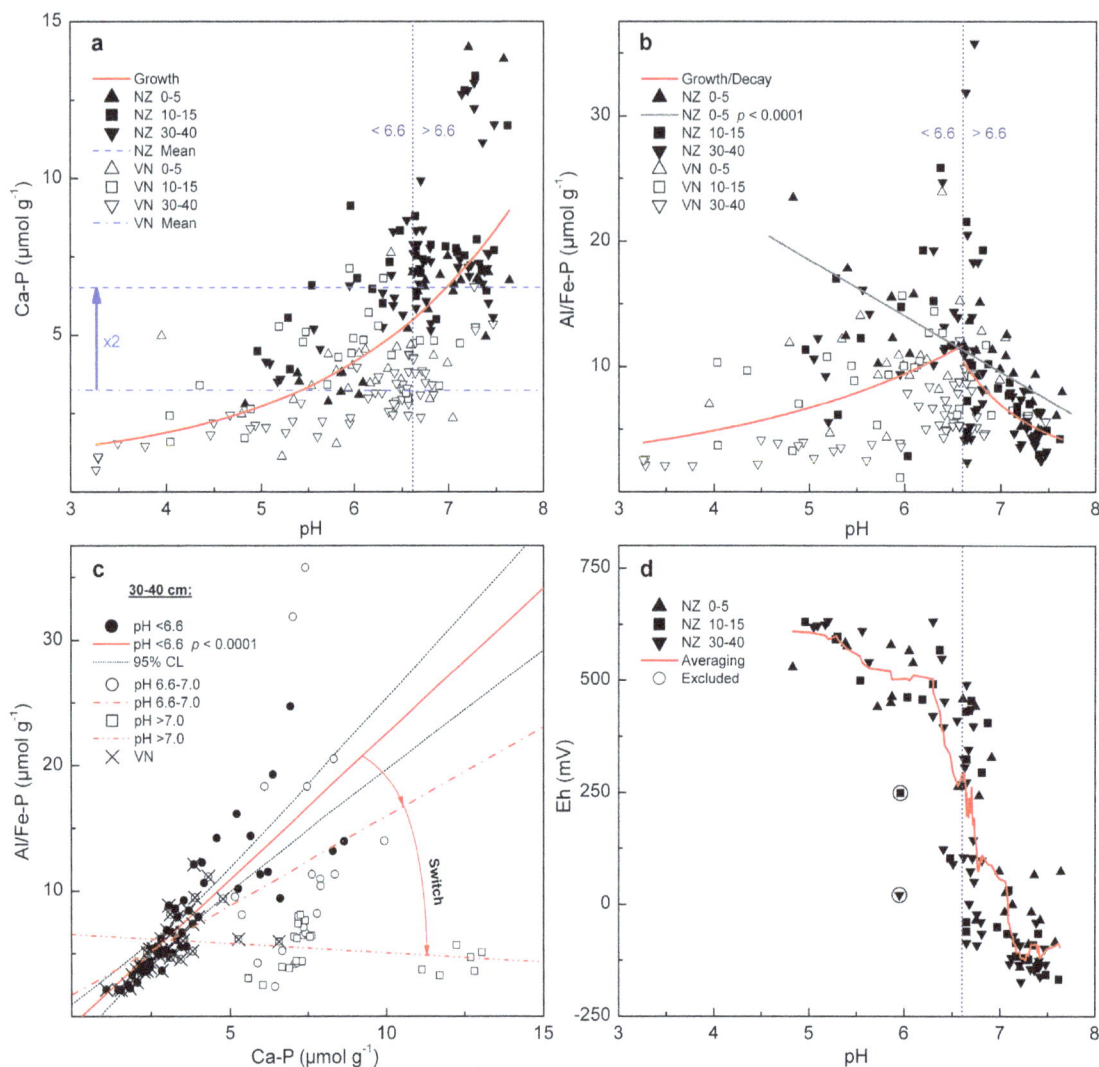

Figure 3. Changes of sediment phosphorus fractions (Al / Fe-P; Ca-P) as a function of pH and Eh variations at transects at the Firth of Thames (NZ) and Saigon River delta (VN) site. (**a**) Ca-P increase with pH (exponential; both sites, all samples). Mean Ca-P concentration of Firth of Thames samples ca. twice that of Saigon River delta samples. (**b**) Al / Fe-P peak at $\sim$ pH 6.6 due to increase below and decrease above that value (exponential; both sites, all samples). (**c**) Linear regressions between Al / Fe-P and Ca-P at deeper depths (30–35, 35–40 cm) for different pH intervals (cf. Fig. 3a, b). Arrows indicate switch from Al / Fe-P to Ca-P with increasing pH (seaward direction). (**d**) Eh vs. pH (NZ; all samples). Different symbols denote surface (0–5 cm), intermediate (10–15 cm) and deeper depth intervals (30–35, 35–40 cm) in (**a**), (**b**) and (**d**). Symbols for VN data marked with cross in (**c**). Smoothing by averaging 10 adjacent Eh values of pH sorted data in (**d**). See text for linear regression in (**b**).

Thames: pH 5.8–7.1, −160–450 mV, 25–50 ‰; Saigon River delta: pH 5.7–7.0, −180–400 mV; 25–40 ‰; Table 3). This adjustment did not change the results. The difference in Ca-P_{meta} concentrations between mangrove plots of the two sites was just as disproportionate when compared to the difference in total Ca-P concentrations between those plots (6-fold vs. 2-fold). The portion of Ca-P_{meta} was still ca. 35 % at the high P site and 10 % at the low P site (Table 3). Moreover, the proportion of Ca-P_{meta} to total Ca-P was equally low for an area of acid sulfate sediments of the low P site (10 %) despite its very different average pH and Eh values (Table 3).

In summary, comparatively large amounts of metastable Ca-P accumulated at the high P site.

3.5 Octacalcium phosphate and authigenic apatite

Distributions of OCP, authigenic apatite and detrital apatite were related to the pH at both sites. Strongly acidic sediments ($\sim$ pH < 4) contained just detrital apatite (FAP), whereas slightly acidic sediments ($\sim$ pH 4–7) contained also authigenic apatite (CFAP). Octacalcium phosphate was additionally present in alkaline mangrove, river, bay and tidal flat

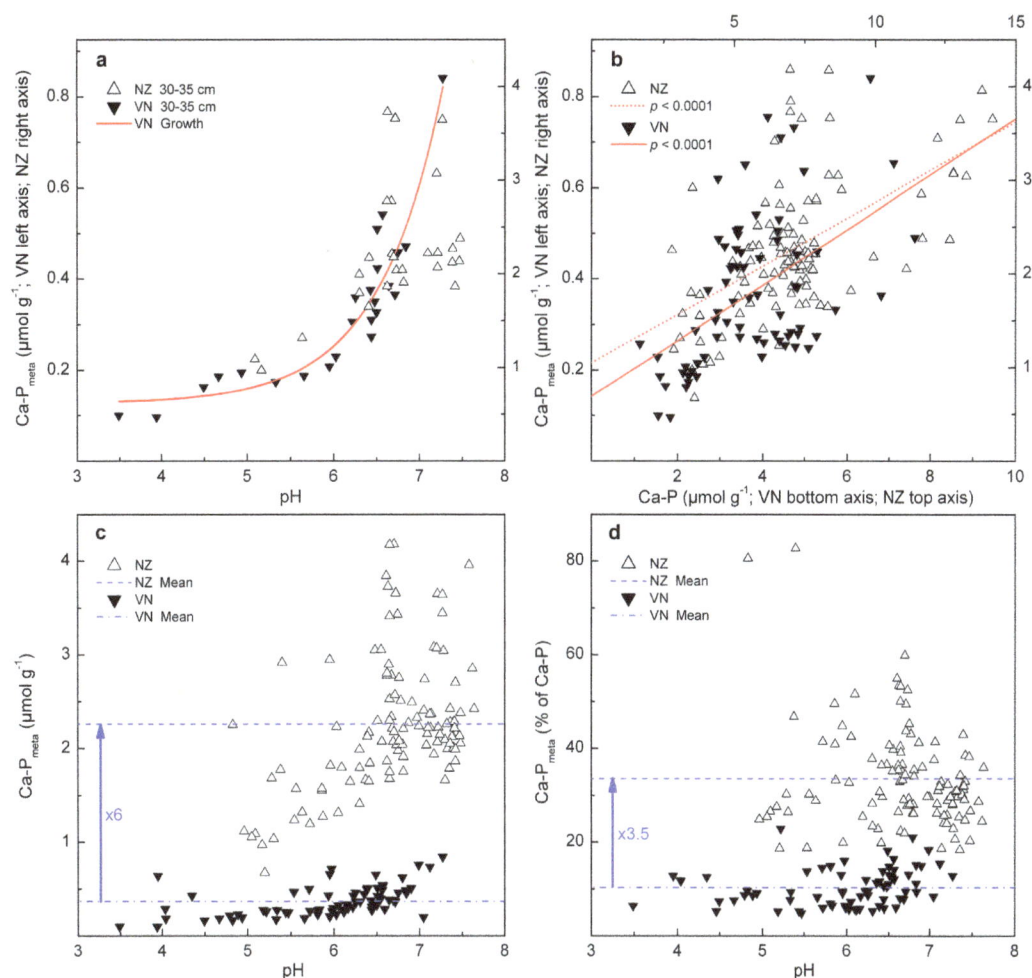

Figure 4. Accumulation and pH dependence of metastable Ca-P (Ca-P$_{meta}$) along transects at the Firth of Thames (NZ) site compared to the Saigon River delta (VN) site. (**a**) Increase of Ca-P$_{meta}$ with pH (exponential; 30–35 cm). (**b**) Linear regressions of Ca-P$_{meta}$ vs. total Ca-P (all plots and depths; NZ: $r = 0.57$, $p < 0.0001$; VN: $r = 0.50$, $p < 0.0001$). (**c**) Ca-P$_{meta}$ vs. pH (all plots and depths). Mean Ca-P$_{meta}$ concentration at the Firth of Thames site ca. 6 times that of the Saigon River delta site. (**d**) Ca-P$_{meta}$ in % of total Ca-P vs. pH (all plots and depths). Mean percentage ca. 3.5 times higher for the Firth of Thames site. Note different axis ranges for the two sites in (**a**) and (**b**).

sediments. Hence, the concentration of more soluble Ca-P$_{OCP+CFAP}$ (hatched area in Fig. 5) significantly increased with pH ($r = 0.88$, $p < 0.0001$; Table 4). However, the portion of Ca-P$_{OCP+CFAP}$ as a percentage of total Ca-P was significantly larger (70.5 ± 17.5 %; numbers above columns in Fig. 5) for sediments of the high P site compared to the low P site (29.5 ± 26.0 %, $t(11) = 3.346$, $p = 0.0065$). This larger portion of Ca-P$_{OCP+CFAP}$ provided supporting evidence for the larger portion of Ca-P$_{meta}$ in sediments of the high P site (cf. Sect. 3.4). Overall, more soluble Ca-P determined by the two independent methods (CONVEX method: Ca-P$_{OCP+CFAP}$; Morgan test: Ca-P$_{meta}$) yielded comparable results. Accordingly, corresponding values obtained by the two methods were significantly correlated ($r = 0.74$, $p < 0.05$; Table 4).

4   Discussion

4.1   Phosphorus status, Firth of Thames

The Firth of Thames sediments were high in P compared to the Saigon River delta site and sediments from other coastal areas (Tables 3 and S3; Figs. 3, 4). Total P concentrations measured along the three transects are classified as enriched ($> 16 \, \mu$mol P g^{-1}) and very enriched ($> 32 \, \mu$mol P g^{-1}) according to the New Zealand classification system (Robertson and Stevens, 2009; Sorensen and Milne, 2009). This is due largely to high P fertilizer application rates, which constitute the main P source (~ 90 %) to the watershed (Waikato region; total input: 41 Gg P yr^{-1}; fertilizer: 37 Gg P yr^{-1}; rate: 28 kg P ha^{-1} yr^{-1}; atmosphere and weathering: 4 Gg P yr^{-1}; Parfitt et al., 2008). A significant increase in TP is correlated with intensification of pastoral farming and contributes to

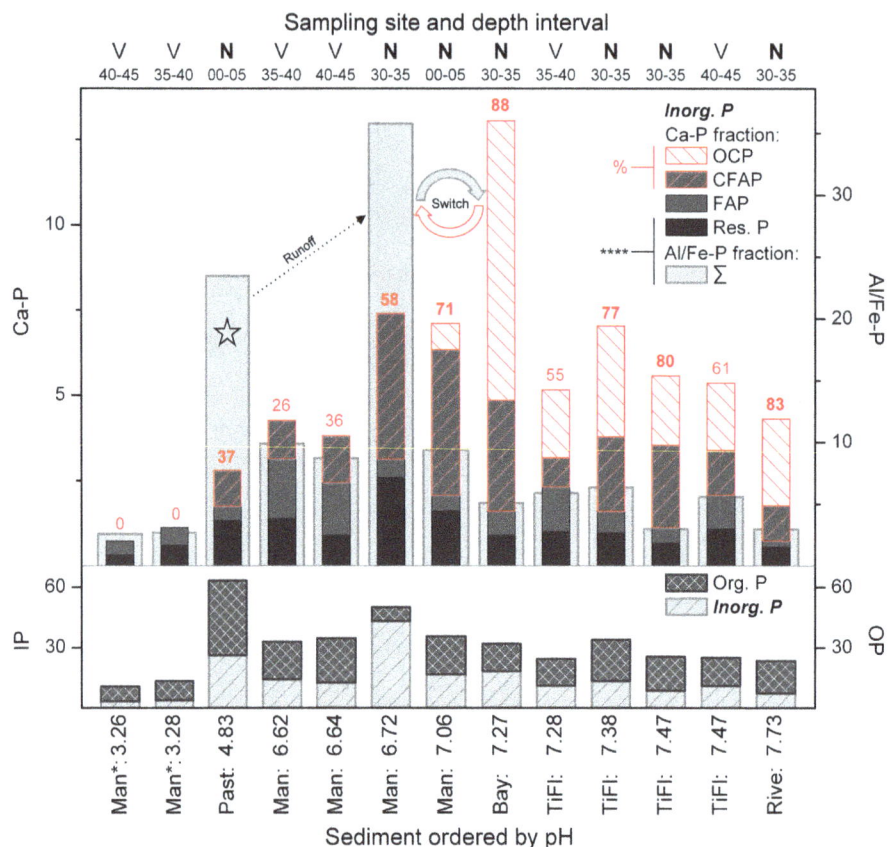

Figure 5. Phosphorus species distributions in sediments of different ecosystems at the Firth of Thames (NZ) and Saigon River delta (VN), ordered by pH (all concentrations in μmol $P\,g^{-1}$). The top axis shows the sampling site (N: NZ, New Zealand; V: VN, Vietnam) and depth interval (cm). The portion of more soluble Ca-$P_{OCP+CFAP}$ (hatched area) is given as a percentage of total Ca-P above columns (NZ values in bold). A typical decrease of adsorbed P and increase of OCP with increasing pH from 6.7 to 7.3 suggests a reversible transformation in that range (arrows; adsorption–precipitation switch). Strong Al / Fe-P predominance in acidic surface sediment of the pasture is denoted (star). Org. P: total organic P; Inorg. P: total inorganic P; OCP: octacalcium phosphate; CFAP: carbonate fluorapatite; FAP: fluorapatite; Res. P: residual P; Man*: strongly acidic mangrove; Man: mangrove; Past: pasture; TiFl: tidal flat; Rive: river. Residual P correlated with Al / Fe-P at a significance level of 0.0001 (asterisks). See Table 4 for all correlations among species distributions.

the deterioration of the river water quality by surface runoff (Vant and Smith, 2004). Further, elevated levels of P and nitrogen in groundwater at coastal farmlands agreed with specific fertilizer application rates (Brownell, 2004). The application rates and previous findings strongly suggest that the P accumulation measured in this study was largely related to P fertilization practices. Hence, the Firth of Thames site is characterized by anthropogenic P enrichment.

4.2   Phosphorus fractions

Phosphorus fractions showed a strong and continuous increase of Ca-P and decrease of Al / Fe-P with increasing inundation duration (Fig. 2). These changes were driven by pH and Eh gradients (Fig. 2). Salinity had no detectable effect on Al / Fe-P (Table 1), similar to findings of Maher and DeVries (1994). Both an increase in Ca-P (decreasing solubility of calcium phosphates with increasing pH; Lindsay et al.,

1989; Hinsinger, 2001) and a decline in Al / Fe-P (desorption of P from metal (hydr)oxides with increasing pH or decreasing Eh; e.g. Mortimer, 1971) commonly occur across estuarine inundation gradients. A similar seaward redistribution of sediment P fractions has been observed, for example, in marsh systems in Portugal (Coelho et al., 2004) and South Carolina (Paludan and Morris, 1999), estuarine zones of a French river (Andrieux-Loyer et al., 2008) and an estuarine transect from the Mississippi River to the Gulf of Mexico (Sutula et al., 2004). In the latter case, the opposing landward-to-seaward changes in sediments (Sutula et al., 2004) were mirrored by similar changes in corresponding samples of surface water particulate matter. Furthermore, the same trends were observed along a continuum from agricultural soils across hard-water stream sediments to lake sediments (Noll et al., 2009).

The pH dependence of the solubility of Ca-P phases was mirrored by the increase in Ca-P with decreasing acidity

(Fig. 3a). In contrast, concentrations of Al / Fe-P showed a maximum amount of adsorbed P at pH 6.6 (Fig. 3b) and therefore agreed with the commonly occurring maximum P availability at pH 6.5 in agricultural soils (e.g. Chapin III et al., 2011). This maximum P availability is caused by the highest P solubility in equilibrium with various P minerals at pH 6.5 (Lindsay et al., 1989). Considering that low amounts of P are precipitated with Ca, Fe and Al at pH 6.5 (e.g. Chapin III et al., 2011), large amounts of soluble reactive P could be available for adsorption on metal (hydr)oxides. Further, Ca-P phases predominated in alkaline downstream environments and may undergo dissolution after upstream transport by tides with an accompanying increase in P adsorption (cf. Fig. 2a, c, d; De Jonge and Villerius; 1989). If these phases are transported to close locations of ∼ pH 6.5 and subsequently dissolve, the phosphate released can be adsorbed on more oxidized sediment, thereby contributing to the elevated Al / Fe-P concentrations at this pH.

The maximum amount of adsorbed P at pH 6.6 also indicated that Eh, which showed the maximum decline at ∼ pH 6.6 (Fig. 3d), did not cause significant desorption of P at this pH. Below this pH, Al / Fe-P did not decline with Eh (cf. Fig. 3b and d) and these parameters were not correlated (Table 1), suggesting that a release of P adsorbed to ferric iron compounds did not occur in the corresponding sediments. Yet, both Eh and Al / Fe-P decreased above pH 6.6 (cf. Fig. 3b and d) and were correlated in this range (Table 1). Because the drop in Al / Fe-P correlated also with an increase in pH (Table 1, Fig. 3b), effects of pH and Eh on P desorption could not be distinguished above pH 6.6. The decreasing amount of adsorbed P at near-neutral to alkaline pH may therefore be due to (i) charge changes of metal (hydr)oxides with pH (Oh et al., 1999; Spiteri et al., 2008; Barrow et al., 1980; Sundareshwar and Morris, 1999), (ii) less efficient sorption by iron in the Fe(II) state compared to the Fe(III) state (e.g. Sundareshwar and Morris, 1999), or (iii) a combination of charge changes and Fe reduction.

Critical redox potentials reported for reduction of ferric iron compounds are around 300 mV at pH 5 and 100 mV at pH 7 (Gotoh and Patrick, 1974; Husson, 2013; Yu et al., 2007). These levels match very well with critical Eh levels for desorption of Fe-(hydr)oxide-bound P, including a similar pH dependence of those levels (compare Delaune et al., 1981 with Gotoh and Patrick, 1974). As Fig. 3d shows, Eh values that did not correlate with Al / Fe-P (sediments with pH < 6.6) were above the critical Eh threshold, whereas Eh values that correlated with Al / Fe-P (sediments with pH > 6.6) were below the critical Eh threshold. This implies that reductive dissolution and related desorption of P could have contributed to the downstream transition from Al / Fe-P to Ca-P. Interestingly, the physicochemically induced P redistribution largely agreed with that from the low P site despite considerable differences of P fraction concentrations between both sites (Fig. 3a, b). This suggests that the effects of physical–chemical sediment characteristics were independent of the site-specific total P status.

4.3 Transformation of phosphorus fractions

The increase in Ca-P along transects, which correlated with an equivalent decrease in Al / Fe-P, strongly suggests that Ca-P formed at the expense of adsorbed P along the salinity gradient. By plotting Al / Fe-P and Ca-P concentrations on a pH scale (Fig. 3a, b) it became evident that the downstream transition from Al / Fe-P to Ca-P was related to a pronounced switch from P adsorption to Ca-P precipitation at ∼ pH 6.6 (see also Fig. 3c). Above this pH, reduction processes are less important for P desorption (Reddy and DeLaune, 2008), P adsorption is usually less pronounced (Murrmann and Peech, 1969) and thermodynamically less stable Ca-P phases such as OCP may form (Bell and Black, 1970; Oxmann and Schwendenmann, 2014).

A similar switch has been suggested for observed fraction changes with increasing sediment depth in non-upwelling continental margin environments, but from organic P to authigenic Ca-P (Ruttenberg and Berner, 1993). This switch was partly explained by the redox state, which could be the controlling parameter for diagenetic redistribution and related downcore changes of P fractions in marine environments. The main difference between the P redistribution in this study and results of Ruttenberg and Berner (1993) relates to the P source to the formation of authigenic Ca-P. Our results show strong interactions between inorganic P forms, with Al / Fe-P being a significant P source for Ca-P. In contrast to these strong interactions between Al / Fe-P and Ca-P all correlations with organic P were not significant.

However, the P redistribution along marine sediment cores may strongly differ from that across intertidal zones. The marked fraction changes suggest that the pH regulates an alternative switch between Al / Fe-P and Ca-P at the coastal sites investigated here. This pH-driven P redistribution could be a common mechanism at coastal pH gradients because it took place along different transects, comprising diverse ecosystems, and it was independent of the site-specific total P status (Fig. 3a, b). Hence, this mechanism could also be important for processes of P accumulation by increased P loads, as discussed next.

4.4 Phosphorus accumulation processes

The anthropogenic P input at the high P site caused two different location-dependent accumulation mechanisms, which mainly resulted in elevated Al / Fe-P at pH < 6.6 (landward) and elevated Ca-P at pH > 6.6 (seaward). The between-site comparison (high vs. low P site) therefore implies that fertilizer-derived P was largely included in the Al / Fe-P fraction (adsorbed P) of acidic landward sediments. Phosphorus inputs by runoff or erosion to downstream areas apparently led to enhanced precipitation of Ca-P by increasing pH.

The accumulation pattern in this site comparison therefore corresponds to location-dependent transformations between Al / Fe-P and Ca-P, which are to be expected from the P redistribution at individual sites (Sect. 4.3).

We hypothesize that more soluble Ca-P minerals accumulate relative to total Ca-P due to anthropogenic P inputs because the formation of sparingly soluble Ca-P minerals is too slow for balancing increased formation rates of thermodynamically less stable Ca-P minerals. This hypothesis is consistent with comparatively large amounts of metastable Ca-P, which apparently accumulated at the high P site due to external factors (Sect. 3.4, Table 3; Fig. 4c, d). Our findings showed that Morgan's weakly acidic acetate–acetic acid solution preferentially extracts metastable Ca-P phases. Because sparingly soluble Ca-P minerals, such as detrital apatite, are unlikely to dissolve in Morgan's solution (pH 4.8; cf. Ruttenberg, 1992), the correlations between Morgan P (Ca-P_{meta}) and Ca-P ($p < 0.0001$ at both sites; Fig. 4b) are attributable to more soluble Ca-P minerals. This conclusion is supported by other studies which have indicated that the Morgan test preferentially extracts more soluble Ca-P phases, whereas most other available P tests preferentially extract adsorbed P (cf. Ahmad et al., 1968; Curran, 1984; Curran and Ballard, 1984; Dabin, 1980; Herlihy and McCarthy, 2006).

Our hypothesis is also consistent with concentrations of OCP, CFAP and FAP, which were separately determined for sediments of both sites (Fig. 5). Sediments of the high P site showed a significantly larger portion of Ca-$P_{OCP+CFAP}$ compared to the low P site. Results of both independent methods, which were significantly correlated (Table 4), therefore provide strong evidence for the proposed accumulation of thermodynamically less stable Ca-P by anthropogenic P inputs. Less stable Ca-P may thus be a useful parameter to monitor anthropogenic accumulations of inorganic P in coastal regions. Because physical–chemical sediment characteristics influence Ca-P formation, an important caveat is the between-site comparability of data. In this study, there was between-site comparability of both the sediment characteristics and the general response of each of the P fractions and P species to the sediment characteristics at different depth intervals along the land–sea continuum (Figs. 3a, b, 4a, b, 5).

A dominant proportion of more soluble Ca-P was contributed by OCP in alkaline sediments (Fig. 5). These OCP concentrations therefore suggest authigenic apatite formation by initial precipitation of OCP in alkaline mangrove, river, bay and tidal flat sediments. Octacalcium phosphate was detected in surface marine sediment using XANES (Oxmann, 2014; XANES spectra of Brandes et al., 2007), which requires minimal sample preparation and is minimally affected by common sample matrices. Results of the CONVEX method, which was validated by the matrix effect-free method of standard addition (Oxmann and Schwendenmann, 2014), therefore agree with field and experimental evidence

for the occurrence of OCP in sediment (see also Gunnars et al., 2004; Jahnke et al., 1983; Krajewski et al., 1994; Morse and Casey, 1988; Nancollas et al., 1989; Schenau et al., 2000; van Cappellen and Berner, 1988).

We conclude that OCP plays a crucial role in the redistribution of sediment P (see arrows in Fig. 5), including the pH-dependent switch from adsorbed P to Ca-P in the landward to seaward direction, the potential reverse transformation after upstream transport, and the pH-dependent accumulation processes. Further, apatite formation by successive crystallization is possibly mainly restricted to alkaline sediments. Octacalcium phosphate is an important intermediate in the formation of apatite in alkaline environments, including calcareous soil (Alt et al., 2013; Beauchemin et al., 2003; Grossl and Inskeep, 1992), lake sediment (e.g. Avnimelech, 1983) and marine sediment (see above references). In agreement with the high OCP concentrations found in this study (Fig. 5), solid-state nuclear magnetic resonance (NMR) and XANES spectroscopy-based studies implied that OCP does belong not only to the most commonly reported but also to the most prevalent inorganic P forms in alkaline environments (Beauchemin et al., 2003; Kizewski et al., 2011; Oxmann, 2014).

In general, the established Ca-P precipitation in sediments across salinity gradients provides some insight into the relevance of factors influencing this precipitation such as changes in salinity, dissolved phosphate and pH. In fact, as the ionic strength increases with increasing salinity for a given phosphate concentration and pH, the apparent Ca-P solubility increases strongly (cf. Atlas, 1975). Yet, increasing Ca-P concentrations imply that the salt effect is usually more than offset by the rise in pH, redox-driven phosphate desorption from iron oxyhydroxides and other potential factors in interstitial waters across salinity gradients. For example, Ca^{2+} concentrations generally increase from land to sea and, hence, increase the saturation state with respect to calcium phosphates (normal seawater and sediment pore water: ca. 10 mM; river water, global average: ca. 0.4 mM; soil pore water, average of temperate region soils: ca. 1.5 mM; Girard, 2004; Lerman and Wu, 2008; Lower et al., 1999; Rengel, 2006; Sun and Turchyn, 2014). Although the correlation of salinity with Al / Fe-P was not significant, the correlation with Ca-P was decreased but still significant (Table 1), indicating that increasing Ca^{2+} concentrations from land to sea may also contribute to Ca-P formation.

Our results imply that when P enters the marine environment, enhanced Ca-P formation takes place in nearshore environments. Given the possibility that CFAP or other less stable Ca-P phases do not readily dissolve in alkaline seawater (Faul et al., 2005; Lyons et al., 2011; Sheldon, 1981; see also Gulbrandsen et al., 1984 as cited in Slomp, 2011), some non-detrital Ca-P at sites further offshore could be derived from Ca-P-generating areas of the lower intertidal zone or even from freshwater environments (see e.g. Raimonet et al., 2013).

5 Conclusions

Our results show a pH-induced switch from P adsorption to Ca-P precipitation at near-neutral pH, which apparently leads to inorganic P accumulation in nearshore sediments. The decrease in Eh and increase in Ca^{2+} concentrations from land to the sea likely contribute to this switch. Further, this P redistribution is apparently driven by OCP formation and enhanced by anthropogenic P inputs. Hence, a significant proportion of authigenic Ca-P may be derived from anthropogenic sources in some coastal regions.

The proposed mechanism, including relatively rapid formation of an apatite precursor, explains several independent observations: the downstream transition from Al / Fe-P to Ca-P at $\sim$ pH 6.6, the Ca-P formation at the expense of adsorbed P, the large increase of Ca-P_{meta} with increasing pH, the dominant proportion of OCP in alkaline sediments, the pH-dependent accumulation mechanisms of Al / Fe-P and Ca-P, and the accumulation of Ca-P_{meta} and Ca-$P_{OCP+CFAP}$ at the high P site. The suggested switch appears to be a very common mechanism because it was observed across different ecosystems and it was independent of the site-specific total P status. Further evidence that this mechanism operates in different environments comes from similar downstream transitions reported by several studies.

Less stable Ca-P is mainly formed and buried during sedimentation rather than being allochthonous material. Hence, CFAP and OCP act as diagenetic sinks for P at the investigated sites and are mainly responsible for the accumulation of inorganic P in the lower intertidal zone and bay. Some authigenic Ca-P, however, could be dissolved when physical–chemical conditions of the sediment change (e.g. altered pH/Eh due to land reclamation) or after upstream transport by tides. Some of it could also be resuspended and transported further offshore, similar to detrital FAP. In general, OCP formation may mitigate a desorption-derived P release from sediment and seems to occur when P adsorption is usually less pronounced – that is, under alkaline conditions.

Acknowledgements. We thank Bharath Thakur for assistance in field work and sample preparation and Peter and Gail Thorburn for assistance with sampling by boat. The project was funded by the German Research Foundation through a research fellowship granted to J. F. Oxmann under the code OX 54/2-1.

Edited by: C. P. Slomp

References

Ahmad, N., Jones, R. L., and Beavers, A. H.: Genesis, mineralogy and related properties of West Indian soils, (i) Montserrat Series, derived from glauconitic sandstone, Central Trinidad, J. Soil Sci., 19, 1-8, 1968.

Alt, F., Oelmann, Y., Schöning, I., and Wilcke, W.: Phosphate release kinetics in calcareous grassland and forest soils in response to H^+ addition, Soil Sci. Soc. Am. J., 77, 2060–2070, 2013.

Andersen, J. M.: An ignition method for determination of total phosphorus in lake sediments, Water Res., 10, 329–331, 1976.

Andrieux-Loyer, F., Philippon, X., Bally, G., Kéroul, R., Youenou, A., and Le Grand, J.: Phosphorus dynamics and bioavailability in sediments of the Penzé Estuary (NW France): in relation to annual P-fluxes and occurrences of *Alexandrium Minutum*, Biogeochemistry, 88, 213–231, 2008.

Anjos, J. T. and Rowell, D. L.: The effect of lime on phosphorus adsorption and barley growth in three acid soils, Plant Soil, 103, 75–82, 1987.

Atlas, E. L.: Phosphate equilibria in seawater and interstitial waters, Ph.D. thesis, Oregon State University, Corvallis, 1975.

Atlas, E. L. and Pytkowicz, R. M.: Solubility behavior of apatites in seawater, Limnol. Oceanogr., 22, 290–300, 1977.

Avnimelech, Y.: Phosphorus and calcium carbonate solubilities in Lake Kinneret, Limnol. Oceanogr., 28, 640–645, 1983.

Barrow, N. J., Bowden, J. W., Posner, A. M., and Quirk, J. P.: Describing the effects of electrolyte on adsorption of phosphate by a variable charge surface, Aust. J. Soil Res., 18, 395–404, 1980.

Baturin, G. N. (Ed.): Principal features of the marine geochemistry of disseminated phosphorus, in: Developments in Sedimentology, Elsevier B. V., Amsterdam, 1981.

Beauchemin, S., Hesterberg, D., Chou, J., Beauchemin, M., Simard, R. R., and Sayers, D. E.: Speciation of phosphorus in phosphorus-enriched agricultural soils using X-ray absorption near-edge structure spectroscopy and chemical fractionation, J. Environ. Qual., 32, 1809–1819, 2003.

Bell, L. C. and Black, C. A.: Transformation of dibasic calcium phosphate dihydrate and octacalcium phosphate in slightly acid and alkaline soils, Soil Sci. Soc. Am. Proc., 34, 583–587, 1970.

Bentor, Y. K. (Ed.): Phosphorites: The unsolved problems, in: Marine Phosphorites: Geochemistry, occurrence, genesis, SEPM Special Publication, 29, 3–18, 1980.

Bolan, N. S., Adriano, D. C., and Curtin, D.: Soil acidification and liming interactions with nutrient and heavy metal transformation and bioavailability, Adv. Agron., 78, 215–272, 2003.

Bowden, J. W., Nagarajah, S., Barrow, N. J., and Quirk, J. P.: Describing the adsorption of phosphate, citrate and selenite on a variable-charge mineral surface, Aust. J. Soil Res., 18, 49–60, 1980.

Brandes, J. A., Ingall, E., and Paterson, D.: Characterization of minerals and organic phosphorus species in marine sediments using soft X-ray fluorescence spectromicroscopy, Mar. Chem., 103, 250–265, 2007.

Brownell, B.: Firth of Thames RAMSAR Site update, EcoQuest Education Foundation, Kaiaua, New Zealand, 2004.

Chapin III, F. S., Matson, P. A., and Vitousek, P. M.: Principles of terrestrial ecosystem ecology, 2nd Edn., Springer-Verlag, New York, 529 pp., 2011.

Clarke, P. J.: Nitrogen pools and soil characteristics of a temperate estuarine wetland in eastern Australia, Aquat. Bot., 23, 275–290, 1985.

Coelho, J. P., Flindt, M. R., Jensen, H. S., Lillebø, A. I., and Pardal, M. A.: Phosphorus speciation and availability in intertidal sediments of a temperate estuary: Relation to eutrophication and annual P-fluxes, Estuar. Coast. Shelf. S., 61, 583–590, 2004.

Curran, M. P.: Soil testing for phosphorus availability to some conifers in British Columbia, B.Sc. thesis, The University of Victoria, B.C., 1984.

Curran, M. P. and Ballard, T. M.: P availability to forest trees in British Columbia, Contract Res. Rep. to B.C. Ministry of Forests, Victoria, B.C., 1984.

Dabin, P.: Phosphorus deficiency in tropical soils as a constraint on agricultural output, priorities for alleviating soil-related constraints to food production in the tropics, IRRI, Los Banos, 217–233, 1980.

De Jonge, V. N. and Villerius, L. A.: Possible role of carbonate dissolution in estuarine phosphate dynamics. Limnol. Oceanogr., 34, 332–340, 1989.

DeLaune, R. D., Reddy, C. N., and Patrick, W. H., Jr.: Effect of pH and redox potential on concentration of dissolved nutrients in an estuarine sediment, J. Environ. Qual., 10, 276–279, 1981.

Eisma, D.: Intertidal deposits: River mouths, tidal flats and coastal lagoons, CRC Press, Boca Raton, Boston, London, New York, 525 pp., 1997.

Faul, K. L., Paytan, A., and Delaney, M. L.: Phosphorus distribution in sinking oceanic particulate matter, Mar. Chem., 97, 307–333, 2005.

Fauzi, A., Skidmore, A. K., Heitkönig, I. M. A., van Gils, H., and Schlerf, M.: Eutrophication of mangroves linked to depletion of foliar and soil base cations, Environ. Monit. Assess., 186, 8487–8498, 2014.

Gao, Y., Cornwell, J. C., Stoecker, D. K., and Owens, M. S.: Effects of cyanobacterial-driven pH increases on sediment nutrient fluxes and coupled nitrification-denitrification in a shallow fresh water estuary, Biogeosciences, 9, 2697–2710, doi:10.5194/bg-9-2697-2012, 2012.

Girard, J.: Principles of environmental chemistry, Jones and Bartlett Publishers, Sudbury, Massachusetts, 2004.

Golubev, S. V., Pokrovsky, O. S., and Savenko, V. S.: Unseeded precipitation of calcium and magnesium phosphates from modified seawater solutions, J. Cryst. Growth, 205, 354–360, 1999.

Gotoh, S. and Patrick, W. H.: Transformation of iron in a waterlogged soil as influenced by redox potential and pH, Soil Sci. Soc. Am. J., 38, 66–71, 1974.

Grossl, P. R. and Inskeep, W. P.: Kinetics of octacalcium phosphate crystal growth in the presence of organic acids, Geochim. Cosmochim. Ac., 56, 1955–1961, 1992.

Gulbrandsen, R. A., Roberson, C. E., and Neil, S. T.: Time and the crystallization of apatite in seawater, Geochim. Cosmochim. Ac., 48, 213–218, 1984.

Gunnars, A., Blomqvist, S., and Martinsson, C.: Inorganic formation of apatite in brackish seawater from the Baltic Sea: an experimental approach, Mar. Chem., 91, 15–26, 2004.

Hartzell, J. L. and Jordan, T. E.: Shifts in the relative availability of phosphorus and nitrogen along salinity gradients, Biogeochemistry, 107, 489–500, 2012.

Haynes, R. J.: Effects of liming on phosphate availability in acid soils – A critical review, Plant Soil, 68, 289–308, 1982.

Heggie, D. T., Skyring, G. W., O'Brien, G. W., Reimers, C., Herczeg, A., Moriarty, D. J. W., Burnett, W. C., and Milnes, A. R.: Organic carbon cycling and modern phosphorite formation on the East Australian continental margin: An overview, in: Phosphorite Research and Development, edited by: Notholt, A. J. G. and Jarvis, I., Geological Society, London, Special Publ., 52, 87–117, 1990.

Herlihy, M. and McCarthy, J.: Association of soil-test phosphorus with phosphorus fractions and adsorption characteristics, Nutr. Cycl. Agroecosys., 75, 79–90, 2006.

Hill, R. and Borman, D.: Estimating pastoral land use change for the Waikato region, in: Adding to the knowledge base for the nutrient manager, edited by: Currie, L. D. and Christensen, C. L., Occasional Report No. 24, Fertilizer and Lime Research Centre, Massey University, Palmerston North, New Zealand, p. 25, 2011.

Hinsinger, P.: Bioavailability of soil inorganic P in the rhizosphere as affected by root-induced chemical changes: A review, Plant Soil, 237, 173–195, 2001.

Hinsinger, P., Bengough, A. G., Vetterlein, D., and Young, I. M.: Rhizosphere: biophysics, biogeochemistry and ecological relevance, Plant Soil, 321, 117–152, 2009.

Howarth, R. W., Jensen, H. S., Marino, R., and Postma, H.: Transport to and processing of P in near-shore and oceanic waters, in: Phosphorus in the Global Environment – Transfers, Cycles and Management (Scope 54), edited by: Tiessen, H., Wiley, New York, NY, 323–345, 1995.

Huang, X. and Morris, J. T.: Distribution of phosphatase activity in marsh sediments along an estuarine salinity gradient, Mar. Ecol. Prog. Ser., 292, 75–83, 2005.

Husson, O.: Redox potential (Eh) and pH as drivers of soil/plant/microorganism systems: a transdisciplinary overview pointing to integrative opportunities for agronomy, Plant Soil, 362, 389–417, 2013.

Jahnke, R. A., Emerson, S. R., Roe, K. K., and Burnett, W. C.: The present day formation of apatite in Mexican continental margin sediments, Geochim. Cosmochim. Ac., 47, 259–266, 1983.

Jordan, T. E., Cornwell, J. C., Boynton, W. R., and Anderson, J. T.: Changes in phosphorus biogeochemistry along an estuarine salinity gradient: The iron conveyer belt, Limnol. Oceanogr., 53, 172–184, 2008.

Kizewski, F., Liu, Y.-T., Morris, A., and Hesterberg, D.: Spectroscopic approaches for phosphorus speciation in soils and other environmental systems, J. Environ. Qual., 40, 751–766, 2011.

Krajewski, K. P., van Cappellen, P., Trichet, J., Kuhn, O., Lucas, J., Martinalgarra, A., Prevot, L., Tewari, V. C., Gaspar, L., Knight, R. I., and Lamboy, M.: Biological processes and apatite formation in sedimentary environments, Eclog. Geol. Helvet., 87, 701–745, 1994.

Kurmies, B.: Zur Fraktionierung der Bodenphosphate, Die Phosphorsäure, 29, 118–149, 1972.

Lerman, A. and Wu, L.: Kinetics of global geochemical cycles, in: Kinetics of water-rock interactions, edited by: Brantley, S. L., Kubiki, J. D., and White, A. F., Springer, New York, 655–736, 2008.

Lindsay, W. L., Vlek, P. L. G., and Chien, S. H.: Phosphate minerals, in: Minerals in soil environments, edited by: Dixon, J. B. and Weed, S. B., Soil Sci. Soc. Am., Madison, 1089–1130, 1989.

Lovelock, C. E., Sorrell, B. K., Hancock, N., Hua, Q., and Swales, A.: Mangrove forest and soil development on a rapidly accreting shore in New Zealand, Ecosystems, 13, 437–451, 2010.

Lower, S. K.: Carbonate equilibria in natural waters – A Chem1 Reference Text, Simon Fraser University, June 1, 1999.

Lyons, G., Benitez-Nelson, C. R., and Thunell, R. C.: Phosphorus composition of sinking particles in the Guaymas Basin, Gulf of California, Limnol. Oceanogr., 56, 1093–1105, 2011.

Maher, W. A. and DeVries, M.: The release of phosphorus from oxygenated estuarine sediments, Chem. Geol., 112, 91–104, 1994.

Martens, C. S. and Harriss, R. C.: Inhibition of apatite precipitation in the marine environment by magnesium ions, Geochim. Cosmochim. Ac., 84, 621–625, 1970.

Morgan, M. F.: Chemical soil diagnosis by the universal soil testing system, Conn. Agric. Exp. Stn. Bull., 450, 579–628, 1941.

Morse, J. W. and Casey, W. H.: Ostwald processes and mineral paragenesis in sediments, Am. J. Sci., 288, 537–560, 1988.

Mortimer, C. H.: Chemical exchanges between sediments and water in Great Lakes – Speculations on probable regulatory mechanisms, Limnol. Oceanogr., 16, 387–404, 1971.

Murphy, J. and Riley, J. P.: A modified single solution method for the determination of phosphate in natural waters, Anal. Chim. Ac., 27, 31–36, 1962.

Murrmann, R. P. and Peech, M.: Effect of pH on labile and soluble phosphate in soils, Soil. Sci. Soc. Am. Proc., 33, 205–210, 1969.

Naidu, R., Syersy, J. K., Tillman, R. W., and Kirkman, J. H.: Effect of liming on phosphate sorption by acid soils, Eur. J. Soil Sci., 41, 165–175, 1990.

Nancollas, G. H., LoRe, M., Perez, L., Richardson, C., and Zawacki, S. J.: Mineral phases of calcium phosphate, Anat. Rec., 224, 234–241, 1989.

Noll, M. R., Szatkowski, A. E., and Magee, E. A.: Phosphorus fractionation in soil and sediments along a continuum from agricultural fields to nearshore lake sediments: Potential ecological impacts, J. Great Lakes Res., 35, 56–63, 2009.

Oh, Y.-M., Hesterberg, D. L., and Nelson, P. V.: Comparison of phosphate adsorption on clay minerals for soilless root media, Commun. Soil Sci. Plan., 30, 747–756, 1999.

Ostrofsky, M. L.: Determination of total phosphorus in lake sediments, Hydrobiologia, 696, 199–203, 2012.

Oxmann, J. F.: Technical Note: An X-ray absorption method for the identification of calcium phosphate species using peak-height ratios, Biogeosciences, 11, 2169–2183, 2014, http://www.biogeosciences.net/11/2169/2014/.

Oxmann, J. F. and Schwendenmann, L.: Quantification of octacalcium phosphate, authigenic apatite and detrital apatite in coastal sediments using differential dissolution and standard addition, Ocean Sci., 10, 571–585, 2014, http://www.ocean-sci.net/10/571/2014/.

Oxmann, J. F., Pham, Q. H., and Lara, R. J.: Quantification of individual phosphorus species in sediment: A sequential conversion and extraction method, Eur. J. Soil Sci., 59, 1177–1190, 2008.

Oxmann, J. F., Pham, Q. H., Schwendenmann, L., Stellman, J. M., and Lara, R. J.: Mangrove reforestation in Vietnam: The effect of sediment physicochemical properties on nutrient cycling, Plant Soil, 326, 225–241, 2010.

Paludan, C. and Morris, J. T.: Distribution and speciation of phosphorus along a salinity gradient in intertidal marsh sediments, Biogeochemistry, 45, 197–221, 1999.

Parfitt, R. L., Baisden, W. T., and Elliott, A. H.: Phosphorus inputs and outputs for New Zealand in 2001 at national and regional scales, J. R. Soc. N. Z., 38, 37–50, 2008.

Raimonet, M., Andrieux-Loyer, F., Ragueneau, O., Michaud, E., Kerouel, R., Philippon, X., Nonent, M., and Mémery, L.: Strong gradient of benthic biogeochemical processes along a macrotidal temperate estuary: Focus on P and Si cycles, Biogeochemistry, 115, 399–417, 2013.

Reddy, K. R. and DeLaune, R. D.: Biogeochemistry of wetlands: Science and applications, CRC Press, Boca Raton, 2008.

Reddy, K. R. and Sacco, P. D.: Decomposition of water hyacinth in agricultural drainage water, J. Environ. Qual., 10, 228–234, 1981.

Rengel, R.: Calcium, in: Encyclopedia of soil science, second edition, edited by: Lal, R., CRC Press, Boca Raton, 198–201, 2005.

Richardson, A. E., Barea, J., McNeill, A. M., and Prigent-Combaret, C.: Acquisition of phosphorus and nitrogen in the rhizosphere and plant growth promotion by microorganisms, Plant Soil, 321, 305–339, 2009.

Robertson, B. and Stevens, L.: Porirua Harbour: Intertidal fine scale monitoring 2008/09, Report prepared for Greater Wellington Regional Council, Wellington, New Zealand, 22 pp., 2009.

Ruttenberg, K. C.: Diagenesis and burial of phosphorus in marine sediments: Implications for the marine phosphorus budget, Ph.D. thesis, Yale Univ., New Haven, Connecticut, 1990.

Ruttenberg, K. C.: Development of a sequential extraction method for different forms of phosphorus in marine sediments, Limnol. Oceanogr., 37, 1460–1482, 1992.

Ruttenberg, K. C.: The global phosphorus cycle, in: Treatise on geochemistry, Vol. 8, edited by: Turekian, K. K. and Holland, D. J., Elsevier B. V., Amsterdam, 585–643, 2003.

Ruttenberg, K. C. and Berner, R. A.: Authigenic apatite formation and burial in sediments from non-upwelling continental margin environments, Geochim. Cosmochim. Ac., 57, 991–1007, 1993.

Schenau, S. J. and De Lange, G. J.: A novel chemical method to quantify fish debris in marine sediments, Limnol. Oceanogr., 45, 963–971, 2000.

Schenau, S. J., Slomp, C. P., and De Lange, G. J.: Phosphogenesis and active phosphorite formation in sediments from the Arabian Sea oxygen minimum zone, Mar. Geol., 169, 1–20, 2000.

Seitzinger, S. P.: The effect of pH on the release of phosphorus from Potomac estuary sediments: Implications for blue-green algal blooms, Estuar. Coast. Shelf. S., 33, 409–418, 1991.

Sharp, J. H., Culberson, C. H., and Church, T. M.: The chemistry of the Delaware estuary. General considerations, Limnol. Oceanogr., 27, 1015–1028, 1982.

Sheldon, R. P.: Ancient marine phosphorites, Annu. Rev. Earth Pl. Sc., 9, 251–284, 1981.

Slomp, C. P.: Phosphorus cycling in the estuarine and coastal zones: Sources, sinks, and Transformations, in: Treatise on estuarine and coastal science, Vol. 5, edited by: Wolanski, E. and McLusky, D. S., Academic Press, Waltham, 201–229, 2011.

Smyth, T. J. and Sanchez, P. A.: Effects of lime, silicate, and phosphorus applications to an Oxisol on phosphorus sorption and iron retention, Soil Sci. Soc. Am. J., 44, 500–505, 1980.

Sorensen, P. G. and Milne, J. R.: Porirua Harbour targeted intertidal sediment quality assessment, Report prepared for Greater
Wellington Regional Council, Wellington, New Zealand, 71 pp.,
2009.

Spiteri, C., Cappellen, P. V., and Regnier, P.: Surface complexa-
tion effects on phosphate adsorption to ferric iron oxyhydroxides
along pH and salinity gradients in estuaries and coastal aquifers,
Geochim. Cosmochim. Ac., 72, 3431–3445, 2008.

Sun, X. and Turchyn, A. V.: Significant contribution of authigenic
carbonate to marine carbon burial, Nature Geosci., 7, 201–204,
2014.

Sundareshwar, P. V. and Morris, J. T.: Phosphorus sorption characteristics of intertidal marsh sediments along an estuarine salinity
gradient, Limnol. Oceanogr., 44, 1693–1701, 1999.

Sutula, M., Bianchi, T. S., and McKee, B. A.: Effect of seasonal
sediment storage in the lower Mississippi River on the flux of
reactive particulate phosphorus to the Gulf of Mexico, Limnol.
Oceanogr., 49, 2223–2235, 2004.

Swales, A., Bentley, S. J., Lovelock, C., and Bell, R. G.: Sediment processes and mangrove-habitat expansion on a rapidly-
prograding muddy coast, New Zealand, Coastal Sediments '07,
New Orleans, Louisiana, May 2007, 1441–1454, 2007.

Van Beusekom, J. E. E. and De Jonge, V. N.: Transformation of
phosphorus in the Wadden Sea: Apatite formation, Deutsche Hy-
drographische Zeitschrift, 49, 297–305, 1997.

Van Cappellen, P. and Berner, R. A.: A mathematical model for the
early diagenesis of phosphorus and fluorine in marine sediments;
apatite precipitation, Am. J. Sci., 288, 289–333, 1988.

Van der Zee, C., Roevros, N., and Chou, L.: Phosphorus speciation,
transformation and retention in the Scheldt estuary (Belgium/The
Netherlands) from the freshwater tidal limits to the North Sea,
Mar. Chem., 106, 76–91, 2007.

Vant, B. and Smith, P.: Trends in river water quality in the Waikato
Region, 1987–2002, EW Technical Report 2004/02, Waikato Regional Council, Hamilton, New Zealand, 32 pp., 2004.

Yu, K., Böhme, F., Rinklebe, J., Neue, H-U, and Delaune, R. D.:
Major biogeochemical processes in soils – A microcosm incubation from reducing to oxidizing conditions, Soil Sci. Soc. Am. J.,
71, 1406–1417, 2007.

10

Secondary calcification and dissolution respond differently to future ocean conditions

N. J. Silbiger and M. J. Donahue

University of Hawaii, at Manoa, Hawaii Institute of Marine Biology, PO Box 1346, Kaneohe, Hawaii

Correspondence to: N. J. Silbiger (silbiger@hawaii.edu)

Abstract. Climate change threatens both the accretion and erosion processes that sustain coral reefs. Secondary calcification, bioerosion, and reef dissolution are integral to the structural complexity and long-term persistence of coral reefs, yet these processes have received less research attention than reef accretion by corals. In this study, we use climate scenarios from RCP 8.5 to examine the combined effects of rising ocean acidity and sea surface temperature (SST) on both secondary calcification and dissolution rates of a natural coral rubble community using a flow-through aquarium system. We found that secondary reef calcification and dissolution responded differently to the combined effect of pCO_2 and temperature. Calcification had a non-linear response to the combined effect of pCO_2 and temperature: the highest calcification rate occurred slightly above ambient conditions and the lowest calcification rate was in the highest temperature–pCO_2 condition. In contrast, dissolution increased linearly with temperature–pCO_2. The rubble community switched from net calcification to net dissolution at +271 μatm pCO_2 and 0.75 °C above ambient conditions, suggesting that rubble reefs may shift from net calcification to net dissolution before the end of the century. Our results indicate that (i) dissolution may be more sensitive to climate change than calcification and (ii) that calcification and dissolution have different functional responses to climate stressors; this highlights the need to study the effects of climate stressors on both calcification and dissolution to predict future changes in coral reefs.

1 Introduction

In 2013, atmospheric carbon dioxide ($CO_{2(atm)}$) reached an unprecedented milestone of 400 ppm (Tans and Keeling, 2013), and this rising $CO_{2(atm)}$ is increasing sea surface temperature (SST) and ocean acidity (Caldeira and Wickett, 2003; Cubasch et al., 2013; Feely et al., 2004). Global SST has increased by 0.78 °C since pre-industrial times (Cubasch et al., 2013), and it is predicted to increase by another 0.8–5.7 °C by the end of this century (Meinshausen et al., 2011; Van Vuuren et al., 2008; Rogelj et al., 2012). The Hawaii Ocean Time-series detected a 0.075 decrease in mean annual pH at station ALOHA over the past 20 years (Doney et al., 2009) and there have been similar trends at stations around the world, including the Bermuda Atlantic Time-series and the European Station for Time-series Observations in the ocean (Bindoff et al., 2007). pH is expected to drop by an additional 0.14–0.35 pH units by the end of the twenty-first century (Bopp et al., 2013). All marine ecosystems are at risk from rising SST and decreasing pH (Doney et al., 2009; Hoegh-Guldberg et al., 2007; Hoegh-Guldberg and Bruno, 2010), but coral reefs are particularly vulnerable to these stressors (reviewed in Hoegh-Guldberg et al., 2007).

Corals create the structurally complex calcium carbonate ($CaCO_3$) foundation of coral reef ecosystems. This structural complexity is at risk from climate-driven shifts from high-complexity, branched coral species to mounding and encrusting growth forms (Fabricius et al., 2011) and from increases in the natural processes of reef destruction, including bioerosion and dissolution (Wisshak et al., 2012, 2013; Tribollet et al., 2006). While substantial research attention has focused on the response of reef-building corals to climate change (reviewed in Hoegh-Guldberg et al., 2007; Fabricius, 2005; Pandolfi et al., 2011), secondary calcification (calcification by non-coral invertebrates and calcareous algae), bioerosion, and reef dissolution that are integral to maintaining the structural complexity and net growth of coral reefs has received less attention (Andersson and Gledhill, 2013; Andersson et al., 2011; Andersson and Mackenzie, 2012). Bioerosion and dissolution break down the reef framework, while secondary calcification helps maintain reef stability by cementing the reef together (Adey, 1998; Camoin and Montaggioni, 1994; Littler, 1973) and producing chemical cues that induce settlement of many invertebrate larvae including several species of corals (Harrington et al., 2004; Price, 2010). Coral reefs will only persist if constructive reef processes (growth by corals and secondary calcifiers) exceed destructive reef processes (bioerosion and dissolution). In this study, we examine the combined effects of rising ocean acidity and SST on both calcification and dissolution rates of a natural community of secondary calcifiers and bioeroders.

Recent laboratory experiments have focused on the response of individual taxa of bioeroders or secondary calcifiers to climate stressors. For example, studies have specifically addressed the effects of rising ocean acidity and/or temperature on bioerosion by a clionid sponge (Wisshak et al., 2012, 2013; Fang et al., 2013) and a community of photosynthesizing microborers (Tribollet et al., 2009; Reyes-Nivia et al., 2013). These studies found that bioerosion increased under future climate change scenarios. Several studies have focused on tropical calcifying algae and have found decreased calcification (Semesi et al., 2009; Johnson et al., 2014; Comeau et al., 2013; Jokiel et al., 2008; Kleypas and Langdon, 2006) and increased dissolution (Diaz-Pulido et al., 2012) with increasing ocean acidity and/or SST. However, the bioeroding community is extremely diverse, and can interact with the surrounding community of secondary calcifiers: for example, crustose coralline algae (CCA) can inhibit internal bioerosion (White, 1980; Tribollet and Payri, 2001). To understand the combined response of bioeroders and secondary calcifiers, we take a community perspective and examine the synergistic effects of rising SST and ocean acidity on a natural community of secondary calcifiers and bioeroders. Using the total alkalinity anomaly technique, we test for net changes in calcification during the day and dissolution (most of which is caused by bioeroders; Andersson and Gledhill, 2013) at night. Our climate change treatments are modeled after the Representative Concentration Pathway (RCP) 8.5 climate scenario (Van Vuuren et al., 2011; Meinshausen et al., 2011), one of the high-emission scenarios used in the most recent Intergovernmental Panel on Climate Change (IPCC) report (Cubasch et al., 2013). The RCP 8.5 scenario predicts an increase in temperature of 3.8–5.7 °C (Rogelj et al., 2012) and an increase in atmospheric CO_2 of 557 ppm by the year 2100 (Meinshausen et al., 2011). We use the RCP 8.5 scenario because the current CO_2 concentrations are tracking just above what this scenario predicts (Sanford et al., 2014). While prior studies have focused on the contributions of individual community members to increased temper-

ature and CO_2, here, we examine the community response to the RCP 8.5 climate scenario and measure calcification, dissolution, and net community production rates.

2 Materials and methods

2.1 Collection site

All collections were made on the windward side of Moku o Loʻe (Coconut Island) in Kaneohe Bay, Hawaii adjacent to the Hawaii Institute of Marine Biology. This fringing reef is dominated by *Porites compressa* and *Montipora capitata*, with occasional colonies of *Pocillopora damicornis*, *Fungia scutaria*, and *Porites lobata*. Kaneohe Bay is a protected, semi-enclosed embayment; the residence time can be more than 1 month long in the protected southern portion of the bay (Lowe et al., 2009a, 2009b) that is coupled with a high daily variance in pH (Guadayol et al., 2014). The wave action is minimal (Smith et al., 1981; Lowe et al., 2009a; Lowe et al., 2009b), and currents are relatively slow (5 cm s^{-1} maximum) and wind driven (Lowe et al., 2009a, 2009b).

2.2 Sample collection

We collected pieces of dead *Porites compressa* coral skeleton (hereafter, referred to as rubble) as representative communities of bioeroders and secondary calcifiers. Rubble was collected with a hammer and chisel from a shallow reef flat (~ 1 m depth) in November 2012. Only pieces of rubble without any live coral were collected. The rubble community in Kaneohe Bay is comprised of secondary calcifiers, including CCA from the genera *Hydrolithon*, *Sporolithon*, and *Peyssonnelia*, and non-coral calcifying invertebrates (e.g., boring bivalves (*Lithophaga fasciola* and *Barbatia divaricate*), oysters (*Crassostrea gigas*), and small crustaceans), filamentous and turf algae, and internal bioeroders, including boring bivalves (*L. fasciola* and *B. divaricate*), sipunculids (*Aspidosiphon elegans*, *Lithacrosiphon cristatus*, *Phascolosoma perlucens*, and *Phascolosoma stephensoni*), phoronids (*Phoronis ovalis*), sponges (*Cliona* spp.) and a diverse assemblage of polychaetes (White, 1980). All rubble pieces were combined after collection and maintained in a 100 L flow-through tank with ambient seawater from Kaneohe Bay until random assignment to treatments.

2.3 Experimental design

The Hawaii Institute of Marine Biology (HIMB) hosts a mesocosm facility with flow-through seawater from Kaneohe Bay and controls for light, temperature, pCO$_2$, and flow rate. The facility is comprised of 24 experimental aquaria split between four racks; each rack has a 150 L header tank that feeds six experimental aquaria, each 50 L in volume (Fig. 1).

Before adding rubble to the experimental aquaria, we collected day and night samples of pH, total alkalinity (TA),

Table 1. Means and standard errors of all measured parameters by rack. pCO_2, HCO_3^-, CO_3^{2-}, DIC, and Ω_{arag} were all calculated from the measured TA and pH samples using CO_2SYS. Each table entry is the mean of 12 water samples: one daytime sample and one nighttime sample for six aquaria within a rack. Data are all from the imposed treatment conditions with no rubble inside the aquaria.

Rack	Pre-industrial	Present day	2050 prediction	2100 prediction
Temp (°C)	23.8 ± 0.07	24.8 ± 0.08	26.2 ± 0.06	27.2 ± 0.08
Salinity	35.65 ± 0.01	35.71 ± 0.02	35.62 ± 0.02	35.71 ± 0.02
TA ($\mu mol\,kg^{-1}$)	2137 ± 1.7	2138 ± 2.3	2139 ± 2.0	2142 ± 1.9
pH_t	8.02 ± 0.02	7.87 ± 0.01	7.74 ± 0.02	7.67 ± 0.02
pCO_2 (μatm)	409 ± 20.0	614 ± 15.6	868 ± 33.0	1047 ± 38.7
HCO_3^- ($\mu mol\,kg^{-1}$)	1692 ± 16.9	1815 ± 7.3	1894 ± 7.8	1939 ± 6.6
CO_3^{2-} ($\mu mol\,kg^{-1}$)	194.20 ± 6.7	147.08 ± 2.8	113.98 ± 3.8	99.24 ± 3.3
DIC ($\mu mol\,kg^{-1}$)	1898 ± 10.9	1980 ± 5.1	2032 ± 5.0	2067 ± 4.5
Ω_{arag}	3.06 ± 0.1	2.32 ± 0.04	1.80 ± 0.06	1.57 ± 0.05
NO_2^- ($\mu mol\,L^{-1}$)	0.082 ± 0.0028	0.078 ± 0.0045	0.074 ± 0.0047	0.070 ± 0.0051
PO_4^{3-} ($\mu mol\,L^{-1}$)	0.017 ± 0.014	0.0097 ± 0.0081	0.033 ± 0.016	0.018 ± 0.0061
$Si(OH)_4$ ($\mu mol\,L^{-1}$)	3.60 ± 0.58	3.64 ± 0.61	3.88 ± 0.49	3.78 ± 0.52
NH_4^+ ($\mu mol\,L^{-1}$)	0.45 ± 0.30	0.19 ± 0.067	0.23 ± 0.15	0.34 ± 0.14
NO_3^- ($\mu mol\,L^{-1}$)	2.13 ± 0.20	2.25 ± 0.21	2.55 ± 0.10	2.48 ± 0.11

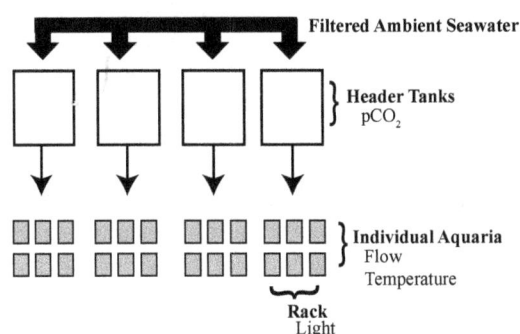

Figure 1. A schematic of the mesocosm system at the Hawaii Institute of Marine Biology. Ambient seawater is pumped into the system from a nearby fringing reef in Kaneohe Bay. The seawater is filtered with a sand trap filter, passed through a water chiller and then fed into one of four header tanks. pCO_2 is manipulated in each header tank by bubbling a mixture of CO_2-free air and pure CO_2 to the desired concentration. The water from one header tank flows into six aquaria (a rack). Light is controlled by a rack with metal-halide lights. There are two metal-halide lights per rack, with each light oscillating over a set of three aquaria. Flow and temperature are controlled in each individual aquarium with flow valves and aquarium heaters and coolers, respectively.

by differences in rubble communities. In the second "treatment experiment", we manipulated pCO_2 and temperature to simulate four climate scenarios (pre-industrial, present day, 2050, and 2100) and tested the response of calcification, dissolution, and net community production. Each experiment used the TA anomaly method (Smith and Key, 1975; Andersson et al., 2009). This method calculates net calcification from changes in TA, and calculates net community production from changes in total dissolved inorganic carbon (DIC) adjusted for changes in carbon due to calcificaiton. Because estimates of calcification are based on changes in TA, this method does not account for mechanical erosion (e.g., small chips of $CaCO_3$ produced by sponge erosion). However, given the short duration of the experiment and the types of bioeroders present, we expect that chemical dissolution captured a significant proportion of the erosion in the system.

Approximately 1.2 L of rubble (3–4 pieces of weight 499 ± 148 g and skeletal density $1.53 \pm 0.1\,g\,cm^{-3}$ (mean $\pm$SD, $n = 85$)) were placed in each of the 24 experimental aquaria and acclimated to tank conditions in ambient seawater for 3 days. On the fourth day, we performed the control experiment, calculating daytime calcification and nighttime dissolution for rubble in ambient seawater conditions using the TA anomaly technique. The next day we manipulated seawater pCO_2 and temperature to replicate four climate scenarios for the treatment experiment: pre-industrial ($-1 \pm 0.057\,°C$ and $-205 \pm 11.9\,\mu atm$), present day (natural Kaneohe Bay seawater $24.8 \pm 0.09\,°C$, $614 \pm 15.6\,\mu atm$), 2050 ($+1.4 \pm 0.09\,°C$ and $+255 \pm 31\,\mu atm$), and 2100 ($+2.4 \pm 0.08$ and $+433 \pm 40\,\mu atm$). Note that all changes in temperature and pCO_2 were made relative to present-day Kaneohe Bay seawater conditions: pCO_2 in Kaneohe Bay

temperature, and salinity from all aquaria to demonstrate the consistency of water conditions across aquaria without any rubble present (Table 1). The long-term temporal stability of the mesocosm system is reported in Putnam (2012). We then conducted "control" and "treatment" experiments to determine how RCP 8.5 predictions affect daytime calcification and nighttime dissolution rates in a natural rubble community. The first "control experiment" characterized baseline calcification and dissolution in each aquarium caused is consistently high relative to the open ocean and can range from 196 to 976 μatm in southern Kaneohe Bay, depending on conditions (Drupp et al., 2013). The yearly average $p\text{CO}_2$ at our collection site ranged from 565 to 675 μatm (Silbiger et al., 2014). After an acclimation time of 7 days, we sampled the treatment experiment, calculating daytime calcification and nighttime dissolution over a 24 h period.

During both experiments, TA, pH, salinity, temperature, and dissolved inorganic nutrient (DIN) samples were collected every 12 h over a 24 h period: just before lights-on in the morning (time 1) and just before lights-off at night (time 2) to capture light conditions, and then again before lights-on the next morning (time 3) to capture dark conditions. Flow into each aquarium was monitored and adjusted every 3 hours to ensure a consistent flow rate over the 24 h experiment. We calculated net ecosystem calcification and net community production using a simple box model (Andersson et al., 2009) and normalized all our calculations to the surface area of the rubble in each tank. The surface area of the rubble was calculated using the wax dipping technique (Stimson and Kinzie III, 1991) at the end of the experiment.

2.4 Mesocosm setup

The mesocosm facility (Fig. 1) is supplied with ambient seawater from Kaneohe Bay, which is filtered through a sand filter, passed through a water chiller (Aqualogic Multi Temp MT-1 model no. 2TTB3024A1000AA), and then fed into one of the four header tanks. $p\text{CO}_2$ was manipulated using a CO_2 gas blending system (see Fangue et al., 2010; Johnson and Carpenter, 2012). Each target $p\text{CO}_2$ concentration was created by mixing CO_2-free atmospheric air with pure CO_2 using mass flow controllers (C100L Sierra Instruments). Output $p\text{CO}_2$ was analyzed using a calibrated infrared CO_2 analyzer (A151, Qubit Systems). CO_2 mixtures were then bubbled into one of the four header tanks and water from each individual header tank fed into the six individual treatment aquaria (Fig. 1). The $p\text{CO}_2$ in each treatment aquarium was estimated with CO_2SYS (Van Heuven et al., 2011) using pH and TA as the parameters.

Temperature was manipulated in each treatment aquarium using dual-stage temperature controllers (Aqualogic TR115DN). The temperature was continuously monitored with temperature loggers (TidbiT v2 Water Temperature Data Logger, sampling every 20 min), and point measurements were taken during every sampling period with a handheld digital thermometer (Traceable Digital Thermometer, Thermo Fisher Scientific; precision = 0.001 °C). Light was controlled by positioning an oscillating pendant metal-halide light (250 W) over a set of three aquaria and was programmed to emit an equal amount of light to each tank ($\sim$ 500 μE of light). Lights were set to a 12 h : 12 h light to dark photoperiod and were monitored using a LI-COR spherical quantum PAR sensor. Flow rate was maintained at 115 ± 1 mL min^{-1}, resulting in a residence time of 7.3 $\pm$ 0.07 h per tank. Each aquarium was equipped with a submersible powerhead pump (Sedra KSP-7000 powerhead) to ensure that the tank was well mixed.

2.5 Seawater chemistry

All sample collection and storage vials were cleaned in a 10 % HCl bath for 24 h and rinsed three times with MilliQ water before use and rinsed three times with sample water during sample collection and processing.

2.5.1 Total alkalinity

Duplicate TA samples were collected in 300 mL borosilicate sample containers with glass stoppers. Each sample was preserved with 100 μL of 50 % saturated HgCl_2 and analyzed within 3 days using open cell potentiometric titrations on a Mettler T50 autotitrator (Dickson et al., 2007). A certified reference material (CRM – Reference Material for Oceanic CO_2 Measurements, A. Dickson, Scripps Institution of Oceanography) was run at the beginning of each sample set. The accuracy of the titrator never deviated more than $\pm$0.8 % from the standard, and TA measurements were corrected for these deviations. The precision was 3.55 μEq (measured as the standard deviation of the duplicate water samples). During the 24 h control experiment, the average changes in TA were 37 μEq over the day and 20 μEq over the night (day and night TA changes were of a larger magnitude in the treatment experiments): these are measurable changes given the precision and accuracy of the TA measurements.

2.5.2 pH$_t$ (total scale)

Duplicate pH$_t$ samples were collected in 20 mL borosilicate glass vials, brought to a constant temperature of 25 °C in a water bath, and immediately analyzed using an m-cresol dye addition spectrophotometric technique (Dickson et al., 2007). The accuracy of the pH was tested against a Tris buffer of known pH$_t$ from the Dickson Lab at the Scripps Institution of Oceanography (Dickson et al., 2007). Our accuracy was better than $\pm$0.04 %, and the precision was 0.004 pH units (measured as the standard deviation of the duplicate water samples). In situ pH and the remaining carbonate parameters were calculated using CO_2SYS (Van Heuven et al., 2011) with the following measured parameters: pH$_t$, TA, temperature, and salinity. The K1K2 apparent equilibrium constants were from Mehrbach (1973) and refit by Dickson and Millero (1987), and HSO_4^- dissociation constants were taken from Uppström (1974) and Dickson (1990).

2.5.3 Salinity

Duplicate salinity samples were analyzed on a Portasal 8410 portable salinometer calibrated with an OSIL IAPSO standard (accuracy = $\pm$0.003, precision = $\pm$0.0003).

2.5.4 Nutrients

Nutrient samples were collected with 60 mL plastic syringes and immediately filtered through combusted 25 mm glass fiber filters (GF/F 0.7 μm) and transferred into 50 mL plastic centrifuge tubes. Nutrient samples were frozen and later analyzed for $Si(OH)_4$, NO_3^-, NO_2^-, NH_4^+, and PO_4^{3-} on a Seal Analytical AA3 HR Nutrient Analyzer at the UH SOEST Lab for Analytical Chemistry.

2.6 Measuring net ecosystem calcification

We assumed that the mesocosms were well-mixed systems; thus, we calculated net ecosystem calcification and net community photosynthesis following the simple box model presented in Andersson et al. (2009). TA was normalized to a constant salinity (35) to account for changes due to evaporation and then corrected for dissolved inorganic nitrogen and phosphate to account for their small contributions to the acid–base system (Wolf-Gladrow et al., 2007). Net ecosystem calcification, or G, was calculated using the following equation:

$$G = \left[F_{\text{TAin}} - F_{\text{TAout}} - \frac{dTA}{dt} \right] / 2, \qquad (1)$$

where F_{TAin} is the rate of TA flowing into an aquarium (the average TA in the header tank times the inflow rate), F_{TAout} is the rate of TA flowing out of an aquarium (the average TA in the aquarium times the outflow rate), and, $\frac{dTA}{dt}$ is the change in TA in an aquarium during the measurement period (change in TA normalized to the volume of water and the surface area of the rubble); specific calculations are given in the supplemental material. The equation is divided by 2 because 1 mole of $CaCO_3$ is precipitated or dissolved for every 2 moles of TA removed or added to the water column. Here, G represents the sum of all the calcification processes minus the sum of all the dissolution processes in mmol $CaCO_3$ m^{-2} h^{-1}; thus, all positive numbers are net calcification, and all negative numbers are negative net calcification (i.e., net dissolution). Net daytime calcification (G_{day}) is calculated from the first 12 h sampling period in the light, net nighttime dissolution (G_{night}) is calculated from the second 12 h sampling period in the dark, and total net calcification (G_{net}) is calculated from the full 24 h cycle ($G_{\text{day}} + G_{\text{night}}$). G_{day}, G_{night}, and G_{net} are converted from hourly to daily rates and presented as mmol $CaCO_3$ m^{-2} d^{-1}.

2.7 Measuring net community production and respiration

Net community production (NCP) was calculated by measuring changes in DIC (Gattuso et al., 1999). DIC was normalized to a constant salinity (35) to account for any evaporation over the 24 h period. We used a simple box model to calculate NCP:

$$\text{NCP} = \left[F_{\text{DICin}} - F_{\text{DICout}} - \frac{dDIC}{dt} \right] - G. \qquad (2)$$

F_{DICin}, F_{DICout}, and $\frac{dDIC}{dt}$ are the rates of DIC flowing into the aquaria, flowing out of the aquaria, and the change in DIC in the aquaria per unit time in mmol C m^{-2} h^{-1}, respectively. To measure NCP, we subtract G to remove any change in carbon due to inorganic processes. NCP represents the sum of all the photosynthetic processes minus the sum of all the respiration processes; thus, all positive numbers are net photosynthesis and all negative numbers are negative net photosynthesis (i.e., net respiration). Net daytime NCP (NCP$_{\text{day}}$) is calculated from the first 12 h sampling period in the light, net nighttime NCP (NCP$_{\text{night}}$) is calculated from the second 12 h sampling period in the dark, and total NCP (NCP$_{\text{net}}$) is calculated from the full 24 h cycle (NCP$_{\text{day}}$ + NCP$_{\text{night}}$). All rates are presented as mmol C m^{-2} d^{-1}.

2.8 Statistical analysis

Each aquarium contained a slightly different rubble community because of the randomization of rubble pieces to each treatment. To ensure there were no systematic differences in rubble communities between racks (rack effects) before the experimental treatments were applied, we tested for differences in calcification and NCP between racks in the control experiment using an ANOVA (Fig. S2 in the Supplement).

In the treatment experiment, we first tested for feedbacks in carbonate chemistry due to the presence of rubble: using a paired t test, we compared the day–night difference in measured pCO_2 in each aquarium with rubble, $(pCO_{2,\,\text{day}} - pCO_{2,\,\text{night}})_{\text{rubble}}$, and without rubble, $(pCO_{2,\,\text{day}} - pCO_{2,\,\text{night}})_{\text{no rubble}}$.

Although we imposed four discrete temperature–pCO_2 scenario treatments on each tank (Table 1), random variation between treatments and the feedback between the rubble communities and the water chemistry resulted in near-continuous variation in temperature–pCO_2 treatments across aquaria (Figs. 2 and S1 in the Supplement). To capture this continuous variation in temperature–pCO_2 in the analysis, we used the measured temperature–pCO_2 seawater condition as a continuous independent variable in a regression rather than the four categorical treatment conditions in an ANOVA (an analysis of G and NCP using the ANOVA approach is included in Figs. S3, S4 and Tables S1, S2 in the Supplement). The regression approach allowed us to capture the quantitative relationships better between net calcification (G) or NCP and the temperature–pCO_2 treatment. We created a single, continuous variable, standardized climate change (SCC), from a linear combination of temperature and pCO_2 values in each aquarium. A simple linear combination was used because pCO_2 increased linearly with temperature (Fig. 2), as imposed by our treatments. We first calculated the relationship between ΔTemp (Eq. 3) and ΔpCO_2 (Eq. 4)

Table 2. Regression results for the treatment experiments: G_{day}, G_{night}, and G_{net} versus standardized climate change (Fig. 3a, c, e) and NCP_{day}, NCP_{night}, and NCP_{net} versus standardized climate change (Fig. 3b, d, f). Bold values indicate a statistically significant p value at $\alpha < 0.05$.

	SS	df	F	p	R^2
G_{day}					
Standardized climate change	3.79	1	1.45	0.06	
(Standardized climate change)^2	23.63	1	9.04	**0.007**	
Error	54.89	21			0.33
G_{night}					
Standardized climate change	67.80	1	39.14	**<0.0001**	
Error	38.11	22			0.64
G_{net}					
Standardized climate change	88.01	1	19.49	**<0.001**	
Error	99.35	22			0.47
NCP_{day}					
Standardized climate change	5687.2	1	57.36	**<0.0001**	
Error	2181.4	22			0.72
NCP_{night}					
Standardized climate change	3816.1	1	52.06	**<0.0001**	
Error	1612.6	22			0.70
NCP_{net}					
Standardized climate change	17925	1	121.47	**<0.0001**	
Error	3246.4	22			0.85

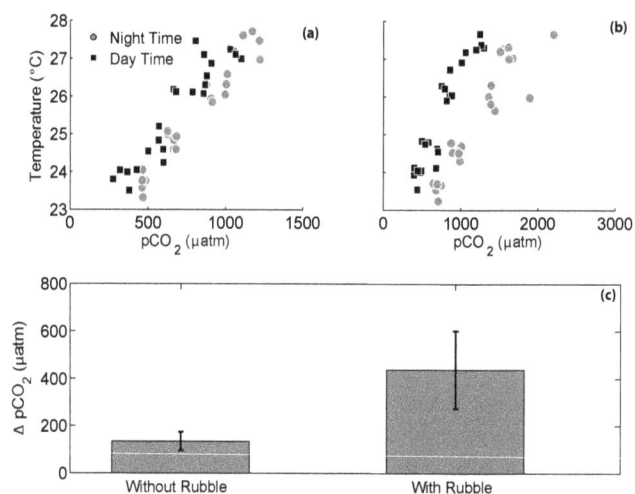

Figure 2. pCO_2 and temperature in each aquarium **(a)** without any rubble present and **(b)** with rubble present. Daily variability in pCO_2 was higher when rubble was present due to feedbacks from the rubble community (note the different x axis scales in panels **(a)** and **(b)**). Panel **(c)** shows the mean difference between day and night pCO_2 with and without rubble present, with observations paired by aquarium (error bars are standard error) ($t_{23} = -7.23$, $p < 0.0001$).

using linear regression. The coefficients from this regression (slope: $\alpha = 0.0031$; y intercept: $\beta = -0.078$) were used to combine pCO_2 and temperature onto the same scale, as a measure of standardized climate change (Eq. 5):

$$\Delta Temp_i = Temp_{trt,i} - Temp_{cont,i}, \qquad (3)$$

$$\Delta pCO_{2i} = pCO_{2trt,i} - pCO_{2cont,i}, \qquad (4)$$

$$SCC_i = \Delta Temp_i + \alpha \cdot \Delta pCO_{2i} + \beta. \qquad (5)$$

This synthetic temperature–pCO_2 axis, SCC, is centered on the ambient (control) conditions such that a value of 0 corresponds to present-day Kaneohe Bay conditions, a negative value corresponds to water that is colder and less acidic (pre-industrial) and a positive value corresponds to water that is warmer and more acidic (future conditions) compared to background seawater. (The independent relationships between G and NCP with $\Delta Temp$ and ΔpCO_2 are shown in Figs. S5 and S6 in the Supplement and are similar to the relationship with SCC.)

With SCC as a continuous, independent variable, we used a regression to test for linear and nonlinear relationships between day, night, and net calcification (G_{day}, G_{night}, and G_{net}) and NCP (NCP_{day}, NCP_{night}, and NCP_{net}) versus SCC. For a simple test of nonlinearity in the response of calcification to SCC, we included a quadratic term (SCC^2) in the model. For G_{day}, we used weighted regression (weight function: $w_i = 1/(1+|r_i|)$), where $w_i =$ weight and $r_i =$ residual, Fair, 1974) to account for heteroscedasticity. All other data met assumptions for a linear regression. Lastly, we used a linear regression to test the relationship between G and NCP.

3 Results

3.1 Control experiment

For rubble in ambient seawater conditions, the average G_{day}, G_{night}, and G_{net} in the control experiment were 3.4 ± 0.16 mmol m^{-2} d^{-1}, -2.4 ± 0.15 mmol m^{-2} d^{-1}, and 0.96 ± 0.20 mmol m^{-2} d^{-1}, respectively. There was no significant difference in G_{day} ($F_{3,23} = 0.68$, $p = 0.58$), G_{night} ($F_{3,23} = 1.52$, $p = 0.24$), or G_{net} ($F_{3,23} = 1.38$, $p = 0.28$) between racks in the control experiment (Fig. S2). NCP rates also did not show any rack effects. Average NCP rates were 23.2 ± 1.4 mmol m^{-2} d^{-1} ($F_{3,23} = 0.07$, $p = 0.94$) during the day, -20.7 ± 1.9 mmol m^{-2} d^{-1} ($F_{3,23} = 1.95$, $p = 0.15$) during the night, and 2.5 ± 2.1 mmol m^{-2} d^{-1} ($F_{3,23} = 1.5$, $p = 0.25$) over the entire 24 h period.

3.2 Treatment experiment

The rubble communities significantly altered the seawater chemistry, with higher pCO_2 than the applied pCO_2 manipulation, particularly at night (Fig. S1). The mean difference between day and night pCO_2 for all treatments was $134.4 \pm 39 \mu$atm without rubble and was $438.5 \pm 163.9 \mu$atm when rubble was present ($t_{23} = -7.23$, $p < 0.0001$; Fig. 2).

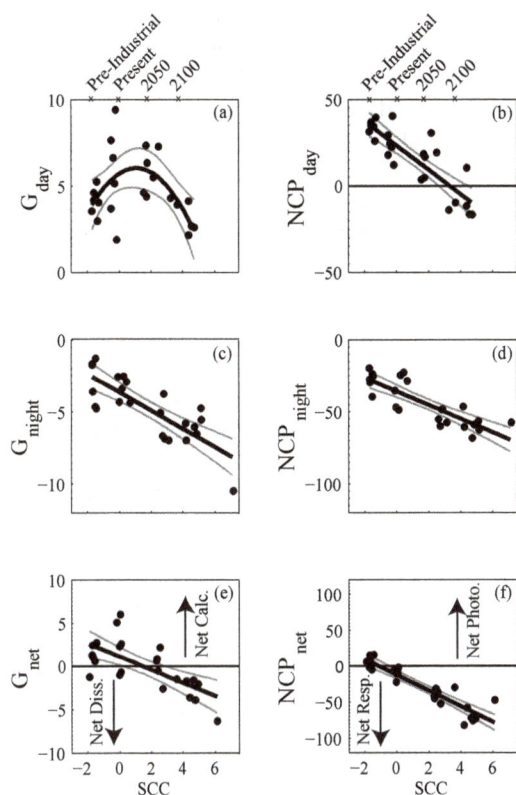

Figure 3. Net ecosystem calcification ((**a**) G_{day}, (**c**) G_{night}, (**e**) and G_{net}) and net community production ((**b**) NCP_{day}, (**d**) NCP_{night}, and (**f**) NCP_{net}) versus standardized climate change (SCC). Each point represents net ecosystem calcification (left panel) or net community production (right panel) calculated from an individual aquarium. Standardized climate change was centered around background seawater conditions such that a value of 0 indicated that there was no change in pCO_2 or temperature. Positive values indicate an elevated pCO_2 and temperature condition relative to background, and negative values represent lower pCO_2 and temperature conditions. G_{day} had a non-linear relationship with standardized climate change ($y = -0.27x^2 + 0.59x + 5.7$), while G_{night} ($y = -0.63x - 3.6$) and G_{net} ($y = -0.76x + 1.1$) each had a negative linear relationship with standardized climate change (Table 2). NCP_{day} ($y = -7.01x + 23.4$), NCP_{night} ($y = -35.76 - 4.74$), and NCP_{net} ($y = -12.07x - 10.85$) all had significant negative relationships with standardized climate change. Black lines are best fit lines for each model with 95 % confidence intervals in gray. The x's in the top panel represent the imposed conditions for pre-industrial, present day, 2050, and 2100. The black horizontal line in panels (**b**), (**e**) and (**f**) shows the point where G and NCP equal 0. Points above the line are net calcifying (**e**) or net photosynthesizing (**f**) and points below the line are net dissolving (**e**) or net respiring (**f**) over the entire 24 h period.

Standardized climate change was a significant predictor for G_{day}, G_{night}, and G_{net} (Table 2; Fig. 3). G_{day} had a non-linear relationship with standardized climate change (Table 2, Fig. 3a), increasing to a threshold and then rapidly declining. G_{night}, however, had a strong linear relationship

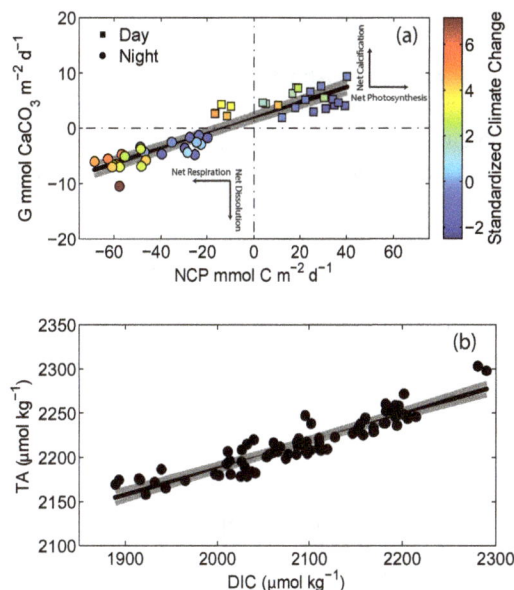

Figure 4. (**a**) Calculated G and NCP rates for all treatment aquaria. Squares are data collected during light (day) conditions and circles represent data collected during dark (night) conditions, and the color represents standardized climate change (color bar). There is a strong positive relationship between G and NCP ($y = 0.14x + 1.9$, $p < 0.0001$, $R^2 = 0.85$). Negative and positive y values are net dissolution and net calcification, respectively; negative and positive x values are net respiration and net photosynthesis, respectively. (**b**) TA versus DIC: There is a strong positive relationship between TA and DIC ($y = 0.31x + 1577.4$, $p < 0.0001$, $R^2 = 0.85$). Black and gray lines represent the best-fit line and 95 % confidence intervals, respectively. As expected, the slope of TA versus DIC (0.31) is approximately twice that of G versus NCP (0.14).

with standardized climate change (Table 2; Fig. 3c), suggesting that joint increases in ocean pCO_2 and temperature will increase nighttime dissolution of coral rubble. Lastly, G_{net} had a strong negative relationship with standardized climate change (Table 2; Fig. 3e), and the rubble community switched from net calcification to net dissolution at an increase in pCO_2 and temperature of 271 μatm and 0.75 °C, respectively. Standardized climate change was also a significant predictor of NCP: day, night, and net NCP rates all declined with standardized climate change (Table 2; Fig. 3b, d, f).

Net ecosystem calcification increased with net community production ($F_{1,46} = 260$, $p < 0.0001$, $R^2 = 0.85$; Fig. 4). In general, communities were net photosynthesizing and net calcifying during the day (Fig. 4a: squares in the upper right quadrant) and were net respiring and net dissolving at night (Fig. 4a: circles in the lower left quadrant). The exceptions were communities in the most extreme temperature–pCO_2 treatment: these communities were net respiring during the day while holding a positive, yet very low, calcification rate (Fig. 4a: squares in the upper left quadrant).

4 Discussion

4.1 Carbonate chemistry feedbacks

The rubble communities in the aquaria significantly altered the seawater chemistry, particularly at night ($t_{23} = -7.23$, $p < 0.0001$; Figs. 2, S1). This day–night difference in seawater chemistry increased under more extreme climate scenarios, as predicted by Jury et al. (2013). This large diel swing in pCO_2 is not uncommon in shallow coral reef environments. pCO_2 ranged from 480 to 975 μatm over 24 h on a shallow reef flat adjacent to our collection site (Silbiger et al., 2014) and from 450 to 742 μatm on a Moloka'i reef flat dominated by coral rubble (Yates and Halley, 2006). Here, pCO_2 had an average difference of 438 μatm between day and night, with a range of 412 μatm in the pre-industrial treatment to 854 μatm in the most extreme temperature–pCO_2 treatments (Fig. 2). In our study, we incorporated these feedbacks into the statistical analysis by using the actual, sampled pCO_2 (and temperature) in each aquarium (Fig. 3) rather than using the intended pCO_2 (and temperature) treatments in an ANOVA (Tables S1, S2 and Figs. S3, S4), better reflecting the pCO_2 experienced by organisms in each aquarium.

4.2 Calcification, dissolution, and net community production in a high CO_2 and temperature environment

Our results suggest that as pCO_2 and temperature increase over time, rubble reefs may shift from net calcification to net dissolution. In our study, this tipping point occurred at a pCO_2 and temperature increase of 271 μatm and 0.75 °C. Furthermore, our results showed that G_{day} and G_{night} in a natural coral rubble community have different functional responses to changing pCO_2 and temperature (Fig. 3). The ranges in G_{day} and G_{night} in our aquaria were similar to in situ rates on Hawaiian rubble reefs. Yates and Halley (2006) saw G_{day} values between 3.3 and 11.7 mmol CaCO$_3$ m^{-2} d^{-1} and G_{night} values between -2.4 and -24 mmol CaCO$_3$ m^{-2} d^{-1} on a Moloka'i reef flat with coral rubble only (note that Yates and Halley calculated G over a 4 h time frame and that the data were multiplied by 3 here to show G in mmol m^{-2} d^{-1}). Also note that we normalized our rates to the surface area of the rubble, while Yates and Halley (2006) normalized their rates to planar surface area.). G_{day} and G_{night} in our experiment ranged from 1.9 to 9.4 and -1.3 to -10.5 mmol CaCO$_3$ m^{-2} d^{-1}, respectively, across all treatment conditions. The higher dissolution rates in the in situ study by Yates and Halley (2006) are likely due to dissolution in the sediment, which was not present in our study.

G_{day} had a non-linear response to standardized climate change. G_{day} increased with temperature–pCO_2 until slightly above ambient conditions, and then decreased under more extreme climate conditions (Fig. 3a). This mixed response, increasing and then decreasing with standardized climate change, is reflected in prior experiments. We suggest three possible mechanisms to explain why calcification increases in slightly higher temperature–pCO_2 than ambient conditions. (1) Some calcifiers can maintain and even increase their calcification rates in acidic conditions (Kamenos et al., 2013; Findlay et al., 2011; Rodolfo-Metalpa et al., 2011; Martin et al., 2013) by either modifying their local pH environment (Hurd et al., 2011) or partitioning their energetic resources towards calcification (Kamenos et al., 2013). For example, in low, stable pH conditions, the coralline algae, *Lithothamnion glaciale*, increased its calcification rate relative to a control treatment, but did not concurrently increase its rate of photosynthesis (Kamenos et al., 2013). Kamenos et al. (2013) suggest that the up-regulation of calcification may limit photosynthetic efficiency. In the present study, the increase in G_{day} coincided with a decrease in net photosynthesis (Fig. 3a, b). Photosynthesizing calcifiers in the community may be partitioning their energetic resources more towards calcification and away from photosynthesis in order to maintain a positive calcification rate (Kamenos et al., 2013). Notably, turf algae likely have a major control over the NCP in this community, which would not have any impact on calcification. (2) An alternative hypothesis is that the calcifiers may be adapted or acclimatized to high pCO_2 conditions (Johnson et al., 2014) and have not yet reached their threshold because the rubble was collected from a naturally high and variable pCO_2 environment (Guadayol et al., 2014; Silbiger et al., 2014). (3) In this study, the calcifiers experienced a combined increase in both pCO_2 and temperature and, thus, the non-linear response in G_{day} may also be due to a metabolic response. In a typical thermal performance curve, organisms increase their metabolism until they have reached a thermal maximum, and then rapidly decline (Huey and Kingsolver, 1989; Pörtner et al., 2006), and we see this response in our results. A recent study found a similar non-linear response to temperature and pCO_2 in the coral *Siderastrea sidera* (Castillo et al., 2014). While they attribute the pCO_2 response to photosynthesis being neutralized (we did not see this response in our non-coral community), they suggest that the thermal response is due to both changes in metabolism and thermally driven changes in the aragonite saturation state (Castillo et al., 2014).

We saw a decline in both calcification and NCP in the extreme temperature–pCO_2 condition (Fig. 3). Calcification has been shown to decline with climate stressors, and the magnitude of decline differs across species (Kroeker et al., 2010; Pandolfi et al., 2011; Ries et al., 2009; Kroeker et al., 2013). The concurrent decline in NCP and calcification (Figs. 3a, b and 4) suggests that non-photosynthesizing invertebrates in the community (such as bivalves) might be dominating the calcification signal in these conditions. This hypothesis would explain the pattern that we see in Fig. 4, where communities in the most extreme pCO_2 and temperature conditions are net respiring during the day while still maintaining a small, positive calcification rate (Fig. 4a: five points in the upper left quadrant).

G_{night} rates are more straightforward, decreasing linearly with pCO_2 and temperature (Figs. 3c and 4). NCP_{night} rates also decreased linearly with pCO_2 and temperature (Fig. 3d). Similarly, Andersson et al. (2009) saw an increase in dissolution under acidic conditions in a community of corals, sand, and CCA. Previous studies on individual bioeroder taxa have also found higher rates of bioerosion or dissolution in more acidic, higher-temperature conditions (Wisshak et al., 2013; Fang et al., 2013; Reyes-Nivia et al., 2013; Tribollet et al., 2009; Wisshak et al., 2012). There are several mechanisms that could be mediating the increased dissolution rates in the high temperature–pCO_2 treatments. (1) Higher temperatures could increase the metabolism of the bioeroder community, thus increasing borer activity (e.g., Davidson et al., 2013). (2) Because many boring organisms excrete acidic compounds to erode the skeletal structure (Hutchings, 1986), reduced pH in the overlaying water column may reduce the metabolic cost to the organisms, making it easier for eroders to break down the $CaCO_3$. (3) Higher dissolution rates could be mediated by an increase in the proportion of dolomite in the skeletal structure of CCA on the rubble. A recent study found a 200 % increase in dolomite in CCA that were exposed to high pCO_2 and temperature conditions; this increase in dolomite resulted in increased bioerosion by endolithic algae (Diaz-Pulido et al., 2014). However, it is unlikely that changes in the mineralogy of the CCA indirectly increased dissolution here, given the short timescale of our study. In the present study, we used the TA anomaly method to calculate chemical dissolution as a proxy for bioerosion. Future studies should also include measures of mechanical breakdown (e.g. the production of sponge chips) in addition to chemical dissolution for a more complete picture of the impacts of climate stress on reef breakdown. Studies, including the present one, which focused on community-level responses, have consistently found that ocean acidification will increase dissolution rates on coral reefs (Andersson and Gledhill, 2013).

Standardized climate change explained more of the variance in dissolution than in calcification in our rubble community ($R^2_{G_{night}} = 0.64 > R^2_{G_{day}} = 0.33$; Table 2). This result is not surprising. Bioerosion, an important driver of dissolution, may be more sensitive to changes in ocean acidity than calcification, leading to net dissolution in high CO_2 waters. Many boring organisms excrete acidic compounds, which may be less metabolically costly in a low pH environment. Erez et al. (2011) hypothesize that increased dissolution, rather than decreased calcification, maybe be the reason that net coral reef calcification is sensitive to ocean acidification. The results of this study support this hypothesis. Although G_{net} declines linearly with temperature–pCO_2, calcification (G_{day}) and dissolution (G_{night}) have distinct responses to standardized climate change: G_{day} had a non-linear response, while G_{night} declined linearly with standardized climate change.

Our results highlight the need to study the effects of climate stressors on both calcification and dissolution.

Author contributions. Conceived and designed the experiments: N. J. Silbiger, M. J. Donahue. Performed the experiments: N. J. Silbiger. Analyzed the data: N. J. Silbiger, M. J. Donahue. Wrote the paper: N. J. Silbiger, M. J. Donahue.

Acknowledgements. Thanks to I. Caldwell, R. Coleman, J. Faith, K. Hurley, J. Miyano, R. Maguire, D. Schar, J. M. Sziklay, and M. M. Walton for help in field collections and lab analyses and to R. Briggs from UH SOEST Lab for Analytical Chemistry. M. J. Atkinson, R. Gates, C. Jury, H. Putnam, and R. Toonen gave thoughtful advice throughout the project. Comments by F. Mackenzie and our two anonymous reviewers improved this manuscript. This project was supported by a NOAA Nancy Foster Scholarship to N. J. Silbiger, a PADI Foundation Grant to N. J. Silbiger, and Hawaii SeaGrant 1847 to M. J. Donahue. This paper is funded in part by a grant/cooperative agreement from the National Oceanic and Atmospheric Administration, Project R/IR-18, which is sponsored by the University of Hawaii Sea Grant College Program, SOEST, under Institutional Grant No. NA09OAR4170060 from the NOAA Sea Grant Office, Department of Commerce. The views expressed herein are those of the author(s) and do not necessarily reflect the views of NOAA or any of its subagencies. This is HIMB contribution no. 1607, Hawaii SeaGrant contribution no. UNIHI-SEAGRANT-JC-12-19, and SOEST contribution no. 9237.

Edited by: J. Middelburg

References

Adey, W. H.: Review – coral reefs: algal structures and mediated ecosystems in shallow turbulent, alkaline waters, J. Phycol., 34, 393–406, 1998.

Andersson, A. J. and Gledhill, D.: Ocean Acidification and Coral Reefs: Effects on Breakdown, Dissolution, and Net Ecosystem Calcification, Ann. Rev. Mar. Sci., 5, 321–348, 2013.

Andersson, A. J. and Mackenzie, F. T.: Revisiting four scientific debates in ocean acidification research, Biogeosciences, 9, 893–905, doi:10.5194/bg-9-893-2012, 2012.

Andersson, A. J., Kuffner, I. B., Mackenzie, F. T., Jokiel, P. L., Rodgers, K. S., and Tan, A.: Net Loss of $CaCO_3$ from a subtropical calcifying community due to seawater acidification: mesocosm-scale experimental evidence, Biogeosciences, 6, 1811–1823, doi:10.5194/bg-6-1811-2009, 2009.

Andersson, A. J., Mackenzie, F. T., and Gattuso, J.-P.: Effects of ocean acidification on benthic processes, organisms, and ecosystems, in: Ocean Acidification, edited by: Gattuso, J.-P. and Hansson, L., Oxford University Press, 122–153, 2011.

Bopp, L., Resplandy, L., Orr, J. C., Doney, S. C., Dunne, J. P., Gehlen, M., Halloran, P., Heinze, C., Ilyina, T., Séférian, R., Tjiputra, J., and Vichi, M.: Multiple stressors of ocean ecosystems in the 21st century: projections with CMIP5 models, Biogeosciences, 10, 6225–6245, doi:10.5194/bg-10-6225-2013, 2013.

Caldeira, K. and Wickett, M. E.: Oceanography: anthropogenic carbon and ocean pH, Nature, 425, 365–365, 2003.

Camoin, G. F. and Montaggioni, L. F.: High energy coralgal-stromatolite frameworks from Holocene reefs (Tahiti, French Polynesia), Sedimentology, 41, 656–676, 1994.

Castillo, K. D., Ries, J. B., Bruno, J. F., and Westfield, I. T.: The reef-building coral *Siderastrea siderea* exhibits parabolic responses to ocean acidification and warming, Proc. R. Soc. B., 281, 20141856, 2014.

Comeau, S., Edmunds, P. J., Spindel, N. B., and Carpenter, R. C.: The responses of eight coral reef calcifiers to increasing partial pressure of CO_2 do not exhibit a tipping point, Limnol. Oceanogr, 58, 388–398, 2013.

Cubasch, U., Wuebbles, D., Chen, D., Facchini, M. C., Frame, D., Mahowald, N., and Winther, J.-G.: Climate Change 2013: The Physical Science Basis, Contribution of Working Group I to the Fifth Assessment Report of the Intergovernmental Panel on Climate Change Cambridge, United Kingdom and New York, NY, USA, 119–158, 2013.

Davidson, T. M., de Rivera, C. E., and Carlton, J. T.: Small increases in temperature exacerbate the erosive effects of a non-native burrowing crustacean, J. Exp. Mar. Biol. Ecol., 446, 115–121, 2013.

Diaz-Pulido, G., Anthony, K., Kline, D. I., Dove, S., and Hoegh-Guldberg, O.: Interactions between ocean acidification and warming on the mortality and dissolution of coralline alge, J. Phycol., 48, 32–39, 2012.

Diaz-Pulido, G., Nash, M. C., Anthony, K. R. N., Bender. D., Opdyke, B. N., Reyed-Nivia, C., and Troitzsch, U.: Greenhouse conditions induce mineralogical changes and dolomite accumulation in coralline algae on tropical reefs, Nature Communications, 5, 3310, doi:10.1038/ncomms4310, 2014.

Dickson, A. G.: Standard potential of the reaction: AgCl (s)+ $12H_2(g) = Ag$ (s)+ HCl (aq), and and the standard acidity constant of the ion HSO_4^- in synthetic sea water from 273.15 to 318.15 K, J. Chem. Thermodyn., 22, 113–127, 1990.

Dickson, A. G. and Millero, F. J.: A comparison of the equilibrium constants for the dissociation of carbonic acid in seawater media, Deep-Sea Res.-Pt. I, 34, 1733–1743, 1987.

Dickson, A. G., Sabine, C. L., and Christian, J. R.: Guide to best practices for ocean CO_2 measurements, Sidney, British Columbia, North Pacific Marine Science Organization, PICES Special Publication 3, 2007.

Doney, S. C., Fabry, V. J., Feely, R. A., and Kleypas, J. A.: Ocean Acidification: The Other CO_2 Problem, Ann. Rev. Mar. Sci., 1, 169–192, 2009.

Drupp, P. S., De Carlo, E. H., Mackenzie, F. T., Sabine, C. L., Feely, R. A., and Shamberger, K. E.: Comparison of CO_2 dynamics and air–sea gas exchange in differing tropical reef environments, Aquat. Geochem., 19, 371–397, 2013.

Erez, J., Reynaud, S., Silverman, J., Schneider, K., and Allemand, D.: Coral calcification under ocean acidification and global change, in: Coral Reefs: an ecosystem in transition, edited by: Dubinski, Z. and Stambler, N., Springer, 151–176, 2011.

Fabricius, K. E.: Effects of terrestrial runoff on the ecology of corals and coral reefs: review and synthesis, Mar. Pollut. Bull., 50, 125–146, 2005.

Fabricius, K., Langdon, C., Uthicke, S., Humphrey, C., Noonan, S., De'ath, G., Okazaki, R., Muehllehner, N., Glas, M., and Lough, J.: Losers and winners in coral reefs acclimatized to elevated carbon dioxide concentrations, Nature Climate Change, 1, 165–169, 2011.

Fair, R. C.: On the robust estimation of econometric models, in: Annals of Economic and Social Measurement, 3, NBER, 117–128, 1974.

Fang, J. K. H., Mello-Athayde, M. A., Schönberg, C. H. L., Kline, D. I., Hoegh-Guldberg, O., and Dove, S.: Sponge biomass and bioerosion rates increase under ocean warming and acidification, Glob. Change Biol., 19, 3581–3591, 2013.

Fangue, N. A., O'Donnell, M. J., Sewell, M. A., Matson, P. G., MacPherson, A. C., and Hofmann, G. E.: A laboratory-based, experimental system for the study of ocean acidification effects on marine invertebrate larvae, Limnol. Oceanogr.-Meth., 8, 441–452, 2010.

Feely, R. A., Sabine, C. L., Lee, K., Berelson, W., Kleypas, J., Fabry, V. J., and Millero, F. J.: Impact of anthropogenic CO_2 on the $CaCO_3$ system in the oceans, Science, 305, 362–366, 2004.

Findlay, H. S., Wood, H. L., Kendall, M. A., Spicer, J. I., Twitchett, R. J., and Widdicombe, S.: Comparing the impact of high CO_2 on calcium carbonate structures in different marine organisms, Mar. Biol. Res., 7, 565–575, 2011.

Gattuso, J.-P., Frankignoulle, M., and Smith, S. V.: Measurement of community metabolism and significance in the coral reef CO_2 source-sink debate, Proc. Natl. Acad. Sci., 96, 13017–13022, 1999.

Guadayol, Ò., Silbiger, N. J., Donahue, M. J., and Thomas, F. I. M.: Patterns in Temporal Variability of Temperature, Oxygen and pH along an Environmental Gradient in a Coral Reef, PloS one, 9, e85213, doi:10.1371/journal.pone.0085213, 2014.

Harrington, L., Fabricius, K., De'ath, G., and Negri, A.: Recognition and selection of settlement substrata determine post-settlement survival in corals, Ecology, 84, 3428–3437, 2004.

Hoegh-Guldberg, O., Mumby, P. J., Hooten, A. J., Steneck, R. S., Greenfield, P., Gomez, E., Harvell, C. D., Sale, P. F., Edwards, A. J., Caldeira, K., Knowlton, N., Eakin, C. M., Iglesias-Prieto, R., Muthiga, N., Bradbury, R. H., Dubi, A., and Hatziolos, M. E.: Coral reefs under rapid climate change and ocean acidification, Science, 318, 1737–1742, 2007.

Huey, R. B. and Kingsolver, J. G.: Evolution of thermal sensitivity of ecotherm performance, Trends Ecol. Evol., 4, 131–135, 1989.

Hutchings, P. A.: Biological destruction of coral reefs, Coral Reefs, 4, 239–252, 1986.

Hoegh-Guldberg, O. and Bruno, J. F.: The impact of climate change on the world's marine ecosystems, Science, 328, 1523–1528, 2010.

Hurd, C. L., Cornwall, C. E., Currie, K., Hepburn, C. D., McGraw, C. M., Hunter, K. A., and Boyd, P. W.: Metabolically induced pH fluctuations by some coastal calcifiers exceed projected 22nd century ocean acidification: a mechanism for differential susceptibility?, Glob. Change Biol., 17, 3254–3262, 2011.

Johnson, M. D. and Carpenter, R. C.: Ocean acidification and warming decrease calcification in the crustose coralline alga *Hy-*

drolithon onkodes and increase susceptibility to grazing, J. Exp. Mar. Biol. Ecol., 434, 94–101, 2012.

Johnson, M. D., Moriarty, V. W., and Carpenter, R. C.: Acclimatization of the Crustose Coralline Alga *Porolithon onkodes* to Variable pCO_2, PLOS ONE, 9, e87678, doi:10.1371/journal.pone.0087678, 2014.

Jokiel, P. L., Rodgers, K. S., Kuffner, I. B., Andersson, A. J., Cox, E. F., and Mackenzie, F. T.: Ocean acidification and calcifying reef organisms: a mesocosm investigation, Coral Reefs, 27, 473–483, 2008.

Jury, C. P., Thomas, F. I. M., Atkinson, M. J., and Toonen, R. J.: Buffer Capacity, Ecosystem Feedbacks, and Seawater Chemistry under Global Change, Water, 5, 1303–1325, 2013.

Kamenos, N. A., Burdett, H. L., Aloisio, E., Findlay, H. S., Martin, S., Longbone, C., Dunn, J., Widdicombe, S., and Calosi, P.: Coralline algal structure is more sensitive to rate, rather than the magnitude, of ocean acidification, Glob. Change Biol., 19, 3621–3628, 2013.

Kleypas, J. and Langdon, C.: Coral reefs and changing seawater chemistry, in: Coral Reefs and Climate Change: Science and Managemen., edited by: Phinney, J., Skirving, W., Kleypas, J., and Hoegh-Guldberg, O., American Geophysical Union, Washington D.C., 73–110, 2006.

Kroeker, K. J., Kordas, R. L., Crim, R. N., and Singh, G. G.: Meta-analysis reveals negative yet variable effects of ocean acidification on marine organisms, Ecol. Lett., 13, 1419–1434, 2010.

Kroeker, K. J., Kordas, R. L., Crim, R., Hendriks, I. E., Ramajo, L., Singh, G. S., Duarte, C. M., and Gattuso, J. P.: Impacts of ocean acidification on marine organisms: quantifying sensitivities and interaction with warming, Glob. Change Biol., 19, 1884–1896, 2013.

Littler, M. M.: The population and community structure of Hawaiian fringing-reef crustose corallinaceae (Rhodophyta, Cryptonemiales), J. Exp. Mar. Biol. Ecol., 11, 103–120, 1973.

Lowe, R. J., Falter, J. L., Monismith, S. G., and Atkinson, M. J.: A numerical study of circulation in a coastal reef-lagoon system, J. Geophys. Res.-Oceans, 114, C06022, doi:10.1029/2008JC005081, 2009a.

Lowe, R. J., Falter, J. L., Monismith, S. G., and Atkinson, M. J.: Wave-driven circulation of a coastal reef-lagoon system, J. Phys. Oceanogr., 39, 873–893, 2009b.

Martin, S., Cohu, S., Vignot, C., Zimmerman, G., and Gattuso, J. P.: One-year experiment on the physiological response of the Mediterranean crustose coralline alga, *Lithophyllum cabiochae*, to elevated pCO_2 and temperature, Ecol. Evol., 3, 676–693, doi:10.1029/2008JC005081, 2013.

Mehrbach, C.: Measurement of the apparent dissociation constants of carbonic acid in seawater at atmospheric pressure, Limnol. Oceanogr., 18, 897–907, 1973.

Meinshausen, M., Smith, S. J., Calvin, K., Daniel, J. S., Kainuma, M. L. T., Lamarque, J. F., Matsumoto, K., Montzka, S. A., Raper, S. C. B., and Riahi, K.: The RCP greenhouse gas concentrations and their extensions from 1765 to 2300, Clim. Change, 109, 213–241, 2011.

Pandolfi, J. M., Connolly, S. R., Marshall, D. J., and Cohen, A. L.: Projecting coral reef futures under global warming and ocean acidification, Science, 333, 418–422, 2011.

Price, N.: Habitat selection, facilitation, and biotic settlement cues affect distribution and performance of coral recruits in French Polynesia, Oecologia, 163, 747–758, 2010.

Pörtner, H. O., Bennet, A. F., Bozinovic, F., Clarke, A., Lardies, M. A., Lucassen, M., Pelster, B., Schiemer, F., and Stillman, J. H.: Trade-offs in thermal adaptation: the need for molecular ecological integration, Phys. Biochem. Zool., 79, 295–313, 2006.

Putnam, H. M.: Resilience and acclimatization potential of reef corals under predicted climate change stressors, PhD, Zoology, University of Hawaii at Manoa, Honolulu, 1–154, 2012.

Reyes-Nivia, C., Diaz-Pulido, G., Kline, D., Guldberg, O.-H., and Dove, S.: Ocean acidification and warming scenarios increase microbioerosion of coral skeletons, Glob. Change Biol., 19, 1919–1929, 2013.

Ries, J. B., Cohen, A. L., and McCorkle, D. C.: Marine calcifiers exhibit mixed responses to CO_2-induced ocean acidification, Geology, 37, 1131–1134, 2009.

Rodolfo-Metalpa, R., Houlbrèque, F., Tambutté, É., Boisson, F., Baggini, C., Patti, F. P., Jeffree, R., Fine, M., Foggo, A., and Gattuso, J. P.: Coral and mollusc resistance to ocean acidification adversely affected by warming, Nature Climate Change, 1, 308–312, 2011.

Rogelj, J., Meinshausen, M., and Knutti, R.: Global warming under old and new scenarios using IPCC climate sensitivity range estimates, Nature Climate Change, 2, 248–253, 2012.

Sanford, T., Frumhoff, P. C., Luers, A., and Gulledge, J.: The climate policy narrative for a dangerously warming world, Nature Climate Change, 4, 164–166, 2014.

Semesi, I. S., Kangwe, J., and Björk, M.: Alterations in seawater pH and CO_2 affect calcification and photosynthesis in the tropical coralline alga, *Hydrolithon* sp. (Rhodophyta), Estuarine, Coast. Shelf Sci., 84, 337–341, 2009.

Silbiger, N., Guadayol, Ò., Thomas, F. I. M., and Donahue, M.: Reefs shift from net accretion to net erosion along a natural environmental gradient, Mar. Ecol. Prog. Ser., 515, 33–44, 2014.

Smith, S. V. and Key, G. S.: Carbon dioxide and metabolism in marine environments, Limnol. Oceanogr., 20, 493–495, 1975.

Smith, S. V., Kimmerer, W. J., Laws, E. A., Brock, R. E., and Walsh, T. W.: Kaneohe Bay sewage diversion experiment – perspectives on ecosystem responses to nutritional perturbation, Pacific Science, 35, 279–402, 1981.

Solomon, S., Qin. D., Manning. M. Chen. Z., Marquis, M., Bindoff, N. L., Willebrand, J., Artale, V., Cazenave, A., Gregory, J., Gulev, S., Hanawa, K., Le Quéré, C., Levitus, S., Nojiri, Y., Shum, C. K., Talley L. D., and Unnikrishnan, A.: Observations: Oceanic Climate Change and Sea Level, in: Climate Change 2007: The Physical Science Basis, Contribution of Working Group I to the Fourth Assessment Report of the Intergovernmental Panel on Climate Change, edited by: Solomon, S., Qin, D., Manning, M., Chen, Z., Marquis, M., Averyt, K. B., Tignor, M., and Miller, H. L., Cambridge University Press, Cambridge, United Kingdom and New York, NY, USA, 387–429, 2007.

Stimson, J. and Kinzie III, R. A.: The temporal pattern and rate of release of zooxanthellae from the reef coral *Pocillopora damicornis* (Linnaeus) under nitrogen-enrichment and control conditions, J. Exp. Mar. Biol. Ecol., 153, 63–74, 1991.

Tans, P. and Keeling, R.: NOAA/ESRL, available at: www.esrl.noaa.gov/gmd/ccgg/trends/ (last access: 8 January 2015), 2013.

Tribollet, A. and Payri, C.: Bioerosion of the coralline alga *Hydrolithon onkodes* by microborers in the coral reefs of Moorea, French Polynesia, Oceanol. Acta, 24, 329–342, 2001.

Tribollet, A., Atkinson, M. J., and Langdon, C.: Effects of elevated $pCO_{(2)}$ on epilithic and endolithic metabolism of reef carbonates, Glob. Change Biol., 12, 2200–2208, 2006.

Tribollet, A., Godinot, C., Atkinson, M., and Langdon, C.: Effects of elevated $pCO_{(2)}$ on dissolution of coral carbonates by microbial euendoliths, Glob. Biogeochem. Cy., 23, GB3008, doi:10.1029/2008GB003286, 2009.

Uppström, L. R.: The boron/chlorinity ratio of deep-sea water from the Pacific Ocean, Deep-Sea Res., 21, 161–162, 1974.

Van Heuven, S., Pierrot, D., Lewis, E., and Wallace, D. W. R.: MATLAB Program developed for CO_2 system calculations, Rep. ORNL/CDIAC-105b, 2009.

van Heuven, S., Pierrot, D., Rae, J. W. B., Lewis, E., and Wallace, D. W. R.: 5 MATLAB program developed for CO_2 system calculations, ORNL/CDIAC-105b, Carbon Dioxide Inf. Anal. Cent., Oak Ridge Natl. Lab., US DOE, Oak Ridge, Tenn., doi:10.3334/CDIAC/otg.CO2SYS_MATLAB_v1.1, 2011.

Van Vuuren, D. P., Meinshausen, M., Plattner, G. K., Joos, F., Strassmann, K. M., Smith, S. J., Wigley, T. M. L., Raper, S. C. B., Riahi, K., and De La Chesnaye, F.: Temperature increase of 21st century mitigation scenarios, Proc. Natl. Acad. Sci., 105, 15258–15262, 2008.

Van Vuuren, D. P., Edmonds, J., Kainuma, M., Riahi, K., Thomson, A., Hibbard, K., Hurtt, G. C., Kram, T., Krey, V., and Lamarque, J.-F.: The representative concentration pathways: an overview, Clim. Change, 109, 5–31, 2011.

White, J.: Distribution, recruitment and development of the borer community in dead coral on shallow Hawaiian reefs, Ph.D., Zoology, University of Hawaii at Manoa, Honolulu, 1980.

Wisshak, M., Schönberg, C. H. L., Form, A., and Freiwald, A.: Ocean acidification accelerates reef bioerosion, Plos One, 7, e45124, doi:10.1371/journal.pone.0045124, 2012.

Wisshak, M., Schönberg, C. H. L., Form, A., and Freiwald, A.: Effects of ocean acidification and global warming on reef bioerosion – lessons from a clionaid sponge, Aquat. Biol., 19, 111–127, 2013.

Wolf-Gladrow, D. A., Zeebe, R. E., Klass, C., Körtzinger, A., and Dickson, A. G.: Total Alkalinity: The explicit conservative expression and its application to biogeochemical processes, Mar. Chem., 106, 287–300, 2007.

Yates, K. K. and Halley, R. B.: CO_3^{2-} concentration and pCO_2 thresholds for calcification and dissolution on the Molokai reef flat, Hawaii, Biogeosciences, 3, 357-369, doi:10.5194/bg-3-357-2006, 2006.

11

Variable C : N : P stoichiometry of dissolved organic matter cycling in the Community Earth System Model

R. T. Letscher, J. K. Moore, Y.-C. Teng, and F. Primeau

Earth System Science, University of California, Irvine, CA, USA

Correspondence to: R. T. Letscher (robert.letscher@uci.edu)

Abstract. Dissolved organic matter (DOM) plays an important role in the ocean's biological carbon pump by providing an advective/mixing pathway for $\sim 20\,\%$ of export production. DOM is known to have a stoichiometry depleted in nitrogen (N) and phosphorus (P) compared to the particulate organic matter pool, a fact that is often omitted from biogeochemical ocean general circulation models. However the variable C : N : P stoichiometry of DOM becomes important when quantifying carbon export from the upper ocean and linking the nutrient cycles of N and P with that of carbon. Here we utilize recent advances in DOM observational data coverage and offline tracer-modeling techniques to objectively constrain the variable production and remineralization rates of the DOM C : N : P pools in a simple biogeochemical-ocean model of DOM cycling. The optimized DOM cycling parameters are then incorporated within the Biogeochemical Elemental Cycling (BEC) component of the Community Earth System Model (CESM) and validated against the compilation of marine DOM observations. The optimized BEC simulation including variable DOM C : N : P cycling was found to better reproduce the observed DOM spatial gradients than simulations that used the canonical Redfield ratio. Global annual average export of dissolved organic C, N, and P below 100 m was found to be $2.28\,\mathrm{Pg\,C\,yr^{-1}}$ ($143\,\mathrm{Tmol\,C\,yr^{-1}}$), $16.4\,\mathrm{Tmol\,N\,yr^{-1}}$, and $1\,\mathrm{Tmol\,P\,yr^{-1}}$, respectively, with an average export C : N : P stoichiometry of 225 : 19 : 1 for the semilabile (degradable) DOM pool. Dissolved organic carbon (DOC) export contributed $\sim 25\,\%$ of the combined organic C export to depths greater than 100 m.

1 Introduction

Dissolved organic matter (DOM) is an important pool linking nutrient cycles of nitrogen (N) and phosphorus (P) to the ocean's carbon cycle. Following its net production in the surface ocean, DOM provides an advective pathway for removal of biologically fixed carbon (C) to the deep ocean, accounting for $\sim 20\,\%$ of the C exported within the ocean's biological pump (Hansell, 2013). Remineralization of DOM in the ocean's interior is carried out by microbial heterotrophs, respiring C while releasing inorganic N and P nutrients back into the water column. The concept of the Redfield ratio (Redfield, 1958; Redfield et al., 1963) has been a unifying paradigm in ocean biogeochemistry linking the stoichiometry of biological production and phytoplankton cellular material to that of the remineralization of detrital organic matter (OM) and inorganic nutrient ratios in the water column. At the global scale, production/decomposition of particulate OM (POM) in the ocean is thought to largely follow the canonical Redfield ratio of 106 : 16 : 1 for C : N : P, however some recent studies have suggested more variable C : N : P ratios (i.e., Martiny et al., 2013a, b) and only recently has variable C : N : P stoichiometry been introduced into Earth system models (e.g., Vichi et al., 2007; Dunne et al., 2013). Large deviations from the Redfield ratio have been documented for DOM (Aminot and Kérouel, 2004; Hopkinson and Vallino, 2005). Hopkinson and Vallino (2005) found DOM production and decomposition to follow a stoichiometry of 199 : 20 : 1, indicating the more efficient export of C within DOM per mole of N and P relative to sinking POM. This finding is significant in light of evidence that future perturbations to the ocean from global climate change may favor enhanced partitioning of production to DOM (Wohlers et al., 2009; Kim et al., 2011). Thus accounting for variable stoichiometry within the DOM pool that deviates from the Redfield ratio requires a re-evaluation of the controls on C export and their response to future perturbations due to climate change.

Here we aim to utilize recent advances in DOM data coverage to incorporate variable production and decomposition stoichiometry within the DOM tracers of the Biogeochemical Elemental Cycling (BEC) model in order to improve representation of this important carbon export flux and associated nutrient cycles. The BEC tracks the cycling of key biogeochemical tracers (e.g., C, N, P and Fe) and runs within the ocean general circulation component of the Community Earth System Model (CESM) (Moore et al., 2004). The current release of CESM v1.2.1 contains five DOM-related tracers: semilabile DOC (dissolved organic carbon), DON (dissolved organic nitrogen), and DOP (dissolved organic phosphorous) pools as well as refractory DON and DOP pools (Moore et al., 2014). Here we have added a sixth DOM tracer, refractory DOC. Our approach is to optimize the BEC DOM parameters using available observations, by applying a fast offline solver based on a direct-matrix inversion (DMI) of a linear model of DOM cycling; an approach similar to previous applications for marine radiocarbon (Khatiwala et al., 2005) and marine organic matter cycling (Kwon and Primeau, 2006; Hansell et al., 2009). The 3-D ocean circulation is obtained from the offline tracer-transport model for the ocean component of the CESM (Bardin et al., 2014). The DMI solver uses a parallel multifrontal sparse matrix inversion approach as implemented in the MUMPS (MUltifrontal Massively Parallel Sparse direct Solver) (Amestoy et al., 2001, 2006) to quickly obtain the equilibrium solutions needed to objectively calibrate the biogeochemical parameters of the DOM cycling model by minimizing the misfit between the model and observations. The DOM cycling parameters from the equilibrium solution of the offline model are then incorporated within the BEC and optimized with only minor additional tuning.

The remainder of this article is organized as follows. Section 2 describes (1) the current representation of DOM cycling in the BEC v1.2.1, (2) the global ocean data set of DOM observations utilized for the optimization, (3) the structure of the offline DOM cycling model and the DMI solver, and (4) the modified BEC model with improved DOM cycling parameters with the metrics employed for optimization. Section 3 details the results of (1) the offline DOM cycling model solution, (2) the reference CESM BEC v1.2.1 simulation, as well as (3) the BEC simulation with optimized DOM cycling, including a comparison of DOM cycling metrics. Sections 3.4 and 3.5 describe a comparison of multiple DOM cycling schema and an evaluation of direct uptake of DOP by phytoplankton in the BEC model, respectively. We conclude with a discussion and summary of our results in Sect. 4.

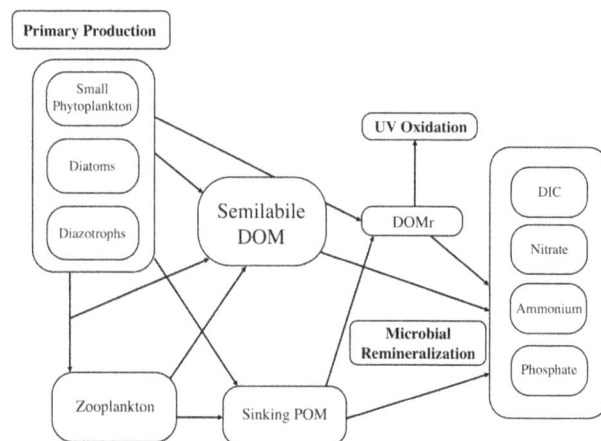

Figure 1. Schematic of organic matter cycling in the CESM BEC. Primary production is carried out by three phytoplankton functional types: small phytoplankton (which also contains a subgroup of calcifying phytoplankton), diatoms, and diazotrophs. Sources to DOM include direct losses from phytoplankton/zooplankton and from zooplankton grazing of phytoplankton. The major sink for DOM is microbial remineralization, parameterized with an assigned lifetime which differs between the euphotic zone and the mesopelagic ocean. A small fraction of phytoplankton production is converted to refractory DOM in the upper ocean with an additional source to DOMr from degradation of sinking POM in the mesopelagic. DOMr is also lost via UV photo-oxidation in the surface layer (< 10 m). The products of organic matter remineralization are dissolved inorganic carbon, nitrate, ammonium, and phosphate.

2 Methods

2.1 DOM cycling in the standard BEC v1.2.1

Model simulations with the optimized DOM parameters are compared against a reference simulation using the standard version of the CESM BEC v1.2.1, which we refer to as REF. The BEC model runs within the ocean physics component of CESM1 (Gent et al., 2011), which is the Parallel Ocean Program v2 (POP2; Smith et al., 2010). Detailed description and evaluation of the ocean general circulation model is given by Danabasoglu et al. (2011). Additional documentation, model output, and model source code are available online (www2.cesm.ucar.edu). The REF simulation has a nominal horizontal resolution of 1° with 60 vertical levels ranging in thickness from 10 m (in the upper 150 m) with increasing layer thickness increasing with depth below 150 m. Results are presented for the final 20 yr annual average from a 310 yr simulation.

A flowchart of organic matter cycling in the BEC is shown in Fig. 1 and a list of DOM parameter values from REF are given in Table 1. Primary production is carried out amongst three phytoplankton groups, which take up available inorganic nutrients and have losses to zooplankton grazing, sinking POM, and semilabile DOM. Organic matter is

Table 1. Optimized DOM parameters from the DMI-enabled linear DOM model (DMI-DOM solver) and the modified DMI model (MOD DMI-DOM solver) as well as the REF and DOM OPT simulations of the CESM BEC. Euphotic zone: 0–100 m for the DMI-DOM models and depths where PAR > 1 % for REF and DOM OPT. The "flux to DOM" represents the fraction of PP that accumulates as DOM while the "fraction of DOM flux" represents the portion of the DOM production flux that accumulates as semilabile (SL) or refractory (R) DOM. Parameters $f_i | i = 1 \cdots 2$, $\kappa_i | i = 1 \cdots 4$ are defined in Eqs. (1)–(4). Surf: surface layer (< 10 m), reminR: remineralization rate, sp: small phytoplankton, diat: diatoms, diaz: diazotrophs, k: half saturation constant for DOP uptake, and NA: not applicable.

Parameter	DMI-DOM solver	MOD DMI-DOM solver	Parameter	REF	DOM OPT
	Flux to DOM (f_1)	Fraction of DOMprod flux	Flux to DOM		
SLDOC	0.099	0.99	f_PP_doc	0.15	0.06
SLDON	0.01	0.9885	f_PP_don	0.15	0.04
SLDOP	0.095	0.997	f_PP_dop	0.15	0.06
			parm_labile_ratio	0.85	0.94
			DOCrefract	NA	0.01
	(f_2)		DONrefract	0.08	0.0115
DOCr	0.006	0.01	DOPrefract	0.03	0.003
DONr	0.004	0.0115	f_to_don	NA	0.66
DOPr	0.0015	0.003			
			DOP uptake		
			sp_kDOP	0.26	0.25
DOP uptake	NA	NA	diat_kDOP	0.9	1.0
			diaz_kDOP	0.09	0.08
DOM lifetimes			DOM lifetimes		
– Euphotic zone			– Euphotic zone		
SLDOC: $1/\kappa_1$	34 yr	15 yr	DOC_reminR	250 d	15 yr
SLDON: $1/\kappa_1$	8.7 yr	15 yr	DON_reminR	160 d	15 yr
SLDOP: $1/\kappa_1$	5.8 yr	62 yr	DOP_reminR	160 d	60 yr
– Layer 1 (> −10 m)			– Layer 1 (> −10 m)		
RDOC: $1/\kappa_4$	NA	15 yr	DOCr_reminR	NA	20 yr
RDON: $1/\kappa_4$	NA	15 yr	DONr_reminR	2.5 yr	20 yr
RDOP: $1/\kappa_4$	NA	15 yr	DOPr_reminR	2.5 yr	20 yr
– Mesopelagic zone			– Mesopelagic zone		
SLDOC: $1/\kappa_2$	2.9 yr	5 yr	DOC_reminR	10 yr	5.5 yr
SLDON: $1/\kappa_2$	1.7 yr	5 yr	DON_reminR	4.4 yr	5 yr
SLDOP: $1/\kappa_2$	0.8 yr	4.5 yr	DOP_reminR	8.8 yr	4 yr
– Layer 2 : 60 (< −10 m)			– Layer 2 : 60 (< −10 m)		
RDOC: $1/\kappa_3$	20 000 yr	15 000 yr	DOCr_reminR	NA	16 000 yr
RDON: $1/\kappa_3$	9000 yr	8000 yr	DONr_reminR	670 yr	9000 yr
RDOP: $1/\kappa_3$	5000 yr	6000 yr	DOPr_reminR	460 yr	5000 yr

produced with a C : N : P stoichiometry set to the slightly modified Redfield ratio of Anderson and Sarmiento (1994), 117 : 16 : 1. Additional sources to semilabile DOM include grazing losses when phytoplankton are grazed by zooplankton as well as direct zooplankton losses. A variable fraction of DOM production is sent to the refractory DOM (DOMr) pool, with different fractions going to the dissolved organic N and P pools. Approximately 15 % of modeled primary production (PP) is sent to the DOM pool via these sources, with the remainder of PP cycling as POM. It is important to note that the BEC does not specifically track the total production/decomposition of DOM, which is estimated to be 30–50 % of net primary production (NPP) (Carlson, 2002; and references therein). Rather, BEC semilabile and refractory DOM tracers track the *accumulated* DOM pools that arise

from the decoupling of DOM production and consumption in time and space and are thus subject to advection by the ocean circulation. These recalcitrant DOM fractions cycle on timescales of years to centuries and represent a smaller portion of NPP, i.e., ∼ 5–10 % (Hansell, 2013). The labile DOM pool, which cycles on timescales of minutes to days (Hansell, 2013) is not explicitly modeled and is instead rapidly converted to inorganic carbon and nutrients at each time step.

Microbial remineralization is the dominant sink for both POM and DOM pools and is parameterized by assigned remineralization rates. POM is remineralized following a prescribed remineralization vs. depth curve, with a length scale that increases with depth (Moore et al., 2014). Semilabile DOM pools are assigned lifetimes (1 / remineralization rate) that depend on the light field with model grid cells where photosynthetically active radiation (PAR) is > 1 % of surface irradiance being assigned a euphotic zone lifetime. Semilabile DOM in model grid cells with PAR < 1 % is assigned a mesopelagic zone lifetime. Remineralization is more rapid for semilabile DOM in the euphotic zone, with lifetimes on the order of 5 months for DON + DOP and ∼8 months for DOC. Longer lifetimes for semilabile DOM are assigned in the mesopelagic zone with the order of remineralization lifetimes following C > P > N. Remineralization of refractory DOM follows a similar light dependence with a faster remineralization rate given to DOMr in euphotic zone grid cells to parameterize a sink via UV oxidation (Carlson, 2002). DOMr below the euphotic zone is remineralized over centennial timescales.

2.2 Database of DOM ocean observations

We compiled publicly available and literature observations of DOM concentrations into a single database for use in both the DMI-enabled linear DOM model as well as to evaluate our BEC DOM optimization model runs. Briefly, the database contains over 34 000 observations of DOC, > 18 000 observations of DON, and > 2000 observations of DOP. Geographic coverage for the five ocean basins is moderately balanced for observations of DOC and DON; however, the Atlantic Ocean dominates available DOP observations with DOP data completely lacking for the Indian, Southern, and Arctic oceans. Semilabile DOM is defined as the total observed DOM concentration less the refractory concentration as determined from the asymptotic concentration of DOM depth profiles. Refractory DOC concentrations vary by ocean basin in the range of 37.7 (South Pacific)–45.0 μM (Arctic). Globally constant concentrations are used for refractory DON (1.8 μM) and refractory DOP (0.03 μM). Full details of this DOM database are given elsewhere (Letscher and Moore, 2014).

2.3 Application of the DMI-enabled solver with a linear DOM cycling model

2.3.1 First iteration – DOM source from BEC PP

The linear DOM cycling model cycles DOM with one source/sink and uses an idealized annual ocean circulation in offline mode from the CESM POP2 ocean circulation model (Bardin et al., 2014); nominal $1° \times 1°$ horizontal resolution with 60 vertical levels, i.e., the same grid as the standard BEC v1.2.1. In this simple model of DOM cycling, two tracers of DOM are simulated for each element, C, N, and P: semilabile (SLDOM) and refractory (RDOM). The source for each DOM tracer is parameterized as some variable fraction, f, of primary production and is formed within the top model grid level with a thickness of 10 m. The sink for each DOM fraction is microbial remineralization parameterized with an assigned remineralization rate, κ, that differs for the euphotic

zone and deep ocean layers in the case of SLDOM. The conservation equations for each DOM tracer are

$$\frac{\partial}{\partial t}\text{SLDOM} + \mathbf{T}\text{SLDOM} = f_1\text{PP} - \begin{cases} \kappa_1\text{SLDOM} \\ \quad \text{if } z > -100\,\text{m} \\ \kappa_2\text{SLDOM} \\ \quad \text{if } z < -100\,\text{m} \end{cases}, \quad (1)$$

$$\frac{\partial}{\partial t}\text{RDOM} + \mathbf{T}\text{RDOM} = f_2\text{PP} - \kappa_3\text{RDOM}, \quad (2)$$

where $\mathbf{T}$ is the advection–diffusion transport operator (a sparse matrix constructed using output from the dynamical CESM POP2 model as described in Bardin et al., 2014) and PP is the annual average 3-D primary production field from the coupled ocean–atmosphere run of the CESM for the 1990s (Moore et al., 2013).

We tested the sensitivity of the linear DOM model results to multiple production functions (CESM PP, DOM production flux from the BEC, satellite estimated PP); however, results suggest the differing source functions do not appreciably alter modeled DOM distributions or parameter values.

Our initial construction of the linear DOM model allowed the sum of $f_1 + f_2$ to vary continuously between 0 and 0.5 and $\kappa_i | i = 1 \cdots 3$ to vary logarithmically between 0.25 and 20 000 yr^{-1} by 24 discrete values. The direct-solver technique makes it possible to objectively calibrate these parameters, $f_i \kappa_i | i = 1 \cdots 3 \{f_1, f_2, \kappa_1, \kappa_2, \kappa_3\}$, by using a numerical optimization algorithm that rapidly tests each permutation of the discretized κi values, scaled by f_i, in order to find the parameter set that minimizes the root mean square difference in the misfit between the model-predicted and observed DOM concentration. A separate linear DOM model (Eqs. 1, 2) is solved for the DOC, DON, and DOP cases. The DMI solver allows us to determine very efficiently the optimal lifetimes for the various DOM pools. It is not practical to determine these using multiple forward simulations of the full CESM BEC, which would require years to decades of computer time.

2.3.2 Optimized DOM parameter incorporation into the BEC model

The optimized parameter values obtained from the DMI-enabled linear DOM model were incorporated within the BEC to improve its representation of DOM cycling. The BEC model has two tracers for each DOM pool, semilabile and refractory, with differing lifetimes for the euphotic vs. mesopelagic zones. Thus the SLDOM lifetimes, κ_1 and κ_2, from the DMI-enabled DOM model were applied to the BEC model semilabile tracers for the euphotic zone and mesopelagic, respectively. The RDOM lifetime from the DMI-enabled DOM model was applied throughout the full water column of the BEC model. Further fine-tuning of DOM lifetimes was carried out to provide the best DOM optimized case, using the mean bias of the modeled concentrations versus the observations and the log-transformed regression correlation coefficient between simulated and observed DOM in the upper ocean, 0–500 m, as comparison metrics. The BEC simulation containing the set of improved DOM cycling parameters following the first iteration of the DMI-enabled linear DOM model is termed DOM DEV.

2.3.3 Second iteration of DMI-enabled linear DOM model – DOM source from BEC DOM production flux

Initial improvements to DOM cycling metrics within the BEC model were large upon incorporation of the DMI-enabled linear DOM model parameter values; however, because of differences between the offline model and the full BEC model, further improvements to the DOM tracer lifetimes were possible. To achieve this the DMI-enabled linear DOM model was modified such that the production for each tracer was held constant allowing only the remineralization rate, κ_i, to be optimized from a choice of 48 discrete tracer lifetimes spanning the range 0.7–$20\,000\,\mathrm{yr}^{-1}$. Rather than using PP to get the production flux of each DOM tracer, the semilabile and refractory DOM production fluxes (SLDOMprod, RDOMprod) were extracted from the DOM DEV simulation and prescribed in the modified DMI-enabled DOM model. The fraction of SLDOMprod and RDOMprod to be applied each $\kappa_i | i = 1 \cdots 4$ was diagnosed from the relative proportions of each tracer residing in the euphotic or deep layers at the end of the DOM DEV simulation of the BEC (see Fig. 2). At this step it was also desired to solve for the remineralization rate associated for a secondary sink for DOMr due to photo-oxidation in the surface layer. Thus Eqs. (1) and (2) were modified to become Eqs. (3) and (4) as follows:

$$\frac{\partial}{\partial t}\mathrm{SLDOM} + \mathbf{T}\mathrm{SLDOM} = \mathrm{SLDOMprod} \qquad (3)$$

$$- \begin{cases} \kappa_1 \mathrm{SLDOM} & \text{if } z > -100\,\mathrm{m} \\ \kappa_2 \mathrm{SLDOM} & \text{if } z < -100\,\mathrm{m} \end{cases},$$

$$\frac{\partial}{\partial t}\mathrm{RDOM} + \mathbf{T}\mathrm{RDOM} = \mathrm{RDOMprod} \qquad (4)$$

$$- \begin{cases} \kappa_4 \mathrm{RDOM} & \text{if } z > -10\,\mathrm{m} \\ \kappa_3 \mathrm{RDOM} & \text{if } z < -10\,\mathrm{m} \end{cases}.$$

The results obtained from the modified DMI-enabled linear DOM model were incorporated into the final DOM optimized (DOM OPT) simulation of the BEC following minor tuning of the κ parameter values.

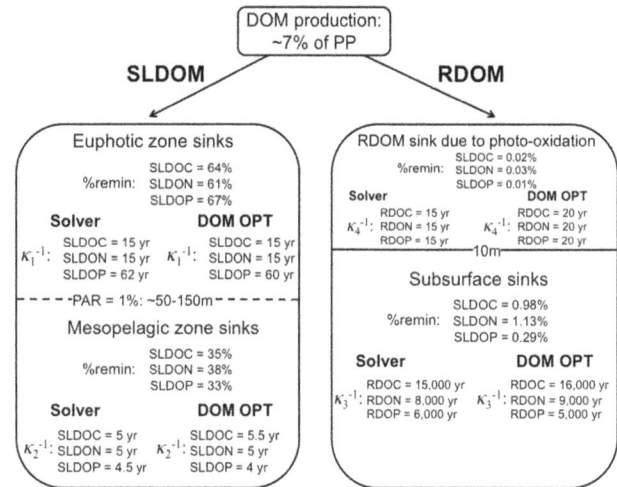

Figure 2. Configuration of the DOM remineralization scheme and parameter values from the modified DMI-enabled DOM model (solver) and the DOM OPT simulation of the CESM BEC. Note the only minor changes to tracer lifetimes, κ_i^{-1}, between the modified DMI-DOM model and the DOM OPT simulation. The value "%remin" represents the percentage of the DOM production flux that is remineralized within each depth horizon on an annual basis and is common to both models.

3 Results

3.1 DOM parameter output from the DMI-enabled linear DOM model

3.1.1 First iteration – DOM source from BEC PP

The objectively optimized DOM parameter values from the solutions to the DMI-enabled linear DOM model (DMI-DOM solver) are shown in Table 1. The fraction of the PP flux that accumulates as DOC, DON, and DOP is $\sim 10\,\%$, with the percentage cycling as refractory DOM: DOCr $= 0.6\,\%$, DONr $= 0.4\,\%$, and DOPr $= 0.15\,\%$. Optimized semilabile DOC exhibited the longest lifetimes with a lifetime of 34 yr in the euphotic zone (EZ) and 2.9 yr in the mesopelagic zone (MZ). Semilabile DON had an intermediate lifetime with respect to DOC and DOP, with an EZ lifetime of 8.7 yr and MZ lifetime of 1.7 yr. Semilabile DOP had the shortest lifetimes, with an EZ lifetime of 5.8 yr and MZ lifetime of 0.8 yr. Optimization of the parameters for the refractory pools yielded lifetimes of 20 000, 9000, and 5000 yr for DOCr, DONr, and DOPr, respectively.

3.1.2 Second iteration of DMI-enabled linear DOM model – DOM source from BEC DOM production flux

Results from the modified DMI-enabled linear DOM model (MOD DMI-DOM solver), which used the BEC DOM production flux from the DOM DEV simulation are shown in

Table 2. DOM production, export, and stoichiometry metrics for the REF and DOM OPT simulations against observational constraints.

Metric	REF	DOM OPT	H09[a]	Metric	REF	DOM OPT	OBS[b]
DOM export 100 m	Tmol (Pg) yr^{-1}	Tmol (Pg) yr^{-1}	Pg yr^{-1}	DOM stoichiometry 100 m			
DOC prod	874 (10.5)	346 (4.16)	3.7	Total pools			
DOC remin	731 (8.78)	157 (1.88)	1.8	C : N	–	15.9	14.0
DOC export	143 (1.72)	189 (2.28)	1.9	N : P	19.4	29.4	40.6
				C : P	–	468.7	580.8
DON prod	120	30.7	–				
DON remin	95.0	14.3	–	Semilabile pools			
DON export	25.0	16.4	–	C : N	7.3	11.9	7.5
				N : P	16.4	18.8	32.2
DOP prod	7.43	2.94	–	C : P	119	223.5	272.7
DOP remin	6.13	1.96	–				
DOP export	1.30	0.98	–				

[a] Hansell et al. (2009) result from a DOC data assimilative biogeochemical/circulation model.
[b] Letscher and Moore (2014) result from analysis of marine DOM database.

Fig. 2 and Table 1. Approximately 7 % of PP is routed to production of DOM, which is divided amongst semilabile (SLDOM) and refractory pools (RDOM). Remineralization lifetimes (κ_i^{-1}) differ for SLDOM depending on location in the water column with longer lifetimes for the euphotic zone (depths where PAR > 1 %) than for the mesopelagic zone. A faster rate of RDOM remineralization is assigned in the surface layer (< 10 m) to parameterize a sink due to photo-oxidation. The parameter, %remin, represents the percentage of the DOM production flux that is remineralized within each depth horizon on an annual basis with the sum equal to 100 % and is diagnosed from the DOM DEV simulation. The relative magnitude of SLDOM remineralization that occurs within the EZ vs. the MZ was found to be ∼ 1.8 : 1 (Fig. 2). Only a small percentage of RDOM remineralization occurs in the surface layer, i.e., 0.01–0.03 % (Fig. 2). The optimal tracer lifetimes from the modified DMI-DOM model were 15 yr for SLDOC in the EZ, 5 yr for SLDOC in MZ, 15 000 yr for RDOC, and 15 yr for RDOC whilst in the surface layer (< 10 m). DON tracer lifetimes were: 15 yr for EZ SLDON, 5 yr for MZ SLDON, 8000 yr for RDON, and 15 yr for RDON at the surface. DOP tracer lifetimes were 62 yr for SLDOP in the EZ, 4.5 yr for MZ SLDOP, 6000 yr for RDOP, and 15 yr for photo-oxidation removal.

3.2 Modeled DOM in the standard CESM BEC v1.2.1 (REF simulation)

A set of metrics were used to assess the performance and improvements to DOM cycling for the CESM BEC simulations including the global integrals of DOM production, export, and C : N : P stoichiometry (Table 2) as well as the mean bias and correlation coefficient (r) of the simulated DOM concentrations against the observational data set in the upper 500 m (Table 3). Results and comparison of DOM cycling metrics from the REF simulation are presented in Tables 2 and 3.

Table 3. DOM mean bias and correlation coefficient in relation to the DOM observations within the upper ocean (0–500 m depth) for the REF and DOM OPT 1° simulations. Observations of semilabile DOM are calculated as the total observed DOM concentration less the asymptotic concentration below 1000 m in each ocean basin.

Metric	REF		DOM OPT	
Total DOM 0–500 m	mean bias	log r	mean bias	log r
DOC	NA	NA	+4 %	0.834
DON	+16 %	0.626	+2 %	0.663
DOP	+32 %	0.362	+7 %	0.439
Semilabile DOM 0–500 m				
DOC	+24 %	0.734	+46 %	0.810
DON	−20 %	0.632	+7 %	0.658
DOP	+4 %	0.388	+4 %	0.431

DOC. Total DOC production in the euphotic zone (upper 100 m) for the REF simulation is 10.5 Pg C yr^{-1} (Table 2). About 85 % of this DOC production is remineralized within the euphotic zone, yielding DOC export from the euphotic zone of ∼ 1.7 Pg C yr^{-1}. Modeled semilabile DOC concentrations from the REF simulation are shown for the surface (Fig. 3a) with observations overlain by the colored dots. The spatial extent of regions with elevated (> 30 μM) semilabile DOC concentrations (i.e., the subtropical gyres) is too large in the REF simulation compared to observations. Large overestimates of simulated DOC are found in the Southern Ocean. Modeled semilabile DOC concentrations for the REF simulation at 200 m are shown in Fig. 3b. Model underestimates (up to ∼ 75 %) are observed in the oxygen-deficient zones in the eastern basins of the equatorial regions. Note that CESM v1.2.1 lacks a DOCr tracer so that simulated DOC is for the semilabile pool only (here we have subtracted the observed deep ocean DOC concentration for each basin from the DOC observations).

DOC

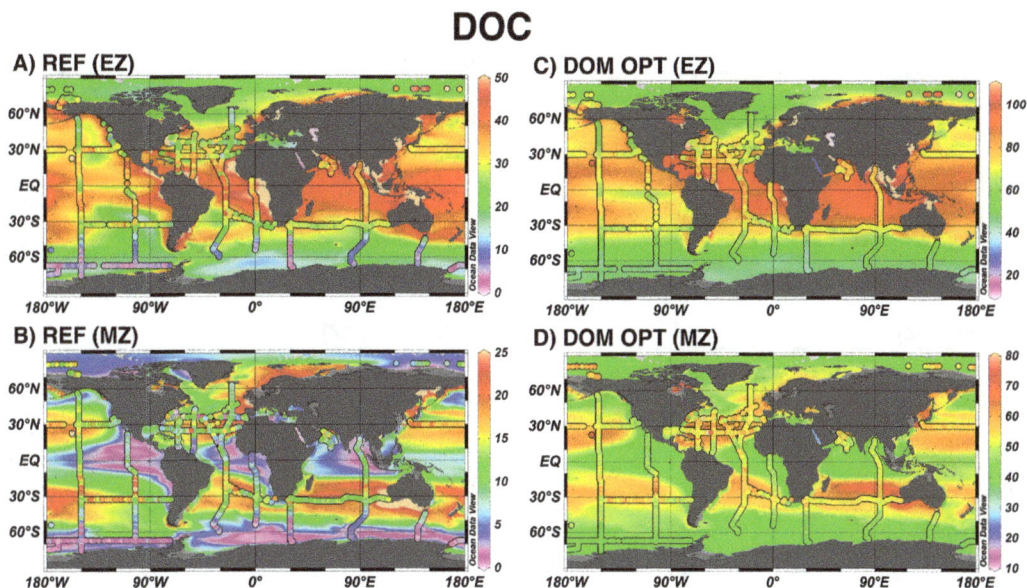

Figure 3. Plots of simulated semilabile [DOC] (μM; colored contours) with observations (colored dots) for the REF simulation at **(a)** the surface (EZ) and **(b)** 200 m (MZ). Total [DOC] (μM; semilabile + refractory) for the DOM OPT simulation is shown for **(c)** the surface (EZ) and **(d)** 200 m (MZ). Note the difference in color scales between plots **(a)** and **(c)**; **(b)** and **(d)** as the REF simulation lacks a DOCr tracer.

DON. Total euphotic zone DON production is 120 Tmol N yr^{-1} with a > 100 m depth export of 25.0 Tmol N yr^{-1} (Table 2). Modeled total DON concentrations (semilabile + refractory) at the surface are similarly overestimated in the REF simulation (Table 3) by up to 100 % within the subtropical gyres of the Pacific and the eastern South Atlantic oceans (Fig. 4a). Model–observation misfit is better at 200 m (Fig. 4b); however, biases of 15–25 % remain (Table 3) in a number of regions (e.g., central equatorial Pacific and southern Indian oceans).

DOP. Total euphotic zone DOP production is 7.43 Tmol P yr^{-1} with export out of the euphotic zone valued at 1.30 Tmol P yr^{-1} (Table 2). Modeled DOP distributions are shown in Fig. 5a (surface), and 5b (200 m), with observations mostly limited to the Atlantic Ocean. The region of elevated simulated DOP (> 0.25 μM) in the eastern South Atlantic surface waters is located further to the east than is observed (Fig. 5a), possibly owing to the snapshot nature of the observations (collected in January–February) compared to the annually averaged simulation. Modeled DOP in the subsurface agrees reasonably well with the Atlantic observations, except for a ∼ 70 % model overestimate in the South Atlantic subtropical gyre (Fig. 5b).

3.3 Modeled DOM in the DOM OPT simulation

Results and comparison of DOM cycling metrics from the DOM OPT simulation against the observational data set and REF simulation are presented in Tables 2 and 3. For a comparison of the set of DOM cycling parameter values between the REF and DOM OPT simulations, see Table 1.

DOC. Total DOC production in the euphotic zone (upper 100 m) for the DOM OPT simulation is 4.16 Pg C yr^{-1} (Table 2). About 45 % of this DOC is remineralized within the euphotic zone, yielding DOC export from the EZ of 2.28 Pg C yr^{-1}, which is ∼ 20 % larger than the result from a separate DOC data assimilative modeling study (Table 2; Hansell et al., 2009). Combined with the particulate organic carbon export from 100 m in the DOM OPT simulation of 7.01 Pg C yr^{-1}, DOC contributes ∼ 25 % to the total 9.29 Pg C yr^{-1} of export production in the CESM BEC. Modeled total DOC concentrations (semilabile + refractory) from the DOM OPT simulation are shown for the surface (Fig. 3c) and at 200 m (Fig. 3d). There is generally good agreement between the simulated fields and observations (colored dots) with the mean bias being < 20 % for the upper ocean (0–500 m; Table 3). Slightly larger model overestimations (up to ∼ 30 %) exist at the surface for certain low-latitude ocean basins (e.g., tropical Atlantic, and Indian oceans).

DON. Total euphotic zone DON production is 30.7 Tmol N yr^{-1} with a > 100 m depth export of 16.4 Tmol N yr^{-1} (Table 2). Modeled total DON concentrations are improved over the REF simulation at 200 m (Fig. 4d); however, overestimations of DON at the surface remain in DOM OPT (Fig. 4c). Simulated surface DON overestimation is largest in the low to mid latitudes, reaching ∼ 30 %. Opposite the pattern obtained for the low latitudes, high-latitude simulated DON is underestimated at the surface in the Southern Ocean (Fig. 4c) by up to ∼ 35 %. However, overall, DON mean biases are small in the DOM OPT simulation, i.e., < 10 % (Table 3).

DON

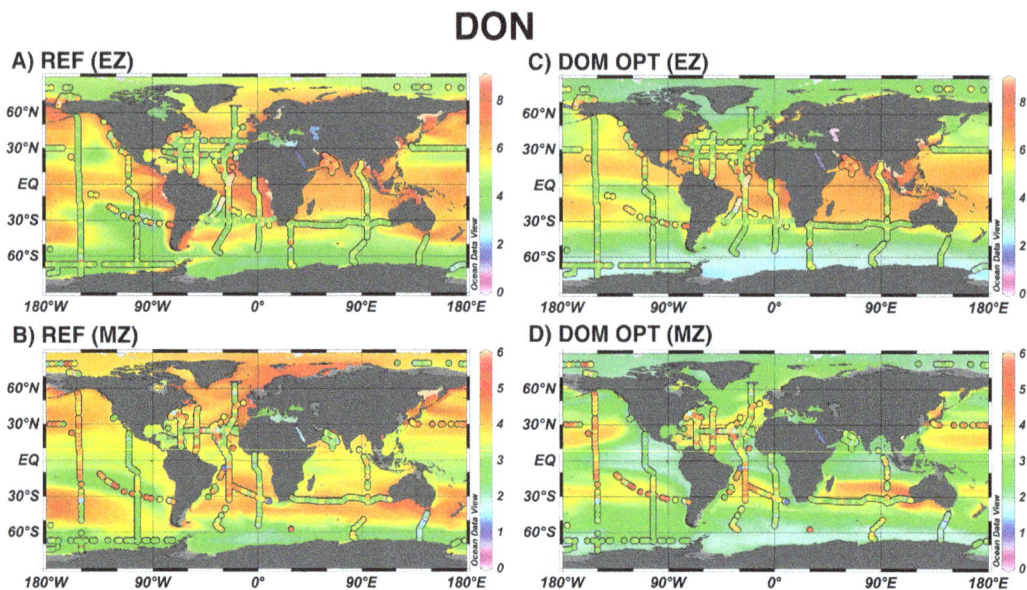

Figure 4. Plots of simulated total [DON] (µM; colored contours) with observations (colored dots) for the REF simulation at (**a**) the surface (EZ) and (**b**) 200 m (MZ), and for the DOM OPT simulation at (**c**) the surface (EZ) and (**d**) 200 m (MZ).

DOP

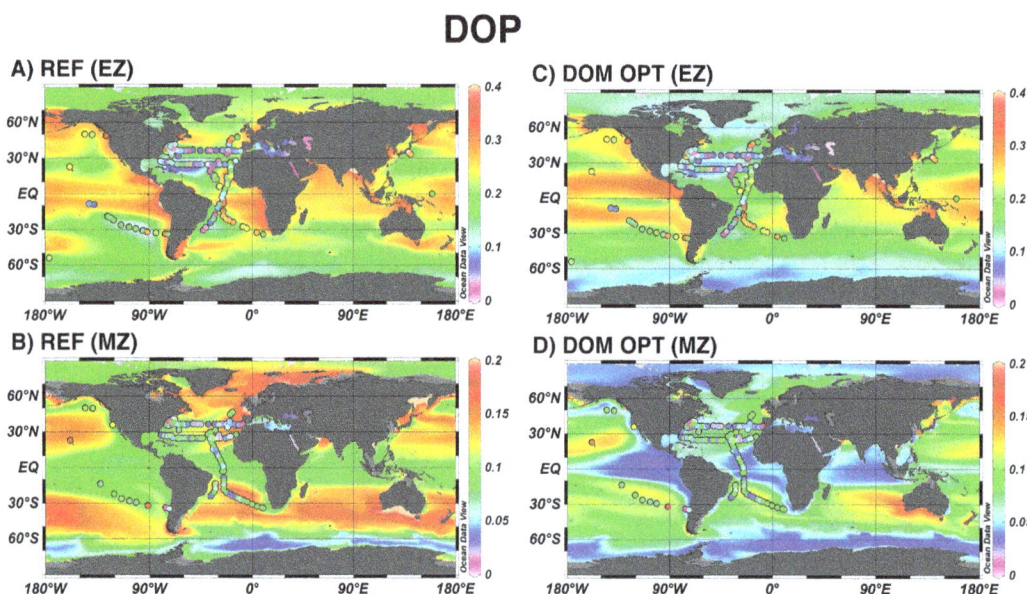

Figure 5. Plots of simulated total [DOP] (µM; colored contours) with observations (colored dots) for the REF simulation at (**a**) the surface (EZ) and (**b**) 200 m (MZ), and for the DOM OPT simulation at (**c**) the surface (EZ) and (**d**) 200 m (MZ).

DOP. Total euphotic zone DOP production is 2.94 Tmol P yr^{-1} with export out of the euphotic zone of ~ 1 Tmol P yr^{-1} (Table 2). Modeled DOP distributions are shown in Fig. 5c (surface), and 5d (200 m). The DOM OPT simulation captures the low DOP concentrations observed in the North Atlantic, largely due to enhanced phytoplankton direct uptake of DOP (see Sect. 3.5). The region of elevated simulated DOP (> 0.25 µM) in the eastern South Atlantic surface waters continues to be located further

to the east than is observed (Fig. 5c) in the DOM OPT simulation as was also the case in the REF simulation. Modeled DOP in the subsurface agrees reasonably well with the Atlantic observations, reducing the large overestimates in the REF simulation (Fig. 5d, b). Overall, mean DOP biases are similarly < 10 % for both the total and semilabile pools (Table 3).

Table 4. DOM mean bias and correlation coefficient in relation to the DOM observations within the upper ocean (0–500 m depth) for the REF, DOM OPT, REDFIELD, and EZRAPID $\sim 3°$ simulations. Observations of semilabile DOM are calculated as the total observed DOM concentration less the asymptotic concentration below 1000 m in each ocean basin.

Metric	REF		DOM OPT		REDFIELD		EZRAPID	
Total DOM 0–500 m	mean bias	$\log r$	mean bias	$\log r$	mean bias	$\log r$	mean bias	$\log r$
DOC	NA	NA	+3 %	0.772	−1 %	0.752	+15 %	0.764
DON	+6 %	0.556	0 %	0.622	+24 %	0.609	+16 %	0.601
DOP	+13 %	0.394	−16 %	0.383	+121 %	0.440	+7 %	0.300
Semilabile DOM 0–500 m								
DOC	+3 %	0.770	+30 %	0.800	+18 %	0.784	+71 %	0.799
DON	−33 %	0.648	−4 %	0.617	+36 %	0.605	+25 %	0.609
DOP	−10 %	0.418	−25 %	0.380	+136 %	0.441	+5 %	0.298

3.4 Comparison of multiple DOM cycling schemes in the CESM BEC

We have also tested other hypotheses for DOM cycling formulations such as non-variable C : N : P cycling stoichiometry (i.e., DOM cycling occurs at the Redfield ratio) as well as more rapid turnover of DOM in the EZ compared to the MZ (the DOM OPT simulation contains more rapid turnover of DOM in the MZ, following the work of Carlson et al., 2004; Letscher et al., 2013a). To test these hypotheses, we performed two additional BEC simulations termed REDFIELD and EZRAPID using a coarser-resolution version of the BEC model with a nominal 3° horizontal resolution. The optimized cycling parameter values obtained for DOC from the DOM OPT simulation were assigned to the DON and DOP pools for the REDFIELD simulation to allow all DOM (C : N : P) to cycle at the same rate and in the same proportions. The ability for phytoplankton to directly utilize DOP is also turned off in the REDFIELD simulation. The optimized EZ and MZ lifetimes for each DOM tracer from the DOM OPT simulation were reversed for the EZRAPID simulation such that the shorter lifetime (more rapid remineralization rate) was assigned to SLDOM in the EZ.

Results from 310 yr simulations of these are compared against $\sim 3°$ simulations of REF and DOM OPT in Table 4. Results are similar for DOC when comparing the DOM OPT and REDFIELD simulations, which is to be expected as the REDFIELD simulation used the same DOC cycling parameters as the DOM OPT simulation. Faster turnover of DOC in the EZ (EZRAPID simulations) had a detrimental effect on DOC mean biases, resulting in large overestimations in the upper 500 m (Table 4) when compared with faster turnover in the MZ (DOM OPT). Large positive mean biases were also found for DON within the REDFIELD and EZRAPID simulations when compared to the DOM OPT (Table 4). Similar positive biases were found for DOP within the REDFIELD and especially for the EZRAPID simulations, i.e., up to ~ 135 % (Table 4).

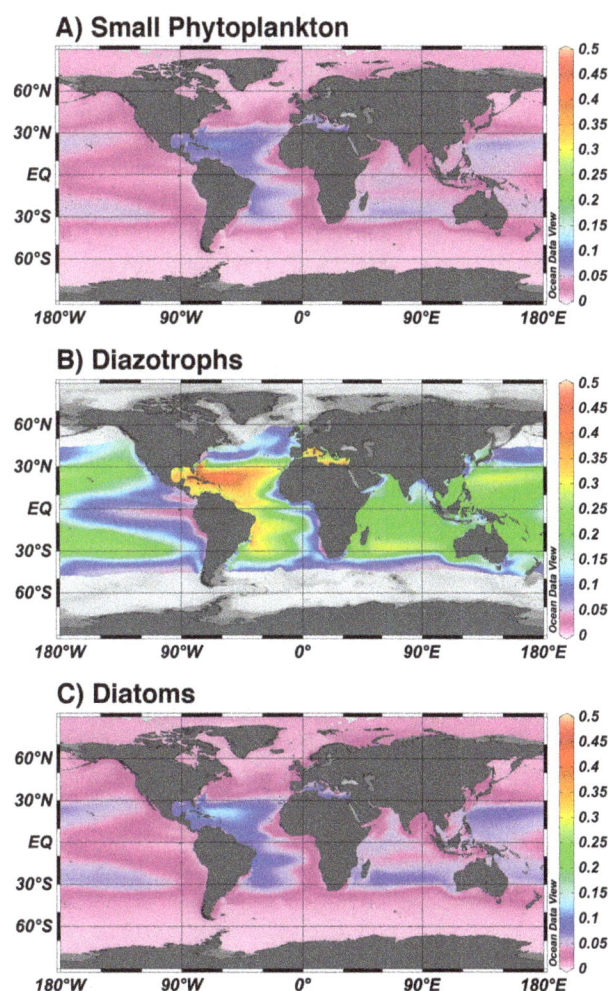

Figure 6. Fraction of total P uptake from DOP integrated over the euphotic zone (upper 100 m) for **(a)** small phytoplankton, **(b)** diazotrophs, and **(c)** diatoms in the DOM OPT simulation.

3.5 Direct DOP uptake by phytoplankton

The longer lifetimes for semilabile DOP in the DOM OPT simulation (on the order of years) allow for significant horizontal advection of DOP from the more productive gyre margins (e.g., the NW African upwelling region) towards the Sargasso Sea, providing an additional phosphorus source to the western North Atlantic. Each phytoplankton group within the BEC model can directly utilize DOP to satisfy their phosphorus requirements when phosphate concentrations are low (Moore et al., 2014). Literature reports of this phenomenon are numerous (e.g., Bjorkman and Karl, 2003; Casey et al., 2009; Lomas et al., 2010; Orchard et al., 2010) whereby phytoplankton make use of extracellular alkaline phosphatases to cleave phosphate groups from DOP moieties such as phosphate mono- and diesters (Dyhrman and Ruttenberg, 2006; Sato et al., 2013) for subsequent uptake of the liberated phosphorus. Sohm and Capone (2006) provide half-saturation constants for DOP uptake by *Trichodesmium* spp. (a diazotroph) and bulk phytoplankton (dominated by nano- and picophytoplankton) from the subtropical North Atlantic, and suggested *Trichodesmium* species obtained much of their required phosphorus from DOP in this region. Based partly on this study, the diazotrophs have been given a lower half-saturation constant for DOP uptake than the other phytoplankton (Moore et al., 2014). Diatoms also exhibit alkaline phosphatase activity albeit at lower rates than other plankton groups (Dyhrman and Ruttenberg, 2006; Nicholson et al., 2006), and were thus assigned a greater half-saturation for DOP uptake than the other phytoplankton groups in the BEC (consistent with their reduced efficiency in taking up dissolved inorganic phosphorus in the model).

The fraction of total phosphorus uptake that is sustained by DOP uptake for each phytoplankton group in the DOM OPT simulation is shown in Fig. 6. DOP uptake is largest by diazotrophs (Fig. 6b), with generally $\sim 20\%$ of P uptake from DOP in the subtropical gyres, increasing to ~ 30–50% in the subtropical North Atlantic, western side of the subtropical South Atlantic, and the eastern Mediterranean Sea. DOP uptake represents a small fraction ($< 5\%$) of P uptake by the small phytoplankton and diatoms (Fig. 6a, c) over much of the ocean, increasing to $\sim 10\%$ in the subtropical ocean gyres.

4 Discussion and summary

This study utilized a rapid solver of a simple linear biogeochemical cycling ocean model, constrained by our compilation of marine DOM observations, to efficiently optimize DOM biogeochemistry in the larger complexity CESM BEC model. This approach allows for a quicker and more quantitatively robust method for optimizing biogeochemical ocean model parameters over the traditional "hand"-tuning approach. Model parameters determined with the modified

DMI-enabled linear DOM model carried over well when implemented in the full CESM BEC (see Fig. 2). The DOM OPT simulation contains reduced mean biases, improved correlation coefficients, and is more consistent with the DOM observational constraints when compared to the REF simulation (Figs. 3, 4, 5; Tables 2, 3).

Our results demonstrate that allowing for non-Redfield stoichiometry in the DOM pools significantly improves the match to observed DOM distributions. The order of lability follows $P > N > C$, diagnosed from the calculated effective tracer lifetimes in DOM OPT which include the net result of the sum of tracer sinks (~ 3.2 vs. 6.3 vs. 6.8 yr for semilabile P, N, C; ~ 4300 vs. 6360 vs. $13\,900$ yr for refractory P, N, and C). The exact values of the DOM lifetimes determined in our study are dependent on the underlying ocean circulation model used; however, Hansell et al. (2012) determined similar values for refractory DOC (16 000 yr) and combined semilabile and semirefractory DOC (~ 7 yr; estimated from their Fig. 5) while using a distinct ocean circulation model than the one employed in the current study. In addition, the DOM lifetimes from the DOM OPT simulation are in general agreement with available estimates from the literature. Semilabile DOC lifetime has been estimated at ~ 1–13 yr in the mesopelagic zone of the Sargasso Sea (Hansell and Carlson, 2001), and ~ 7–22 yr in the mesopelagic zone of the North Pacific subtropical gyre (Abell et al., 2000). Semilabile DON lifetimes have been estimated at ~ 3–12 yr (Letscher et al., 2013a) or ~ 11–20 yr (Abell et al., 2000) for marine DON and ~ 4–14 yr for terrigenous-derived DON in the Arctic Ocean (Letscher et al., 2013b).

Our DOM OPT simulation-estimated export $C : N : P$ ratio of $225 : 19 : 1$ for the semilabile DOM is in excellent agreement with the estimate of $199 : 20 : 1$ by Hopkinson and Vallino (2005) and strongly supports the idea that DOC is exported efficiently relative to DOP compared with the canonical Redfield ratio. The calculated export efficiencies, that is the fraction of euphotic zone DOM production that is exported below 100 m, are 55, 53, and 17.5 % for DOC, DON, and DOP, respectively.

We found that best agreement with observed DOM distributions required a more rapid degradation of semilabile DOM in the mesopelagic than in the euphotic zone. This result is consistent with some incubation studies of DOM degradation (Carlson et al., 2004; Letscher et al., 2013a). Possible hypotheses for this depth dependence on DOM lifetimes in the real ocean are numerous, including differences in DOM composition/quality (Skoog and Benner, 1997; Aluwihare et al., 2005; Goldberg et al., 2011), microbial community structure (Giovannoni et al., 1996; Delong et al., 2006; Treusch et al., 2009; Carlson et al., 2004, 2009; Morris et al., 2012), availability of inorganic nutrients for heterotrophic utilization (Cotner et al., 1997; Rivkin and Anderson, 1997; Caron et al., 2000), abundance of bacterial grazers (Caron et al., 2000), and the presence or specific affinity of microbial cell membrane nutrient transporters (Azam and Malfatti,

2007; Morris et al., 2010). However, the relative importance of each of these mechanisms is not well constrained, nor are any considered in the BEC model formulation, and thus require further investigation.

Direct uptake of DOP by phytoplankton seemed necessary in our simulations to capture the observed very low surface DOP concentrations in the Sargasso Sea. Yet there are large uncertainties in the preference and uptake efficiencies for dissolved inorganic phosphorus versus dissolved organic phosphorus by different phytoplankton groups. Future field and lab studies are needed to reduce these uncertainties and to better quantify the role of DOP in determining spatial patterns of nitrogen fixation. There is also a great need for additional DOP measurements in every basin except the North Atlantic, along with improved quality control and the development of a DOP standard reference material.

Acknowledgements. The authors thank Ann Bardin and Keith Lindsay for their contribution to the development of the CESM POP2 offline transport matrix that was used for the DMI solver. We would also like to thank the British Oceanographic Data Centre and all of the principal investigators involved with production and sharing of DOM data from the Atlantic including Claire Mahaffey, Sinhue Torres-Valdés, Xi Pan, Malcolm Woodward, Rhiannon Mather, Angela Landolfi, and Richard Sanders. This work was supported by grant ER65358 from the U.S. Department of Energy Office of Biological and Environmental Research to F. Primeau and J. K. Moore. J. K. Moore also acknowledges support from the National Science Foundation grants AGS-1021776 and AGS-1048890.

Edited by: K. Thonicke

References

Abell, J., Emerson, S., and Renaud, P.: Distributions of TOP, TON, and TOC in the North Pacific subtropical gyre: Implications for nutrient supply in the surface ocean and remineralization in the upper thermocline, J. Mar. Res., 58, 203–222, 2000.

Aluwihare, L. I., Repeta, D. J., Pantoja, S., and Johnson, C. G.: Two chemically distinct pools of organic nitrogen accumulate in the ocean, Science, 308, 1007–1010, 2005.

Amestoy, P. R., Duff, I. S., Koster, J., and L'Excellent, J.-Y.: A fully asynchronous multifrontal solver using distributed dynamic scheduling, SIAM Journal of Matrix Analysis and Applications, 23, 15–41, 2001.

Amestoy, P. R., Guermouche, A., L'Excellent, J.-Y., and Pralet, S.: Hybrid scheduling for the parallel solution of linear systems, Parallel Computing, 32, 136–156, 2006.

Aminot, A. and Kérouel, R.: Dissolved organic carbon, nitrogen and phosphorus in the NE Atlantic and the NW Mediterranean with particular reference to non-refractory fractions and degradation, Deep Sea Res. I, 51, 1975–1999, 2004.

Anderson, L. A. and Sarmiento, J. L.: Redfield ratios of remineralization determined by nutrient data analysis, Global Biogeochem. Cy., 8, 65–80, 1994.

Azam, F. and Malfatti, F.: Microbial structuring of marine ecosystems, Nature Rev. Microbiol., 5, 782–791, 2007.

Bardin, A., Primeau, F., and Lindsay, K.: An offline implicit solver for simulating prebomb radiocarbon, Ocean Model., 73, 45–58, 2014.

Bjorkman, K. M. and Karl, D. M.: Bioavailability of dissolved organic phosphorus in the euphotic zone at Station ALOHA, North Pacific Subtropical Gyre, Limnol. Oceanogr., 48, 1049–1057, 2003.

Carlson, C. A.: Production and removal processes, in: Biogeochemistry of marine dissolved organic matter, edited by: Hansell, D. A. and Carlson, C. A., Academic Press, San Diego, 91–151, 2002.

Carlson, C. A., Giovannoni, S. J., Hansell, D. A., Goldberg, S. J., Parsons, R., and Vergin, K.: Interactions between DOC, microbial processes, and community structure in the mesopelagic zone of the northwestern Sargasso Sea, Limnol. Oceanogr., 49, 1073–1083, 2004.

Carlson, C. A., Morris, R., Parsons, R., Treusch, A. H., Giovannoni, S. J., and Vergin, K. (2008): Seasonal dynamics of SAR11 populations in the euphotic and mesopelagic zones of the northwestern Sargasso Sea, The ISME Journal, 3, 283–295, 2009.

Caron, D. A., Lim, E. L., Sanders, R. W., Dennett, M. R., and Berninger, U. G.: Responses of bacterioplankton and phytoplankton to organic carbon and inorganic nutrient additions in contrasting oceanic ecosystems, Aquat. Microbial. Ecol., 22, 175–184, 2000.

Casey, J. R., Lomas, M. W., Michelou, V. K., Dyhrman, S. T., Orchard, E. D., Ammerman, J. W., and Sylvan, J. B.: Phytoplankton taxon-specific orthophosphate (Pi) and ATP utilization in the western subtropical North Atlantic, Aquat. Microbial. Ecol., 58, 31, 31–44, 2009.

Cotner, J. B., Ammerman, J. W., Peele, E. R., and Bentzen, E.: Phosphorus-limited bacterioplankton growth in the Sargasso Sea, Aquat. Microbial. Ecol., 13, 141–149, 1997.

Danabasoglu, G., Bates, S. C., Briegleb, B. P., Jayne, S. R., Jochum, M., Large, W. G., and Yeager, S. G.: The CCSM4 ocean component, J. Climate, 25, 1361–1389, 2011.

DeLong, E. F., Preston, C. M., Mincer, T., Rich, V., Hallam, S. J., Frigaard, N. U., and Karl, D. M.: Community genomics among stratified microbial assemblages in the ocean's interior, Science, 311, 496–503, 2006.

Dunne, J. P., John, J. G., Shevliakova, E., Stouffer, R. J., Krasting, J. P., Malyshev, S. L., Milly, P. C. D., Sentman, L. T., Adcroft, A. J., Cooke, W., Dunne, K. A., Griffies, S. M., Hallberg, R. W., Harrison, M. J., Levy, H., Wittenberg, A. T., Phillips, P. J., and Zadeh, N.: GFDL's ESM2 global coupled climate-carbon Earth System Models Part II: Carbon system formulation and baseline simulation characteristics, J. Climate, 26, 2247–2267, 2013.

Dyhrman, S. T. and Ruttenberg, K. C.: Presence and regulation of alkaline phosphatase activity in eukaryotic phytoplankton from the coastal ocean: Implications for dissolved organic phosphorus remineralization, Limnol. Oceanogr., 51, 1381–1390, 2006.

Gent, P. R., Danabasoglu, G., Donner, L. J., Holland, M. M., Hunke, E. C., Jayne, S. R., and Zhang, M.: The community climate system model version 4, J. Climate, 24, 4973–4991, 2011.

Giovannoni, S. J., Rappé, M. S., Vergin, K. L., and Adair, N. L.: 16S rRNA genes reveal stratified open ocean bacterioplankton populations related to the green non-sulfur bacteria, P. Natl. Acad. Sci., 93, 7979–7984, 1996.

Goldberg, S. J., Carlson, C. A., Brzezinski, M., Nelson, N. B., and Siegel, D. A.: Systematic removal of neutral sugars within dissolved organic matter across ocean basins, Geophys. Res. Lett., 38, L17606, doi:10.1029/2011GL048620, 2011.

Hansell, D. A.: Recalcitrant dissolved organic carbon fractions, Ann. Rev. Mar. Sci., 5, 421–445, 2013.

Hansell, D. A. and Carlson, C. A.: Biogeochemistry of total organic carbon and nitrogen in the Sargasso Sea: control by convective overturn, Deep Sea Res. II, 48, 1649–1667, 2001.

Hansell, D. A., Carlson, C. A., Repeta, D. J., and Schlitzer, R.: Dissolved organic matter in the ocean: A controversy stimulates new insights, Oceanography, 22, 202–211, 2009.

Hansell, D. A., Carlson, C. A., and Schlitzer, R.: Net removal of major marine dissolved organic carbon fractions in the subsurface ocean, Global Biogeochem. Cy., 26, GB1016, doi:10.1029/2011GB004069, 2012.

Hopkinson, C. S. and Vallino, J. J.: Efficient export of carbon to the deep ocean through dissolved organic matter, Nature, 433, 142–145, 2005.

Khatiwala, S., Visbeck, M., and Cane, M. A.: Accelerated simulation of passive tracers in ocean circulation models, Ocean Modell., 9, 51–69, 2005.

Kim, J. M., Lee, K., Shin, K., Yang, E. J., Engel, A., Karl, D. M., and Kim, H. C.: Shifts in biogenic carbon flow from particulate to dissolved forms under high carbon dioxide and warm ocean conditions, Geophys. Res. Lett., 38, L08612, doi:10.1029/2011GL047346, 2011.

Kwon, E. Y. and Primeau, F.: Optimization and sensitivity study of a biogeochemistry model using an implicit solver and in situ phosphate data, Global Biogeochem. Cy., 20, GB4009, doi:10.1029/2005GB002631, 2006.

Letscher, R. T. and Moore, J. K.: Preferential remineralization of dissolved organic phosphorus and non-Redfield DOM dynamics in the global ocean, Global Biogeochem. Cy., under review, 2014.

Letscher, R. T., Hansell, D. A., Carlson, C. A., Lumpkin, R., and Knapp, A. N.: Dissolved organic nitrogen in the global surface ocean: Distribution and fate, Global Biogeochem. Cy., 27, 141–153, 2013a.

Letscher, R. T., Hansell, D. A., Kadko, D., and Bates, N. R.: Dissolved organic nitrogen dynamics in the Arctic Ocean, Mar. Chem., 148, 1–9, 2013b.

Lomas, M. W., Burke, A. L., Lomas, D. A., Bell, D. W., Shen, C., Dyhrman, S. T., and Ammerman, J. W.: Sargasso Sea phosphorus biogeochemistry: an important role for dissolved organic phosphorus (DOP), Biogeosciences, 7, 695–710, doi:10.5194/bg-7-695-2010, 2010.

Martiny, A. C., Pham, C. T. A., Primeau, F. W., Vrugt, J. A., Moore, J. K., Levin, S. A., and Lomas, M. W.: Strong latitudinal patterns in the elemental ratios of marine plankton and organic matter, Nature Geosci., 6, 279–283, 2013a.

Martiny, A. C., Vrugt, J. A., Primeau, F. W., and Lomas, M. W.: Regional variation in the particulate organic carbon to nitrogen ratio in the surface ocean, Global Biogeochem. Cy., 27, 723–731, 2013b.

Moore, J. K., Doney, S. C., and Lindsay, K.: Upper ocean ecosystem dynamics and iron cycling in a global three-dimensional model, Global Biogeochem. Cy., 18, GB4028, doi:10.1029/2004GB002220, 2004.

Moore, J. K., Lindsay, K., Doney, S. C., Long, M. C., and Misumi, K.: Ecosystem Dynamics and biogeochemical cycling in the Community Earth System Model [CESM1(BGC)]: Comparison of the 1990s with the 2090s under the RCP4.5 and RCP8.5 Scenarios, J. Climate, 26, 9291–9312, 2013.

Moore, J. K., Lindsay, K., Letscher, R. T., and Mayorga, E.: Improving ocean biogeochemistry simulation in the Community Earth System Model, Geosci. Model Dev. Discuss., in preparation, 2014.

Morris, R. M., Nunn, B. L., Frazar, C., Goodlett, D. R., Ting, Y. S., and Rocap, G.: Comparative metaproteomics reveals ocean-scale shifts in microbial nutrient utilization and energy transduction, The ISME Journal, 4, 673–685, 2010.

Morris, R. M., Frazar, C. D., and Carlson, C. A.: Basin-scale patterns in the abundance of SAR11 subclades, marine Actinobacteria (OM1), members of the Roseobacter clade and OCS116 in the South Atlantic, Environ. Microbiol., 14, 1133–1144, 2012.

Nicholson, D., Dyhrman, S., Chavez, F., and Paytan, A.: Alkaline phosphatase activity in the phytoplankton communities of Monterey Bay and San Francisco Bay, Limnol. Oceanogr., 51, 874–883, 2006.

Orchard, E. D., Ammerman, J. W., Lomas, M. W., and Dyhrman, S. T.: Dissolved inorganic and organic phosphorus uptake in Trichodesmium and the microbial community: The importance of phosphorus ester in the Sargasso Sea, Limnol. Oceanogr., 55, 1390–1399, 2010.

Redfield, A. C.: The biological control of chemical factors in the environment, American Scientist, 46, 205–221, 1958.

Redfield, A. C., Ketchum, B. H., and Richards, F. A.: The influence of organisms on the composition of seawater, in: The sea: ideas and observations on progress in the study of the seas, Wiley, 2, 1963.

Rivkin, R. B. and Anderson, M. R.: Inorganic nutrient limitation of oceanic bacterioplankton, Limnol. Oceanogr., 730–740, 1997.

Sato, M., Sakuraba, R., and Hashihama, F.: Phosphate monoesterase and diesterase activities in the North and South Pacific Ocean, Biogeosciences, 10, 7677–7688, doi:10.5194/bg-10-7677-2013, 2013.

Skoog, A. and Benner, R.: Aldoses in various size fractions of marine organic matter: Implications for carbon cycling, Limnol. Oceanogr., 42, 1803–1813, 1997.

Smith, R. D., Jones, P. W., Briegleb, B., Bryan, F., Danabasoglu, G., Dennis, J., and Yeager, S.: The parallel ocean program (POP) reference manual: ocean component of the community climate system model (CCSM), Los Alamos National Laboratory, LAUR-10-01853, 2010.

Sohm, J. A. and Capone, D. G.: Phosphorus dynamics of the tropical and subtropical north Atlantic: Trichodesmium spp. versus bulk plankton, Mar. Ecol. Prog. Ser., 317, 21, 21–28, 2006.

Treusch, A. H., Vergin, K. L., Finlay, L. A., Donatz, M. G., Burton, R. M., Carlson, C. A., and Giovannoni, S. J.: Seasonality and vertical structure of microbial communities in an ocean gyre, The ISME Journal, 3, 1148–1163, 2009.

Vichi, M., Pinardi, N., and Masina, S.: A generalized model of pelagic biogeochemistry for the global ocean ecosystem, Part I: Theory, J. Mar. Syst., 64, 89–109, 2007.

Wohlers, J., Engel, A., Zöllner, E., Breithaupt, P., Jürgens, K., Hoppe, H. G., and Riebesell, U.: Changes in biogenic carbon flow in response to sea surface warming, P. Natl. Acad. Sci., 106, 7067–7072, 2009.

12

Long-term spatial and temporal variation of CO_2 partial pressure in the Yellow River, China

L. Ran[1], X. X. Lu[1,2], J. E. Richey[3], H. Sun[4], J. Han[4], R. Yu[2], S. Liao[5], and Q. Yi[6]

[1]Department of Geography, National University of Singapore, 117570, Singapore
[2]College of Environment & Resources, Inner Mongolia University, Hohhot, 010021, China
[3]School of Oceanography, University of Washington, Box 355351, Seattle, WA 98195-5351, USA
[4]Institute of Geology and Geophysics, Chinese Academy of Sciences, Beijing, 100029, China
[5]Tongguan Hydrographic Station, Yellow River Conservancy Commission, Tongguan, 714399, China
[6]Toudaoguai Hydrographic Station, Yellow River Conservancy Commission, Baotou, 014014, China

Correspondence to: X. X. Lu (geoluxx@nus.edu.sg)

Abstract. Carbon transport in river systems is an important component of the global carbon cycle. Most rivers of the world act as atmospheric CO_2 sources due to high riverine CO_2 partial pressure (pCO_2). By determining the pCO_2 from alkalinity and pH, we investigated its spatial and temporal variation in the Yellow River watershed using historical water chemistry records (1950s–1984) and recent sampling along the mainstem (2011–2012). Except the headwater region where the pCO_2 was lower than the atmospheric equilibrium (i.e. $380\,\mu$atm), river waters in the remaining watershed were supersaturated with CO_2. The average pCO_2 for the watershed was estimated at $2810 \pm 1985\,\mu$atm, which is 7-fold the atmospheric equilibrium. As a result of severe soil erosion and dry climate, waters from the Loess Plateau in the middle reaches had higher pCO_2 than that from the upper and lower reaches. From a seasonal perspective, the pCO_2 varied from about $200\,\mu$atm to $> 30\,000\,\mu$atm with higher pCO_2 usually occurring in the dry season and lower pCO_2 in the wet season (at 73 % of the sampling sites), suggesting the dilution effect of water. While the pCO_2 responded exponentially to total suspended solids (TSS) export when the TSS concentration was less than $100\,\mathrm{kg\,m^{-3}}$, it decreased slightly and remained stable if the TSS concentration exceeded $100\,\mathrm{kg\,m^{-3}}$. This stable pCO_2 is largely due to gully erosion that mobilizes subsoils characterized by low organic carbon for decomposition. In addition, human activities have changed the pCO_2 dynamics. Particularly, flow regulation by dams can diversely affect the temporal changes of pCO_2, depending on the physiochemical properties of the regulated waters and adopted operation scheme. Given the high pCO_2 in the Yellow River waters, large potential for CO_2 evasion is expected and warrants further investigation.

1 Introduction

Rivers play a crucial role in the global carbon cycle, because they can modulate the carbon dynamics not only of the watersheds but also of the coastal systems into which river waters are discharged (Aufdenkampe et al., 2011). Fluvial carbon export represents an important pathway linking land and the ocean. Approximately 0.9 Gt of carbon is delivered into the oceans per year via inland waters (Cole et al., 2007; Battin et al., 2009). However, rivers are not merely passive conduits. Evidence is accruing to indicate that, while only a small portion of carbon that enters a river network finally reaches the ocean, a considerable fraction would be buried within the river network or returned to the atmosphere en route (Yao et al., 2007; Wallin et al., 2013). Consequently, rivers are viewed as sources of atmospheric carbon dioxide (CO_2) (Cole et al., 2007; Butman and Raymond, 2011). Recent estimates show that global inland waters can transfer $0.75–2.1\,\mathrm{Gt\,C\,yr^{-1}}$ into the atmosphere (Cole et al., 2007; Tranvik et al., 2009; Raymond et al., 2013). Comparative studies associated with lateral carbon fluxes have highlighted the significance of CO_2 evasion in assessing global carbon

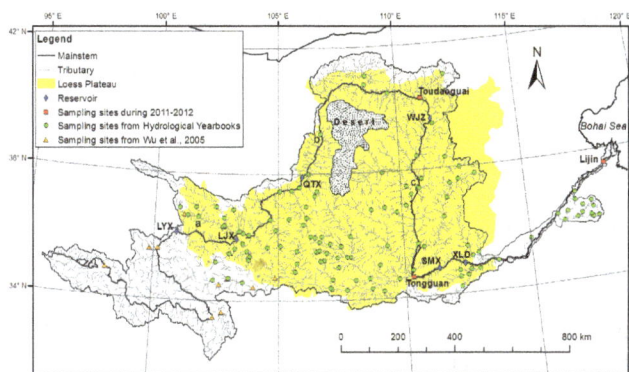

Figure 1. Location map of the sampling sites in the Yellow River watershed. Acronyms for the mainstem dams: LYX – Longyangxia since 1986; LJX – Liujiaxia since 1968; QTX – Qingtongxia since 1968; WJZ – Wanjiazhai since 1998; SMX – Sanmenxia since 1960; and XLD – Xiaolangdi since 2000.

budget (Melack, 2011). For example, Richey et al. (2002) show that CO_2 emission in the Amazon River basin is an order of magnitude greater than fluvial export of organic carbon to the ocean.

Decomposition of terrestrially derived organic carbon and aquatic respiration are the primary sources of riverine CO_2 (Humborg et al., 2010). As an important parameter in estimating CO_2 outgassing, partial pressure of riverine CO_2 (pCO_2) indicates the CO_2 concentration in rivers and the gradient relative to the atmospheric equilibrium (i.e. $380\,\mu atm$). Most rivers of the world have higher pCO_2 than the overlying atmosphere, suggesting a great emission potential (Cole et al., 2007; Striegl et al., 2012). While the riverine pCO_2 of mainstem or estuary waters has been widely recognized, such as the Amazon (Richey et al., 2002), Pearl (Yao et al., 2007), and Columbia (Evans et al., 2013), a holistic assessment concerning a complete river network is rare. This is largely caused by the constraints of time and logistics to conduct spatial sampling covering not only the mainstem but also the lower stream-order tributaries. Indeed, tributaries are physically and biogeochemically more active because they have stronger turbulence and more rapid mixing with the benthic substrate and the atmosphere than the mainstem (Alin et al., 2011; Butman and Raymond, 2011; Benstead and Leigh, 2012). For instance, Aufdenkampe et al. (2011) found that the CO_2 outgassing fluxes from small streams could be 2–3 times higher than from larger rivers. Thus, estimating CO_2 evasion based only on mainstem waters will underestimate the total efflux of a specific river system. Analysing pCO_2 at space- and timescales by high-resolution sampling is a prerequisite for precisely evaluating CO_2 outgassing and its implications for the carbon cycle.

The Yellow River is characterized by high sediment and total dissolved solids (TDS) among the world's large rivers, primarily because of severe soil erosion and intensive chemical weathering and human activity. Its TDS concentration

of $452\,mg\,L^{-1}$ is about four times the world median value (Chen et al., 2005). Based on measurements at hydrological gauges or in specific river reaches, prior studies have investigated its chemical weathering and carbon transport (e.g. Zhang et al., 1995; Wu et al., 2008; Wang et al., 2012; Ran et al., 2013). Soil respiration in terrestrial ecosystems and impact of land use change on carbon storage have also been analysed (Zhao et al., 2008; Li et al., 2010). By contrast, few studies have examined its carbon dynamics in river waters and how riverine pCO_2 has responded to catchment features (Wang et al., 2012; Ran et al., 2013). Using historical records across the watershed during the period 1950s–1984 and recent sampling along the mainstem, we calculated the riverine pCO_2 from alkalinity and pH. This study aimed to investigate the spatial and temporal variation of pCO_2 and its responses to natural and human factors. The results will provide insights into the coupling between soil erosion and riverine pCO_2 and the impact of dam operation on downstream riverine pCO_2 changes.

2 Materials and methods

2.1 The Yellow River

The Yellow River drains $752\,000\,km^2$ of north China, originating in the Tibetan Plateau and flowing eastward into the Bohai Sea (Fig. 1). Located in a semiarid–arid climate, its precipitation is spatially highly variable, decreasing from $700\,mm\,yr^{-1}$ in the southeast to $250\,mm\,yr^{-1}$ in the northwest (Zhao, 1996). Likewise, temperature changes significantly, with the mean temperature in the upper (above Toudaoguai), middle (approximately between Toudaoguai and the Xiaolangdi Dam), and lower (below the Xiaolangdi Dam) reaches being 1–8, 8–14 and 12–14 °C, respectively (Chen et al., 2005). Because the Yellow River basin is in large part surrounded by the Loess Plateau that has typically accumulated huge erodible loess deposits (Fig. 1), it suffers from severe soil erosion. Approximately 1.6 Gt of sediment was transported to the ocean per year prior to the 1970s (Syvitski et al., 2005). For comparison, the mean water discharge was only $49\,km^3\,yr^{-1}$ over the same period (Zhao, 1996).

Both hydrological regime and landscape within the watershed have been greatly altered due to intensive human activity (Ran and Lu, 2012). While the water discharge has dropped to $15\,km^3\,yr^{-1}$ during the recent decade, the sediment flux has decreased to about $0.14\,Gt\,yr^{-1}$ as a result of massive soil conservation and sediment trapping by dams. Among the numerous dams, these constructed on the mainstem channel play fundamental roles in regulating delivery of water, sediment and dissolved solids (Ran and Lu, 2012), especially the joint operation of the Sanmenxia and Xiaolangdi dams since 2000. With about 140 million people currently residing within the watershed, the population density is 180 person km^{-2} (Chen et al., 2005), and it exceeds

Table 1. pH at Luokou station during the period 1980–1984: a comparison of different data sources (arithmetic mean ± standard deviation).

Data source	Year				
	1980	1981	1982	1983	1984
GEMS/Water Programme	8.27 ± 0.11	8.21 ± 0.21	8.10 ± 0.20	8.14 ± 0.10	8.25 ± 0.09
This study	8.11 ± 0.13	8.14 ± 0.07	8.13 ± 0.07	8.10 ± 0.03	8.09 ± 0.05
% of variation	1.93	0.85	−0.37	0.49	1.94

300 person km^{-2} in some agricultural areas in the middle reaches. Consequently, land use has become increasingly both extensive and intensive.

The Yellow River basin was mainly developed on the Sino-Korean Shield with Quaternary loess deposits overlying the vast middle reaches and Archean to Tertiary granites and metamorphic rocks in areas near the basin boundaries and in the lower reaches (Chen et al., 2005). Chemical analyses of loess samples show that feldspar, micas and quartz are the most common detrital minerals with carbonates accounting for 10–20 % (Zhang et al., 1995). Because the loess deposits cover about 46 % of the total drainage area, the river presents high alkalinity and intense rock weathering. With exceptionally high TDS concentration the Yellow River delivers around 11 Mt of dissolved solids per year to the Bohai Sea (Gaillardet et al., 1999).

2.2 Historical records of water chemistry

Historical records of major ions (e.g. Ca^{2+}, Mg^{2+}, Na^+, K^+, Cl^-, HCO_3^- and SO_4^{2-}) measured from a hydrological monitoring network were extracted from the Yellow River Hydrological Yearbooks, which are yearly produced by the Yellow River Conservancy Commission (YRCC). Other variables concurrently measured at each sampling event, including pH, water temperature, water discharge and total suspended solids (TSS), were also retrieved from the yearbooks for this study. The water samples for pH and temperature were taken in the same time period as these for ion analysis. Over the period from the 1950s to 1984, the sampling frequency ranged from 1 to 5 times per month, depending on hydrological regime. Sampling at some stations during the period 1966–1975 were suspended or completely stopped. Post-1984 records are not in the public domain. Given the discontinuity in sampling, only the stations with at least 6 samples in a year were analysed. A total of 129 stations with 15 029 water chemistry measurements were compiled (Fig. 1).

Chemical analyses of the collected water samples were performed under the authority of the YRCC following the standard procedures and methods described by Alekin et al. (1973) and the American Public Health Association (1985). The pH and temperature were measured in field, and total alkalinity (TAlk) was determined using a fixed endpoint titration method. Detailed description of the sampling and analysis procedures can be found in Chen et al. (2002a). The results are summarized in the Supplement (Table S1).

Use of historical records always raises the issue of data reliability. No detailed information on quality assurance and quality control is available in the hydrological reports. Extensive efforts have been made to assess the data quality by analysing the parameter differences measured at the same station but by different agencies. The Luokou station on the lower Yellow River mainstem has been monitored under the United Nations Global Environment Monitoring System (GEMS) Water Programme since 1980 (only yearly means available at http://www.unep.org/gemswater). As pCO$_2$ is considerably sensitive to pH changes (Li et al., 2012), the pH values from the two sources were compared (Table 1), which showed that the data set from the Hydrological Yearbooks agreed well with the GEMS/Water Programme data set with differences of < 2 %. High data quality of the hydrological reports can also be confirmed from the concentration comparison of major ions in the two data sets (see Chen et al., 2005).

Given the data paucity for the upper Yellow River, data collected at 17 sites in the headwater region were retrieved from Wu et al. (2005) (Fig. 1 and Table S1 in the Supplement). They measured pH and temperature along the mainstem and major tributaries and determined the TAlk through Gran titration. Comparison with previous sampling results (Zhang et al., 1995) showed their data agreed well.

2.3 Recent field sampling

From July 2011 to July 2012, weekly sampling on the mainstem was undertaken at Toudaoguai, Tongguan and Lijin stations (Fig. 1). The frequency increased (i.e. daily) when large floods occurred. Water column samples were collected ∼ 0.5 m below the surface water and kept in acid-washed, but carefully neutralized, high-density polyethylene containers. Concomitant determination of pH and water temperature was performed in situ using a Hanna HI9125 pH meter on the NBS scale, which was calibrated prior to each measurement against pH7.01 and pH10.01 buffers. Replicate measurements showed the precision for pH and temperature were ±0.04 units and ±0.1 °C, respectively. The TAlk was determined by titrating 50 mL filtered water through 47 mm Whatman GF/F filters (0.7 µm pore size) with 0.02 M HCl solution within 5 h after sampling. Three parallel titrations showed the analytical error was below 3 %. The parallel alkalinity results were then averaged. In total, 163 samples were collected. Ancillary data, including daily water discharge and TSS, were acquired from the YRCC. Generally, the sampling results at Toudaoguai and Tongguan reflect the TAlk and pCO$_2$ changes on the Loess Plateau, while the Lijin measurements represent seaward export as it is located 110 km upstream of the river mouth and free of tidal influences.

2.4 Calculations of DIC species and pCO$_2$

Total dissolved inorganic carbon (DIC) species in river systems include HCO$_3^-$, CO$_3^{2-}$, H$_2$CO$_3$ and aqueous CO$_2$ (CO$_{2aq}$). Their relative concentration is a function of temperature and pH (Li et al., 2012). DIC species can be determined by Henry's Law, from which the pCO$_2$ can be calculated using the CO2SYS program (Lewis and Wallace, 1998):

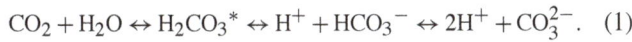

$$CO_2 + H_2O \leftrightarrow H_2CO_3{}^* \leftrightarrow H^+ + HCO_3{}^- \leftrightarrow 2H^+ + CO_3^{2-}. \quad (1)$$

At chemical equilibrium, the activities of the reactants and products are determined from the thermodynamic reaction constants (K_i) that are temperature (T) dependent:

$$K_{CO_2} = [H_2CO_3{}^*]/[pCO_2] \quad (2)$$

$$K_1 = [H^+][HCO_3{}^-]/[H_2CO_3{}^*] \quad (3)$$

$$K_2 = [H^+][CO_3{}^{2-}]/[HCO_3{}^-], \quad (4)$$

where H$_2$CO$_3^*$ is the sum of CO$_{2aq}$ and the true H$_2$CO$_3$. The pK_i values (negative log of K_i) can be calculated by the following equations (Clark and Fritz, 1997):

$$pK_{CO_2} = -7 \times 10^{-5}T^2 + 0.016T + 1.11 \quad (5)$$

$$pK_1 = 1.1 \times 10^{-4}T^2 - 0.012T + 6.58 \quad (6)$$

$$pK_2 = 9 \times 10^{-5}T^2 - 0.0137T + 10.62 \quad (7)$$

Then, the pCO$_2$ can be simply expressed as:

$$pCO_2 = [H_2CO_3{}^*]/K_{CO_2} = [H^+][HCO_3{}^-]/K_{CO_2}K_1. \quad (8)$$

With the pH mostly ranging from 7.4 to 8.6 indicative of natural processes for the Yellow River (Chen et al., 2005), HCO$_3^-$ is considered equivalent to alkalinity because it represents > 96 % of the TAlk. This approach has been frequently used and has demonstrated high pCO$_2$ in Chinese river systems (e.g. Yao et al., 2007; Li et al., 2012). To validate the simplification, we also estimated the pCO$_2$ using the program PHREEQC (Hunt et al., 2011). The pCO$_2$ result derived by PHREEQC are very close to that by CO2SYS with < 3 % differences. However, the calculated pCO$_2$ results may have slightly overestimated the actual values (Cole and Caraco, 1998; Abril et al., 2015).

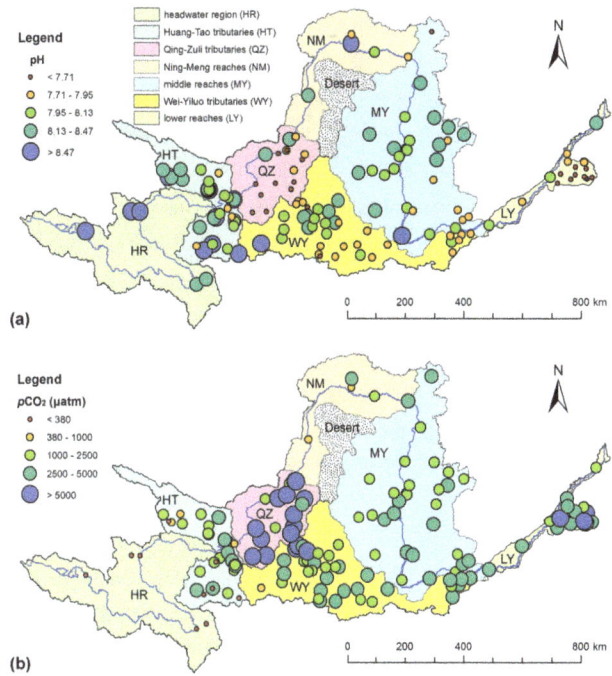

Figure 2. Spatial variations of pH (**a**) and pCO$_2$ (**b**) in the Yellow River watershed.

3 Results

3.1 Characteristics of hydrochemical setting

To better investigate the spatial changes of hydrochemical variables, the watershed was divided into seven sub-basins: the headwater region (HR), the Huang–Tao tributaries (HT), the Qing–Zuli tributaries (QZ), the Ning–Meng reaches (NM), the Wei–Yiluo tributaries (WY), the middle reaches (MY), and the lower reaches (LY) (Fig. 2). The Yellow River waters were characterized by high alkalinity with the pH presenting significant spatial variations (Fig. 2a). While high pH values were mostly observed in the HR sub-basin where the highest was 9.1, relatively low pH (i.e. < 7.71) was recorded at the QZ tributary sites with the lowest being 6.4. For the waters from the Loess Plateau, the pH ranged from 7.71 to 8.47. Towards the river mouth, it showed a downward trend in the lower reaches (LY). With one exception at Lijin (Fig. 1), the pH values were all below 8.13 and even below 7 at some tributary sites. In addition to spatial variations, it showed considerable seasonal changes. As exemplified in Fig. 3a, the waters were generally more alkaline in the dry season (October–May) than in the wet season (June–September).

Similarly, with a range of 855–8633 μmol L^{-1}, the TAlk presented complex spatial variability throughout the watershed. While the HR and LY sub-basins showed the lowest TAlk (< 2600 μmol L^{-1}), the sub-basins on the Loess Plateau had considerably high alkalinity with a mean TAlk of

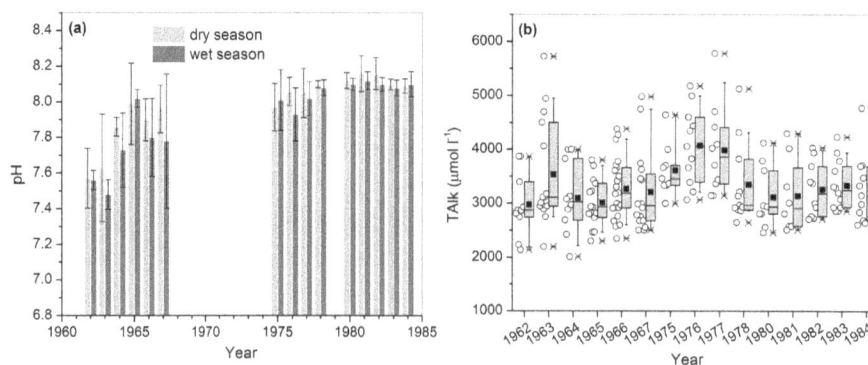

Figure 3. Seasonal comparison of pH (**a**) and box-and-whisker plot of TAlk (**b**) at Luokou station. The horizontal line is the median value, the black square is the mean value, the boxes represent the 25th to 75th percentile, the whiskers represent the 10th and 90th percentile, and the asterisks represent the maximum and minimum. Raw TAlk data are added to the left in (**b**).

$3850 \pm 1000 \, \mu\text{mol L}^{-1}$. The highest TAlk ($8633 \, \mu\text{mol L}^{-1}$) was measured in the QZ sub-basin. It is evident that the TAlk and pH showed similar spatial variations, but in the reverse direction with high TAlk coinciding with low pH (Fig. S1). With regard to all the sampling results, about 58 % of the TAlk values fell into the range of 3000–4000 $\mu\text{mol L}^{-1}$ and 92 % into the range of 2000–5000 $\mu\text{mol L}^{-1}$. For the whole Yellow River watershed, its mean TAlk was $3665 \pm 988 \, \mu\text{mol L}^{-1}$. In addition, its TAlk remained largely stable during the sampling period. Figure 3b shows an example of the TAlk changes at Luokou. Despite the discontinuous measurement from 1968 to 1974, the TAlk did not change significantly over time ($p = 0.48$).

3.2 Spatial and temporal variability of pCO_2

The pCO_2 varied significantly throughout the watershed with 2 orders of magnitude from $\sim 200 \, \mu\text{atm}$ to more than 30 000 μatm. Except the headwater region that showed lower pCO_2 than the overlying atmosphere, the remaining watershed had a considerably high pCO_2 (Fig. 2b). The highest pCO_2 of 36 790 μatm was estimated on a tributary in the QZ sub-basin resulting from low pH and high TAlk. For the middle Yellow River, including the MY and WY sub-basins, the waters were considerably supersaturated in CO$_2$ with the pCO_2 ranging from 1000 to 5000 μatm (Fig. 2b). Moreover, the pCO_2 level in the lower Yellow River reaches (LY) was much higher and can exceed 10 000 μatm. On average, the pCO_2 in the Yellow River watershed was $2810 \pm 1985 \, \mu\text{atm}$, 7-fold the atmospheric CO$_2$ equilibrium. However, it must be recognized that, unlike the historical data set that was monthly measured, sampling in the HR sub-basin by Wu et al. (2005) was conducted only during the late May and June of 1999 and 2000 when the wet season had barely started. Given the flushing effect of infiltrating rainfall and snowmelt flows at the beginning of the wet season (Clow and Drever, 1996; Melack, 2011), the resultant TAlk and pCO_2 are expected to be close to highest.

Similar to TAlk, the pCO_2 at most sites also presented strong seasonal variations. At 73 % of the sampling sites, higher pCO_2 occurred in the months before the onset of the wet season. During the wet season, it decreased to a relatively low level before going up from October onwards. The seasonal ratio of pCO_2, defined as the ratio of pCO_2 in the dry season over that in the wet season, ranged from 0.8 to 2.3. To more clearly show its spatial and temporal changes, Fig. 4 shows the high temporal-resolution results at Toudaoguai, Tongguan and Lijin. Both the TAlk and pCO_2 exhibited large spatial differences among the three sites. The mean TAlk at Tongguan ($4075 \pm 796 \, \mu\text{mol L}^{-1}$) was higher than at the upstream Toudaoguai and the downstream Lijin (3664 ± 399 and $3622 \pm 292 \, \mu\text{mol L}^{-1}$, respectively). Likewise, the mean pCO_2 at Tongguan ($4770 \pm 3470 \, \mu\text{atm}$) was about 3 and 3.5 times that at Toudaoguai ($1624 \pm 778 \, \mu\text{atm}$) and Lijin ($1348 \pm 689 \, \mu\text{atm}$), respectively. The highest pCO_2 of 26 318 μatm was estimated at Tongguan in early May.

Compared with tributary streams showing pronounced seasonal variation, the mainstem exhibited more complicated seasonal patterns (Fig. 4). The TAlk was higher in the dry season than in the wet season, in particular for Tongguan located downstream of the Loess Plateau (Table 2; Fig. 1). The pCO_2 showed similar seasonal cycles. A contrast to the weak seasonal changes at Toudaoguai and Lijin, the pCO_2 at Tongguan in the dry season (6016 μatm on average) was twofold that in the wet season. It is clear the pCO_2 increased substantially in both seasons as waters from the Loess Plateau entered the mainstem, and then decreased along the channel course towards the ocean (Table 2). Furthermore, the pCO_2 presented complex relationships with water discharge. While the pCO_2 changed synchronously with water at Toudaoguai, it decreased with increasing water in the wet season at Tongguan and Lijin (Fig. 4). The pCO_2 at all three stations was significantly higher than the atmospheric equilibrium, though the gradient varied substantially between different stations or different seasons.

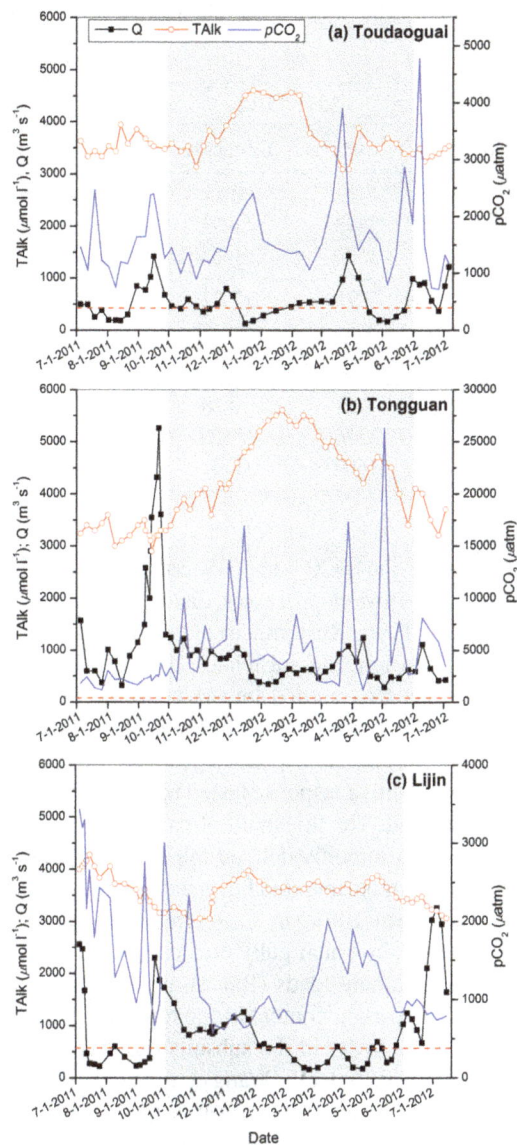

Figure 4. Weekly variations in water discharge (Q), TAlk, and $p\mathrm{CO_2}$ at (**a**) Toudaoguai, (**b**) Tongguan and (**c**) Lijin from July 2011 to July 2012. The dotted line denotes the atmospheric CO_2 equilibrium (i.e. 380 μatm) and the shaded grey represents the dry season.

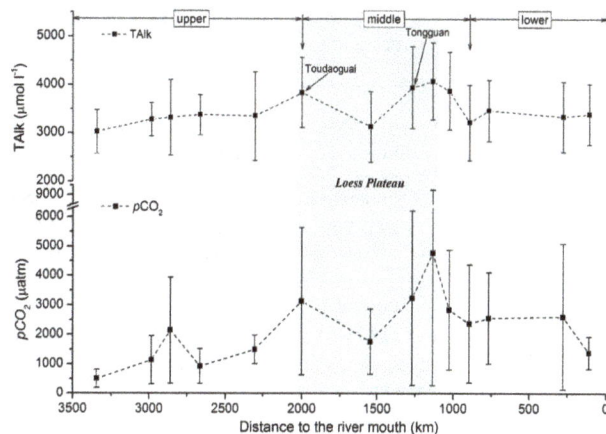

Figure 5. Longitudinal variations of TAlk and $p\mathrm{CO_2}$ along the mainstem channel. The shaded region approximately represents the Loess Plateau. Whiskers indicate the standard deviation.

$100\,\mathrm{kg\,m^{-3}}$, however, the $p\mathrm{CO_2}$ decreased slightly and remained stable thereafter (Fig. 6).

4 Discussion

4.1 Environmental controls on riverine $p\mathrm{CO_2}$

The alkalinity of river water reveals its buffering capacity in a carbonate system to neutralize acids and bases. Due to abundant carbonate outcrops, groundwater in the Yellow River basin was highly alkaline (Chen et al., 2002b), which directly led to higher TAlk in the dry season when baseflow constituted 90 % of the river runoff. High TAlk on the Loess Plateau was probably the result of chemical weathering. With widespread carbonates, chemical weathering in the loess deposits has generated high dissolved solids with $\mathrm{HCO_3^-}$ being the dominant ion (Zhang et al., 1995; Chen et al., 2005). Plotting TAlk against flow showed that they were negatively correlated (Fig. 7). However, the TAlk did not change synchronously with water in the wet season. It decreased more slowly as revealed by the exponents of the fitted equations. Compared with the flow changes, a narrower TAlk fluctuation suggested the coupling results of enhanced alkalinity export in the wet season and the dilution effect of water (Piñol and Avila, 1992; Raymond and Cole, 2003). Analysing the temporal variations of major ions during 1958–2000, Chen et al. (2005) found that they persistently increased due largely to human impacts. In contrast, the long-term stable TAlk (Fig. 3b) indicates that it is not significantly affected. Natural weathering processes must have played a more important role in controlling the export of DIC species and TAlk.

Carbon in river waters is largely derived from biogeochemical processes occurring in terrestrial ecosystems. Changes in terrestrial ecosystems will thus affect riverine carbon cycle. Because soil respiration and CO_2 production

Longitudinal variations of TAlk and $p\mathrm{CO_2}$ along the mainstem indicated that the waters from the Loess Plateau had higher TAlk and were more supersaturated in CO_2 than the upper and lower Yellow River waters (Fig. 5). Both the TAlk and $p\mathrm{CO_2}$ decreased remarkably downstream of the Loess Plateau. In addition, with extremely high suspended solids, the Yellow River provides an excellent case study for understanding the responses of $p\mathrm{CO_2}$ to TSS export (Fig. 6). Based on measurements in the sediment-yielding areas on the Loess Plateau, the $p\mathrm{CO_2}$ increased exponentially with increasing TSS concentration under low TSS scenarios (i.e. $100\,\mathrm{kg\,m^{-3}}$). When the TSS concentration was higher than

Table 2. Inter-annual and seasonal differences of pH, TAlk, and $p\text{CO}_2$ at the three stations. The number below the station name denotes the channel length to the river mouth.

Station	Variable	1950s–1984		2011–2012	
		Wet season	Dry season	Wet season	Dry season
Toudaoguai	pH	7.89	8.01	8.19	8.11
(2002 km)	TAlk (μmol L^{-1})	3595	4091	3513	3771
	$p\text{CO}_2$ (μatm)	3716	2708	1580	1655
Tongguan	pH	–	–	7.91	7.72
(1147 km)	TAlk (μmol L^{-1})	–	–	3356	4562
	$p\text{CO}_2$ (μatm)	–	–	2927	6016
Lijin	pH	8.18	8.19	8.23	8.28
(110 km)	TAlk (μmol L^{-1})	2942	3789	3576	3584
	$p\text{CO}_2$ (μatm)	1344	1349	1609	1132

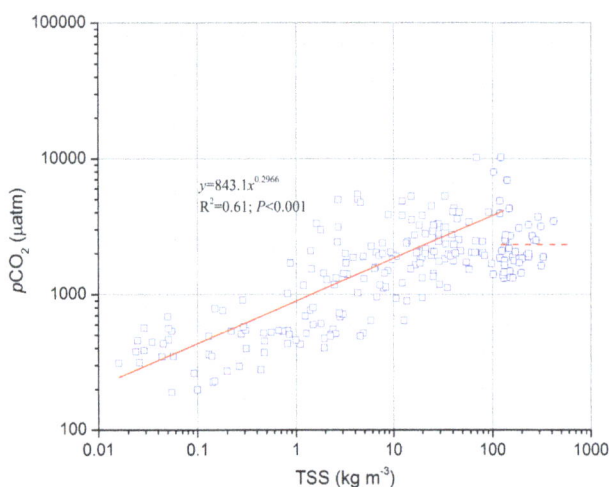

Figure 6. Relationship between total suspended solids (TSS) and $p\text{CO}_2$ based on measurements on the Loess Plateau. The solid line denotes the fitted line for the TSS concentration ranging from 0 to 100 kg m^{-3}, and the dashed line indicates the stable trend of $p\text{CO}_2$ when the TSS concentration is higher than 100 kg m^{-3}.

are highly dependent on temperature and rainfall (Epron et al., 1999; Hope et al., 2004; Shi et al., 2011), higher riverine $p\text{CO}_2$ is expected in the wet season due to soil CO_2 flushing. This is in contrast to the observed seasonal variations in the Yellow River. A unique precipitation distribution and hydrological regime may have contributed to these anomalous observations. The Yellow River basin is characteristic of high-intensity rainfalls; several storms in the wet season can account for > 70 % of the annual precipitation (Zhao, 1996). Coupled with its distinct soil surface microtopography with texture consisted mainly of silt and clay, Hortonian overland flow is the dominant runoff process (Liu and Singh, 2004). As a result of greatly reduced soil infiltration capacity, the generated overland flow by high-intensity rainfalls may have diluted the TAlk and caused the lowered riverine $p\text{CO}_2$.

Responses of the $p\text{CO}_2$ to TSS concentration (Fig. 6) reflect the soil erosion processes distinctive to the Loess Plateau (Zhao, 1996; Rustomji et al., 2008). At the initial stage of soil erosion, the surficial soils with abundant organic carbon are first eroded into river water. Decomposition of the labile organic carbon in the eroded soils will increase the $p\text{CO}_2$. Thus, it responded positively to soil erosion and TSS. This positive response lasted until the topsoils were completely eroded. The threshold of ca. 100 kg m^{-3} is consistent with the commonly defined hyperconcentrated flows (Xu, 2002). Hyperconcentrated flows indicating TSS concentration greater than 100 kg m^{-3} are frequently recorded in the Yellow River, in which gully erosion contributes > 50 % of the fluvial sediment loads (Ran et al., 2014). Compared with the organic carbon content in the topsoils (usually 0.5–1.5 %), it is much lower in the subsoils (i.e. 0.2–0.3 %) and shows uniformity with depth (Wang et al., 2010; Zhang et al., 2012). The mobilized subsoils through gully erosion therefore have lower organic carbon quantity for decomposition, resulting in the reduced and stable riverine $p\text{CO}_2$ regardless of the increasing TSS concentration.

Lower $p\text{CO}_2$ in the HR sub-basin was caused by relatively low TAlk and high pH. Statistical analyses showed that its TAlk was 25 % lower than the basin average while the pH was 7 % higher. Compared to other sub-basins, the HR sub-basin is covered by an alpine meadow ecosystem with soils being slightly eroded, which may have constrained the leaching of organic matter. Moreover, this sub-basin is in cold environments and its temperature falls below zero from October to March. Microbial decomposition of organic matter and ecosystem respiration are kinetically inhibited as affected by the low temperature (Kato et al., 2004), resulting in the low $p\text{CO}_2$. Unique climate also denotes the seasonal patterns of $p\text{CO}_2$ in the upper and middle Yellow River. Occurrence of the highest $p\text{CO}_2$ at Toudaoguai and Tongguan in March through May is likely controlled by ice-melt floods (Fig. 4a and b). In the coldest months, water surface

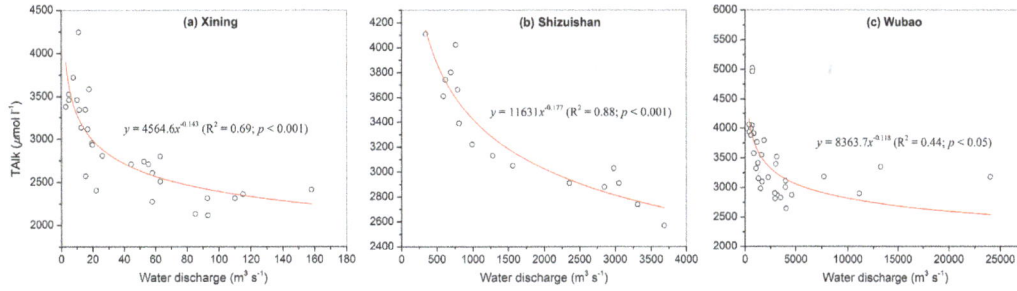

Figure 7. Dependence of TAlk on natural water discharge for different water discharge scales at typical sampling sites: **(a)** is from a tributary and **(b)** and **(c)** are from the Yellow River mainstem (see Fig. 1 for locations).

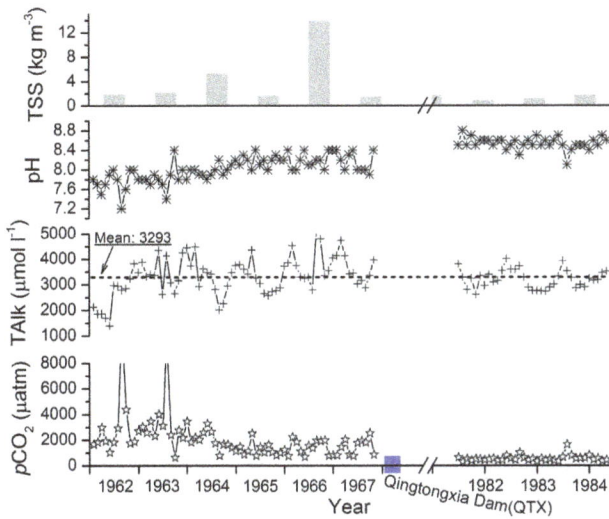

Figure 8. Impacts of Qingtongxia (QTX) dam on riverine TSS, pH, TAlk and $p\mathrm{CO_2}$. Measurements were conducted ~ 850 m downstream of the dam that was built in 1968 (see Fig. 1 for its location).

in the northernmost reaches (between QTX and Toudaoguai; Fig. 1) will freeze up (Chen and Ji, 2005). Aqueous CO_2 could not be efficiently released due to ice protection and typically accumulates below the ice cover. Starting from early spring, the ice begins to thaw and CO_2-laden waters are exported downstream to the sampling sites, probably causing the sharply increased $p\mathrm{CO_2}$. Also, the lower temperature in the dry season would be responsible for the higher $p\mathrm{CO_2}$ as the solubility of CO_2 increases with decreasing water temperature.

High $p\mathrm{CO_2}$ in the QZ sub-basin (Fig. 2b) was primarily the result of high TAlk due to its geological background. Its major rock types are carbonates, detritus (quartz and feldspar) and red-beds (gypsum and halite) (Zhang et al., 1995). These rocks are highly vulnerable to weathering, thus producing TAlk (mainly HCO_3^-) into river. In fact, its mean TDS was 8–14 times the basin average (Chen et al., 2005). Further, this sub-basin has high drought index with its annual evaporation being > 8 times the annual precipitation (Chen et

al., 2005). Such strong evaporation will result in not only the precipitation of minerals with low solubility, but also the elevated concentrations of solutes not removed during the crystallization process. Another possible cause is the severe erosion due to sparse vegetation cover. In addition to mobilization of organic matter, soil erosion is able to enhance chemical weathering by increasing the exposure surface of fresh minerals to atmosphere (Millot et al., 2002). This would also contribute to greatly condensed DIC species in river waters and thus high $p\mathrm{CO_2}$.

As for the longitudinal variations (Fig. 5), severe soil erosion on the Loess Plateau may be the major reason as discussed above. In addition, low groundwater table in the arid climate allows deeper soil horizons to adequately interact with the atmosphere, which could also facilitate the exposure of mineral surfaces to weathering and generate huge quantities of alkalinity. Increasing $p\mathrm{CO_2}$ until Tongguan suggested the integrated responses of $p\mathrm{CO_2}$ to these factors. Without large tributaries joining the lower Yellow River (Fig. 1), the decreasing TAlk and $p\mathrm{CO_2}$ along the mainstem revealed reduced TAlk input. Overall, the spatial changes of TAlk and $p\mathrm{CO_2}$ were the combined results of differences in soil property, hydrological regime, climate and landform development.

4.2 Anthropogenic impacts on riverine $p\mathrm{CO_2}$

Agricultural activity within a watershed can affect its riverine $p\mathrm{CO_2}$. Tilling practices can not only expand the exposure area of soil materials, but also alter the hydrology of surficial soils, increasing the contact rate between water and minerals and thus the alkalinity export (Raymond and Cole, 2003). With a history of more than 2000 years, agriculture in the Yellow River basin is possibly an important reason for the observed high TAlk (Chen et al., 2005). Further, significant decreases in pH in the middle and lower Yellow River basin have been widely detected and are hypothesized to result from acid rain that is likely caused by anthropogenic sulfur emissions to the atmosphere (Guo et al., 2010). The reduced pH may have been partially responsible for the elevated $p\mathrm{CO}_2$ in these regions relative to the headwater region that had higher pH (Fig. 2).

Differences in the hydrochemical parameters between historical records and recent sampling clearly reveal the temporal changes over the period. Significant increase in pH at Toudaoguai was largely caused by widespread salinization of agricultural soils. There are two large irrigation zones upstream of Toudaoguai; large quantities of water is diverted for desalination and irrigation (Chen et al., 2005). The diverted water volume has gradually increased since 1960 due to growing demand (Ran et al., 2014). When washed out from irrigated farmlands, the return water characterized by high pH caused the riverine pH to increase, leading further to greatly reduced $p\mathrm{CO}_2$ despite the roughly stable TAlk (Fig. 3b). Particularly, it is worth noting that the magnitude of reduction was much higher in the wet season when the high-pH return water reached the mainstem with floods (Table 2).

Trapping of water and suspended solids by dams will alter river-borne carbon dynamics (Cole et al., 2007). Extended residence time combined with sufficient organic matter availability may enhance CO_2 production, causing a higher $p\mathrm{CO}_2$. This is particularly true for tropical reservoirs into which organic matter inputs are sufficient, especially in the initial years after impoundment (Roland et al., 2010; Raymond et al., 2013). On the other hand, reduced flow turbulence and increases in water residence time would promote photosynthesis of aquatic plants and reduce aqueous CO_2 concentration (Teodoru et al., 2009; Wang et al., 2011). An example of the impact of dams on downstream $p\mathrm{CO}_2$ changes is presented in Fig. 8. Located on the upper mainstem channel (Fig. 1), operation of the Qingtongxia Dam since 1968 has substantially affected the $p\mathrm{CO}_2$. Despite insignificant changes of the TAlk between the pre- and post-dam periods (Fig. 8), enhanced aquatic photosynthesis after dam operation owing to reduced TSS concentration may have absorbed aqueous CO_2 and resulted in increased pH by shifting the chemical equilibrium of Eq. (1). Accordingly, the riverine $p\mathrm{CO}_2$ declined during the post-dam period with the elevated pH and roughly stable TAlk.

For the dams on the Loess Plateau constructed mostly in 1960–2000, however, aqueous photosynthesis appears to be at a low level owing to its extremely high TSS concentration and limited light availability (Chen et al., 2005). In contrast, flow regulation plays a more important role in controlling the seasonal patterns of downstream $p\mathrm{CO}_2$. Man-made floods have been regularly released from the Xiaolangdi Dam sluice gates since 2000 to flush sediment deposition in the lower Yellow River, usually from late June (Wang et al., 2012; Ran et al., 2014). The deep waters supersaturated with CO_2 are first discharged, resulting in the high $p\mathrm{CO}_2$ at Lijin in the wet season (Fig. 4c). Unlike the seasonal variations at Toudaoguai and Tongguan as mentioned above, duration of the high $p\mathrm{CO}_2$ at Lijin coincided well with the sediment flushing period, indicating the impact of flow reg-

ulation on $p\mathrm{CO}_2$ dynamics. Operation of dam cascade has also modified the TAlk and $p\mathrm{CO}_2$ levels at the inter-annual scale. Affected by upstream dams (see Fig. 1), both the TAlk and $p\mathrm{CO}_2$ at Lijin in the wet season were elevated during the period 2011–2012, by 22 and 20 %, respectively, relative to the baseline period 1950s–1984 (Table 2). Furthermore, soil conservation and vegetation restoration conducted on the Loess Plateau since the 1970s have contributed to the inter-annual changes. More organic carbon has been sequestrated as a result of these land management practices (Chen et al., 2007). Given the strong flushing and leaching effects of high-intensity rainfalls, riverine organic matter export tends to increase in the wet season, and the accompanying decomposition can elevate $p\mathrm{CO}_2$.

4.3 Implications for CO_2 outgassing

CO_2 outgassing from river waters into the atmosphere during the carbon transport processes from land to the ocean has not been fully realized until recent years (Richey et al., 2002; Cole et al., 2007; Battin et al., 2009). Because riverine $p\mathrm{CO}_2$ demonstrates the CO_2 concentration in surface water, a higher riverine $p\mathrm{CO}_2$ usually represents stronger CO_2 outgassing under favourable environmental conditions, forming a carbon source for the atmosphere. However, accurate estimates of global CO_2 outgassing have been hampered by the absence of a spatially resolved $p\mathrm{CO}_2$ database. Previous estimates from rivers alone involve large uncertainties, varying from 0.23 to 0.56 Gt C yr^{-1} (Cole et al., 2007; Aufdenkampe et al., 2011). A recent study has even concluded that up to 1.8 Gt of carbon is annually emitted from global rivers (Raymond et al., 2013), considerably higher than was previously thought. This estimate accounts for about 32 % of the annual carbon flux transferred from terrestrial systems to inland waters (Wehrli, 2013). Given the existing uncertainties, quantifying $p\mathrm{CO}_2$ in different orders of streams of a complete river network is critical to resolve a robust estimate of riverine CO_2 evasion.

With respect to the Yellow River, the lower riverine $p\mathrm{CO}_2$ in the HR sub-basin relative to the atmospheric equilibrium indicates a potential CO_2 drawdown. In comparison, the river waters in the remaining watershed are generally supersaturated with CO_2, mostly greater than 1000 µatm (Fig. 2b). With an average $p\mathrm{CO}_2$ of 2810 ± 1985 µatm for the whole watershed comparable to the median of global rivers (i.e. 1300–4300 µatm; Aufdenkampe et al., 2011), the Yellow River waters tend to act as a net carbon source for the atmosphere. As stated earlier, despite the uncertainties associated with outgassing calculation, recent studies on watershed-scale carbon delivery demonstrate that CO_2 efflux from rivers can substantially exceed lateral carbon export (Richey et al., 2002). The Yellow River has experienced abrupt reductions in flow and TSS fluxes over the past decades and these reductions will continue in future. Its carbon fluxes reaching the ocean will therefore further decrease, and more carbon is likely to be emitted as CO_2 into the atmosphere. In view of the severe soil erosion and high TSS transport (Syvitski et al., 2005), interpretation of these fluxes in the context of climate change is of great importance for understanding the role of Yellow River in the global carbon cycle.

5 Conclusions

The Yellow River was characterized by high alkalinity with a mean TAlk of $3665 \pm 988\,\mu\text{mol}\,\text{L}^{-1}$. Although with significant spatial variations, the TAlk remained largely stable over the study period. However, it showed seasonal variability and decreased in the wet season, suggesting the dilution effect of water discharge. Except for the HR sub-basin where the $p\text{CO}_2$ was lower than the atmospheric equilibrium, river waters in the remaining watershed were supersaturated with CO_2. The basin-wide mean $p\text{CO}_2$ was estimated at $2810 \pm 1985\,\mu\text{atm}$. Similar to the pH and TAlk, the $p\text{CO}_2$ also presented significant spatial and seasonal variations. The middle reaches, mainly the Loess Plateau, showed higher $p\text{CO}_2$ than the upper and lower reaches, which were principally resulting from severe soil erosion and the unique hydrological regime. The $p\text{CO}_2$ correlated exponentially with TSS transport when the erosion intensity was low and only the topsoils rich in organic carbon were eroded. When the TSS concentration exceeded $100\,\text{kg}\,\text{m}^{-3}$ indicating the predominance of gully erosion, the subsoils with low organic carbon content were mobilized. Owing to the reduced organic carbon available for decomposition, the $p\text{CO}_2$ slightly decreased and remained stable thereafter, regardless of the increasing TSS concentration.

The observed spatial and temporal variations of riverine $p\text{CO}_2$ were collectively controlled by natural processes and human activities. High $p\text{CO}_2$ in the upper and middle reaches was usually estimated from March through May when ice-melt floods transported the accumulated CO_2-laden waters in winter. Human activities, especially flow regulation, have significantly changed its seasonal patterns by altering hydrological regime and riverine carbon delivery processes. While reduced turbidity and extended residence time due to dam trapping has enhanced aquatic photosynthesis and resulted in a decreased $p\text{CO}_2$, man-made floods through flow regulation would increase downstream $p\text{CO}_2$. Other anthropogenic perturbations, such as acidification, soil conservation and irrigation, have also affected $p\text{CO}_2$. The accelerating human activity within the watershed is likely to expand the role of anthropogenic over natural factors on the $p\text{CO}_2$ dynamics, because stronger anthropogenic impacts are certain to occur concerning present economic development. Considerably high riverine $p\text{CO}_2$ in the Yellow River waters with respect to the overlying atmosphere indicates that substantial amounts of CO_2 are emitted into the atmosphere. Given the huge human impacts on flow, TSS and carbon fluxes, future efforts to estimate CO_2 evasion and assess its importance in the global carbon cycle are urgently needed.

Acknowledgements. This work was supported by the Ministry of Education, Singapore (MOE2011-T2-1-101). We are grateful to the YRCC for access to the hydrological data. We thank the staff at Toudaoguai, Tongguan and Lijin gauges for their assistance in the field. Comments from two anonymous reviewers were instrumental in improving this paper.

Edited by: S. Bouillon

References

Abril, G., Bouillon, S., Darchambeau, F., Teodoru, C. R., Marwick, T. R., Tamooh, F., Ochieng Omengo, F., Geeraert, N., Deirmendjian, L., Polsenaere, P., and Borges, A. V.: Technical Note: Large overestimation of $p\text{CO}_2$ calculated from pH and alkalinity in acidic, organic-rich freshwaters, Biogeosciences, 12, 67–78, doi:10.5194/bg-12-67-2015, 2015.

Alekin, O. A., Semenov, A. D., and Skopintsev, B. A.: Handbook of Chemical Analysis of Land Waters, Gidrometeoizdat, St. Petersburg, Russia, 269 pp., 1973.

Alin, S. R., Rasera, M. d. F. F. L., Salimon, C. I., Richey, J. E., Holtgrieve, G. W., Krusche, A. V., and Snidvongs, A.: Physical controls on carbon dioxide transfer velocity and flux in low-gradient river systems and implications for regional carbon budgets, J. Geophys. Res., 116, G01009, doi:10.1029/2010jg001398, 2011.

American Public Health Association (APHA): Standard Methods for the Examination of Water and Wastewater, 16th edition, American Public Health Association, Washington, DC, 1268 pp., 1985.

Aufdenkampe, A. K., Mayorga, E., Raymond, P. A., Melack, J. M., Doney, S. C., Alin, S. R., Aalto, R. E., and Yoo, K.: Riverine coupling of biogeochemical cycles between land, oceans, and atmosphere, Front. Ecol. Environ., 9, 53–60, 2011.

Battin, T. J., Luyssaert, S., Kaplan, L. A., Aufdenkampe, A. K., Richter, A., and Tranvik, L. J.: The boundless carbon cycle, Nat. Geosci., 2, 598–600, 2009.

Benstead, J. P. and Leigh, D. S.: An expanded role for river networks, Nat. Geosci., 5, 678–679, 2012.

Butman, D. and Raymond, P. A.: Significant efflux of carbon dioxide from streams and rivers in the United States, Nat. Geosci., 4, 839–842, 2011.

Chen, L. D., Gong, J., Fu, B. J., Huang, Z. L., Huang, Y. L., and Gui, L. D.: Effect of land use conversion on soil organic carbon sequestration in the loess hilly area, loess plateau of China, Ecol. Res., 22, 641–648, 2007.

Chen, S. and Ji, H.: Fuzzy optimization neural network approach for ice forecast in the Inner Mongolia Reach of the Yellow River, Hydrolog. Sci. J., 50, 319–330, 2005.

Chen, J., Wang, F. Y., Xia, X. H., and Zhang, L. T.: Major element chemistry of the Changjiang (Yangtze River), Chem. Geol., 187, 231–255, 2002a.

Chen, J., Tang, C. T., Sakura, Y., Kondoh, A., and Shen, Y.: Groundwater flow and geochemistry in the lower reaches of the Yellow River: a case study in Shandang Province, China, Hydrogeol. J., 10, 587–599, 2002b.

Chen, J., Wang, F., Meybeck, M., He, D., Xia, X., and Zhang, L.: Spatial and temporal analysis of water chemistry records (1958–2000) in the Huanghe (Yellow River) basin, Global Biogeochem. Cy., 19, GB3016, doi:10.1029/2004gb002325, 2005.

Clark, I. D. and Fritz, P.: Environmental isotopes in hydrogeology, CRC Press/Lewis Publishers, New York, 328 pp., 1997.

Clow, D. W. and Drever, J. I.: Weathering rates as a function of flow through an alpine soil, Chem. Geol., 132, 131–141, 1996.

Cole, J. J. and Caraco, N. F.: Atmospheric exchange of carbon dioxide in a low-wind oligotrophic lake measured by the addition of SF_6, Limnol. Oceanogr., 43, 647–656, 1998.

Cole, J. J., Prairie, Y. T., Caraco, N. F., McDowell, W. H., Tranvik, L. J., Striegl, R. G., Duarte, C. M., Kortelainen, P., Downing, J. A., Middelburg, J. J., and Melack, J.: Plumbing the global carbon cycle: Integrating inland waters into the terrestrial carbon budget, Ecosystems, 10, 171–184, 2007.

Epron, D., Farque, L., Lucot, E., and Badot, P. M.: Soil CO_2 efflux in a beech forest: dependence on soil temperature and soil water content, Ann. For. Sci., 56, 221–226, 1999.

Evans, W., Hales, B., and Strutton, P. G.: pCO_2 distributions and air-water CO_2 fluxes in the Columbia River estuary, Estuar. Coastal Shelf S., 117, 260–272, 2013.

Gaillardet, J., Dupre, B., Louvat, P., and Allegre, C. J.: Global silicate weathering and CO_2 consumption rates deduced from the chemistry of large rivers, Chem. Geol., 159, 3–30, 1999.

Guo, J. H., Liu, X. J., Zhang, Y., Shen, J. L., Han, W. X., Zhang, W. F., Christie, P., Goulding, K. W. T., Vitousek, P. M., and Zhang, F. S.: Significant acidification in major Chinese croplands, Science, 327, 1008–1010, 2010.

Hope, D., Palmer, S. M., Billett, M. F., and Dawson, J. J. C.: Variations in dissolved CO_2 and CH_4 in a first-order stream and catchment: an investigation of soil-stream linkages, Hydrol. Process., 18, 3255–3275, 2004.

Humborg, C., Mörth, C., Sundbom, M., Borg, H., Blenckner, T., Giesler, R., and Ittekkot, V.: CO_2 supersaturation along the aquatic conduit in Swedish watersheds as constrained by terrestrial respiration, aquatic respiration and weathering, Glob. Change Biol., 16, 1966–1978, 2010.

Hunt, C. W., Salisbury, J. E., and Vandemark, D.: Contribution of non-carbonate anions to total alkalinity and overestimation of pCO_2 in New England and New Brunswick rivers, Biogeosciences, 8, 3069–3076, doi:10.5194/bg-8-3069-2011, 2011.

Kato, T., Tang, Y., Gu, S., Cui, X., Hirota, M., Du, M., Li, Y., Zhao, X., and Oikawa, T.: Carbon dioxide exchange between the atmosphere and an alpine meadow ecosystem on the Qinghai-Tibetan Plateau, China, Agr. Forest Meteorol., 124, 121–134, 2004.

Lewis, E. and Wallace, D. W. R.: Program developed for CO_2 system calculations, ORNL/CDIAC-105, Carbon dioxide Information Analysis Center, Oak Ridge National Laboratory, Oak Ridge, TN, 1998.

Li, S., Lu, X. X., He, M., Zhou, Y., Li, L., and Ziegler, A. D.: Daily CO_2 partial pressure and CO_2 outgassing in the upper Yangtze River basin: A case study of the Longchuan River, China, J. Hydrol., 466–467, 141–150, 2012.

Li, X. D., Fu, H., Guo, D., Li, X. D., and Wan, C. G.: Partitioning soil respiration and assessing the carbon balance in a *Setaria italica* (L.) Beauv. Cropland on the Loess Plateau, Northern China, Soil Biol. Biochem., 42, 337–346, 2010.

Liu, Q. and Singh, V.: Effect of microtopography, slope length and gradient, and vegetative cover on overland flow through simulation, J. Hydrol. Eng., 9, 375–382, 2004.

Melack, J.: Riverine carbon dioxide release, Nat. Geosci., 4, 821–822, 2011.

Millot, R., Gaillardet, J., Dupre, B., and Allegre, C. J.: The global control of silicate weathering rates and the coupling with physical erosion: new insights from rivers of the Canadian Shield, Earth Planet. Sci. Lett., 196, 83–98, 2002.

Piñol, J. and Avila, A.: Streamwater pH, alkalinity, pCO_2 and discharge relationships in some forested Mediterranean catchments, J. Hydrol., 131, 205–225, 1992.

Ran, L. and Lu, X. X.: Delineation of reservoirs using remote sensing and their storage estimate: an example of the Yellow River basin, China, Hydrol. Process., 26, 1215–1229, 2012.

Ran, L., Lu, X. X., Sun, H., Han, J., Li, R., and Zhang, J.: Spatial and seasonal variability of organic carbon transport in the Yellow River, China, J. Hydrol., 498, 76–88, 2013.

Ran, L., Lu, X. X., and Xin, Z.: Erosion-induced massive organic carbon burial and carbon emission in the Yellow River basin, China, Biogeosciences, 11, 945–959, doi:10.5194/bg-11-945-2014, 2014.

Raymond, P. A. and Cole, J. J.: Increase in the export of alkalinity from North America's largest river, Science, 301, 88–91, 2003.

Raymond, P. A., Hartmann, J., Lauerwald, R., Sobek, S., McDonald, C., Hoover, M., Butman, D., Striegl, R., Mayorga, E., and Humborg, C.: Global carbon dioxide emissions from inland waters, Nature, 503, 355–359, 2013.

Richey, J. E., Melack, J. M., Aufdenkampe, A. K., Ballester, V. M., and Hess, L. L.: Outgassing from Amazonian rivers and wetlands as a large tropical source of atmospheric CO_2, Nature, 416, 617–620, 2002.

Roland, F., Vidal, L. O., Pacheco, F. S., Barros, N. O., Assireu, A., Ometto, J. P., Cimbleris, A. C., and Cole, J. J.: Variability of carbon dioxide flux from tropical (Cerrado) hydroelectric reservoirs, Aquat. Sci., 72, 283–293, 2010.

Rustomji, P., Zhang, X., Hairsine, P., Zhang, L., and Zhao, J.: River sediment load and concentration responses to changes in hydrology and catchment management in the Loess Plateau region of China, Water Resour. Res., 44, W00A04, doi:10.1029/2007WR006656, 2008.

Shi, W.-Y., Tateno, R., Zhang, J.-G., Wang, Y.-L., Yamanaka, N., and Du, S.: Response of soil respiration to precipitation during the dry season in two typical forest stands in the forest-grassland transition zone of the Loess Plateau, Agr. Forest Meteorol., 151, 854–863, 2011.

Striegl, R. G., Dornblaser, M., McDonald, C., Rover, J., and Stets, E.: Carbon dioxide and methane emissions from the Yukon River system, Global Biogeochem. Cy., 26, GB0E05, doi:10.1029/2012GB004306, 2012.

Syvitski, J. P. M., Vorosmarty, C. J., Kettner, A. J., and Green, P.: Impact of humans on the flux of terrestrial sediment to the global coastal ocean, Science, 308, 376–380, 2005.

Teodoru, C. R., del Giorgio, P. A., Prairie, Y. T., and Camire, M.: Patterns in pCO$_2$ in boreal streams and rivers of northern Quebec, Canada, Global Biogeochem. Cy., 23, GB2012 doi:10.1029/2008gb003404, 2009.

Tranvik, L. J., Downing, J. A., Cotner, J. B., Loiselle, S. A., Striegl, R. G., Ballatore, T. J., Dillon, P., Finlay, K., Fortino, K., Knoll, L. B., Kortelainen, P. L., Kutser, T., Larsen, S., Laurion, I., Leech, D. M., McCallister, S. L., McKnight, D. M., Melack, J. M., Overholt, E., Porter, J. A., Prairie, Y., Renwick, W. H., Roland, F., Sherman, B. S., Schindler, D. W., Sobek, S., Tremblay, A., Vanni, M. J., Verschoor, A. M., von Wachenfeldt, E., and Weyhenmeyer, G. A.: Lakes and reservoirs as regulators of carbon cycling and climate, Limnol. Oceanogr., 54, 2298–2314, 2009.

Wallin, M. B., Grabs, T., Buffam, I., Laudon, H., Ågren, A., Öquist, M. G., and Bishop, K.: Evasion of CO$_2$ from streams–The dominant component of the carbon export through the aquatic conduit in a boreal landscape, Glob. Change Biol., 19, 785–797, 2013.

Wang, F. S., Wang, B. L., Liu, C. Q., Wang, Y. C., Guan, J., Liu, X. L., and Yu, Y. X.: Carbon dioxide emission from surface water in cascade reservoirs-river system on the Maotiao River, southwest of China, Atmos. Environ., 45, 3827–3834, 2011.

Wang, X., Ma, H., Li, R., Song, Z., and Wu, J.: Seasonal fluxes and source variation of organic carbon transported by two major Chinese Rivers: The Yellow River and Changjiang (Yangtze) River, Global Biogeochem. Cy., 26, GB2025, doi:10.1029/2011gb004130, 2012.

Wang, Z., Liu, G., and Xu, M.: Effect of revegetation on soil organic carbon concentration in deep soil layers in the hilly Loess Plateau of China, Acta Ecologica Sinica, 30, 3947–3952, 2010 (in Chinese).

Wehrli, B.: Biogeochemistry: Conduits of the carbon cycle, Nature, 503, 346–347, 2013.

Wu, L. L., Huh, Y., Qin, J. H., Du, G., and van der Lee, S.: Chemical weathering in the Upper Huang He (Yellow River) draining the eastern Qinghai-Tibet Plateau, Geochimi. Cosmochim. Ac., 69, 5279–5294, 2005.

Wu, W., Xu, S., Yang, J., and Yin, H.: Silicate weathering and CO$_2$ consumption deduced from the seven Chinese rivers originating in the Qinghai-Tibet Plateau, Chem. Geol., 249, 307–320, 2008.

Xu, J.: Implication of relationships among suspended sediment size, water discharge and suspended sediment concentration: the Yellow River basin, China, CATENA, 49, 289–307, 2002.

Yao, G. R., Gao, Q. Z., Wang, Z. G., Huang, X. K., He, T., Zhang, Y. L., Jiao, S. L., and Ding, J.: Dynamics of CO$_2$ partial pressure and CO$_2$ outgassing in the lower reaches of the Xijiang River, a subtropical monsoon river in China, Sci. Total Environ., 376, 255–266, 2007.

Zhang, J., Huang, W. W., Létolle, R., and Jusserand, C.: Major element chemistry of the Huanghe (Yellow River), China: Weathering processes and chemical fluxes, J. Hydrol., 168, 173–203, 1995.

Zhang, J., Xu, M., Wang, Z., Ma, X., and Qiu, Y.: Effects of revegetation on organic carbon storage in deep soils in hilly Loess Plateau region of Northwest China, Chinese Journal of Applied Ecology, 23, 2721–2727, 2012 (in Chinese).

Zhao, M. X., Zhou, J. B., and Kalbitz, K.: Carbon mineralization and properties of water-extractable organic carbon in soils of the south Loess Plateau in China, Eur. J. Soil Biol., 44, 158–165, 2008.

Zhao, W.: The Yellow River's Sediment, Yellow River Conservancy Press, Zhengzhou, China, 807 pp., 1996.

13

Fluxes of carbon and nutrients to the Iceland Sea surface layer and inferred primary productivity and stoichiometry

E. Jeansson[1], R. G. J. Bellerby[2,1], I. Skjelvan[1,3], H. Frigstad[4], S. R. Ólafsdóttir[5], and J. Olafsson[5,6]

[1]Uni Research Climate, Bergen, Norway
[2]Norwegian Institute for Water Research (NIVA), Bergen, Norway
[3]Geophysical Institute, University of Bergen, Bergen, Norway
[4]Norwegian Environment Agency, Oslo, Norway
[5]Marine Research Institute, Reykjavik, Iceland
[6]Institute of Earth Sciences, University of Iceland, Reykjavik, Iceland

Correspondence to: E. Jeansson (emil.jeansson@uni.no)

Abstract. This study evaluates long-term mean fluxes of carbon and nutrients to the upper 100 m of the Iceland Sea. The study utilises hydro-chemical data from the Iceland Sea time series station (68.00° N, 12.67° W), for the years between 1993 and 2006. By comparing data of dissolved inorganic carbon (DIC) and nutrients in the surface layer (upper 100 m), and a sub-surface layer (100–200 m), we calculate monthly deficits in the surface, and use these to deduce the long-term mean surface layer fluxes that affect the deficits: vertical mixing, horizontal advection, air–sea exchange, and biological activity. The deficits show a clear seasonality with a minimum in winter, when the mixed layer is at the deepest, and a maximum in early autumn, when biological uptake has removed much of the nutrients. The annual vertical fluxes of DIC and nitrate amounts to 2.9 ± 0.5 and $0.45 \pm 0.09 \, \mathrm{mol \, m^{-2} \, yr^{-1}}$, respectively, and the annual air–sea uptake of atmospheric CO_2 is $4.4 \pm 1.1 \, \mathrm{mol \, C \, m^{-2} \, yr^{-1}}$. The biologically driven changes in DIC during the year relates to net community production (NCP), and the net annual NCP corresponds to export production, and is here calculated as $7.3 \pm 1.0 \, \mathrm{mol \, C \, m^{-2} \, yr^{-1}}$. The typical, median C : N ratio during the period of net community uptake is 9.0, and clearly higher than the Redfield ratio, but is varying during the season.

1 Introduction

Increasing our knowledge of the oceanic cycles of carbon and nutrients, and how they are linked, is crucial for improving ocean biogeochemical models and, thus, producing better projections of oceanic response and feedback to a changing climate.

The biological carbon pump (i.e. the biologically driven transport of carbon from the surface waters to the deep ocean) is a pathway that can sequester atmospheric CO_2 on long timescales (Falkowski et al., 1998; Sabine et al., 2004). With the present increase in atmospheric CO_2 (http://www.esrl.noaa.gov/gmd/ccgg/trends/global.html) the strength of the future biological carbon pump is very uncertain, and warrants further investigation (e.g. Passow and Carlson, 2012). To be able to reveal changes in the oceans, we need repeated measurements and long-term time series stations, such as the Hawaii Ocean Time series (HOT) and the Bermuda Atlantic Time series Study (BATS) (e.g. Church et al., 2013). In the Nordic Seas, the time series stations in the Norwegian Sea (ocean weather station Mike) and the Iceland Sea, have greatly increased our knowledge of the carbon cycle in this region (e.g. Skjelvan et al., 2008; Ólafsson et al., 2009). In this paper, we focus on the Iceland Sea, which is the shallowest of the main basins in the Nordic Seas. The Iceland Sea (Fig. 1) is most often defined as the waters delimited by Greenland in the west; the Denmark Strait and the continental shelf break south of Iceland to the south; by Jan Mayen

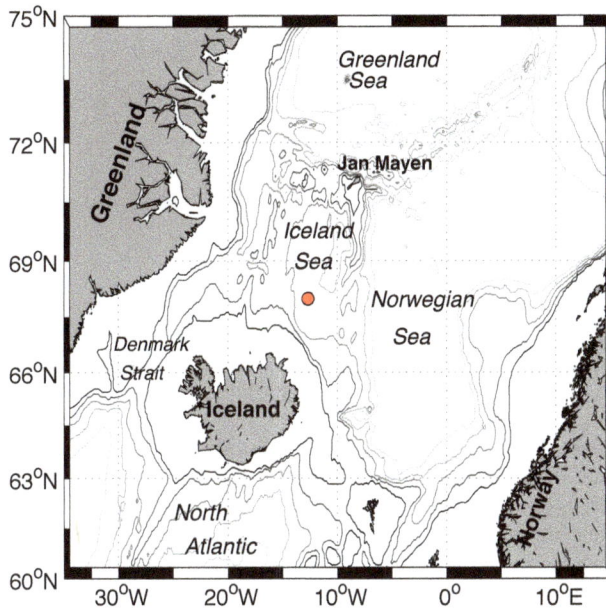

Figure 1. Map of the Nordic Seas region. The red filled circle marks the position of the time series station.

and the Jan Mayen Fracture Zone to the north and by the Jan Mayen Ridge to the east (Pálsson et al., 2012). The hydrographic properties of the Iceland Sea can generally be described as Arctic Intermediate Water overlying Arctic Deep Water (e.g. Swift and Aagaard, 1981). See Assthorsson et al. (2007) for a more detailed description.

The biological carbon pump in the Nordic Seas has not been studied in great detail, and we need to improve our understanding of the driving processes. Until now there are few estimates of the primary productivity in the relatively cold and low-salinity Arctic waters that dominate the upper water column of the Iceland Sea. Production estimates in this Arctic domain are in the range 75–179 g C m^{-2} yr^{-1}, based on data and remote sensing (Thordardottir, 1984; Zhai et al., 2012).

There are several production terms used in the literature, illustrating somewhat different fluxes. New production, as defined by Dugdale and Goering (1967), is the production that results from allochthonous (new) nitrate added to the surface layer by vertical or horizontal advection, or via air–sea exchange. This is different from total production, which also includes nitrogen regenerated within the surface layer (see Dugdale and Goering, 1967). Net community production (NCP) is defined as net primary production minus community respiration (e.g. Platt et al., 1989). Estimates of NCP have traditionally been based on bottle oxygen incubations (Gaarder and Gran, 1927), but are often based on oxygen budgets (e.g. Falck and Gade, 1999) or seasonal mixed-layer changes in oxygen or inorganic carbon, corrected for the air–sea fluxes (e.g. Körtzinger et al., 2008; Frigstad et al., in preparation), or oxygen-to-argon ($O_2 Ar^{-1}$) ratios (e.g.

Reuer et al., 2007; Quay et al., 2012). Export production is the excess organic matter produced in the euphotic zone, on top of the production needed to sustain the productive system (Dugdale and Goering, 1967; Eppley and Peterson, 1979). Thus, the export production cannot exceed the rate of added nutrients (i.e. new production), and these fluxes have been assumed to be equivalent on an annual average (Eppley and Peterson, 1979).

An issue under debate over the last few decades, is the universal validity of the so-called Redfield ratio, describing the stoichiometry between carbon and inorganic nutrients in marine plankton, where the average C : N : P ratios are 106 : 16 : 1 (Redfield et al., 1963). Observations of deviations from this relationship are numerous (e.g. Takahashi et al., 1993; Anderson and Sarmiento, 1994; Daly et al., 1999; Körtzinger et al., 2001; Koeve, 2006, Tamelander et al., 2013; Frigstad et al., 2014). It is common practise to use the traditional Redfield ratio to convert changes of nutrients into production of organic matter, both in observational and model studies (e.g. Skjelvan et al., 2001; Falck and Anderson, 2005; Skogen et al., 2007), so any significant variability or deviations of these ratios could have a marked impact on estimated primary production.

In this study we use observational data of inorganic nutrients (nitrate, phosphate, and silicate) and inorganic carbon (total dissolved inorganic carbon (DIC) and $p CO_2$) from the upper layers of the Iceland Sea to evaluate annual fluxes of carbon and nutrients into the surface layer, which we here define as the upper 100 m of the water column. From these fluxes we estimate the long-term mean in primary production in the Iceland Sea, and the related stoichiometric relationships.

2 Data

The study utilises data from the Iceland Sea time series station, located at 68.00° N, 12.67° W (Fig. 1). Surface sampling of DIC and $p CO_2$ started in 1983, and water column sampling for DIC and $p CO_2$ started in 1991 and 1993, respectively (Ólafsson et al., 2010). Here we include data of inorganic carbon, nutrient and hydrography between 1993 and 2006. For details of analytical methods and data quality, see Olsen (2009), Olafsson and Olsen (2010) and Olafsson et al. (2010). The data are available via the CARINA database (http://cdiac.ornl.gov/oceans/CARINA/).

Monthly long-term surface wind speed data are from the NCEP/NCAR reanalysis project (Kalnay et al., 1996), provided by the NOAA/OAR/ESRL PSD, Boulder, Colorado, USA, from their web site at http://www.esrl.noaa.gov/psd/.

For the atmospheric CO_2 near Iceland we use Globalview data from Vestmannaeyjar, south of Iceland, ICE_01DO (GLOBALVIEW-CO2, 2012), and the barometric pressure are monthly means of sea level pressure (SLP) obtained from NOWW Fisheries Service, Environmental Research Division (http://www.pfeg.noaa.gov/products/las.html).

Table 1. Monthly computed median mixed layer depths (MLDs) and entrainment velocities (v_{mix}). These are used when calculating the vertical fluxes. The values in italic are interpolated from surrounding monthly data. See text for details.

Month	MLD median (m)	v_{mix}[a] (m month^{-1})	Number of sampled months[b]
1	*118*	−29	2
2	147	−29	16
3	168	−21	3
4	*116*	−3	1
5	65	−3	14
6	30	−3	8
7	*25*	−3	1
8	21	−3	16
9	32	−11	4
10	37	−5	4
11	59	−22	14
12	*89*	−30	2

[a] v_{mix} is defined as negative to get a negative flux into the surface layer.
[b] This is the number of sampled months in the data set. For months sampled less than three times, interpolated numbers have been used.

3 Methods

This study is based on the climatology (long-term means) of the hydrographical and chemical properties observed in the Iceland Sea. We calculated long-term monthly mean profiles by averaging all data for every month, for the chosen depths (every 10 m in the upper 300 m, every 50 m between 300 and 500 m, and then every 100 m from 500 down to the bottom (1900 m)) and further interpolated to the chosen depth intervals, using piecewise cubic Hermite interpolation in Matlab® (e.g. Fritsch and Carlson, 1980).

The sampling frequency for the different months during the course of the time series sampling is shown in Table 1. The sampling program of the time series station is largely quarterly (February, May, August, and November), which is clearly seen in Table 1. Four months (January, April, July, and December) have been sampled less than three times, and for these months we use interpolated values.

The wintertime mixed layer in the Iceland Sea typically reaches down to 200 m at the end of the winter mixing (Ólafsson, 2003), which is supported by our calculated mean mixed layer depth (MLD) (Fig. 2). We tested several criteria for the MLD, based on either a difference in temperature ($\Delta T = 0.2\,°C$), or density ($\Delta\sigma_\theta = 0.01, 0.03, 0.05$, and $0.125\,kg\,m^{-3}$), all referenced to a near-surface value at 10 m (e.g. de Boyer Montégut et al., 2004). The temperature criteria gave unreasonably deep winter convection, with median values of 600–800 m. All density criteria were shallower, however, the $0.125\,kg\,m^{-3}$ criterion gave a median winter MLD of nearly 400 m, which is not supported by depth profiles of hydrography or biogeochemical parameters (Fig. 3), or by previous estimates (e.g. Ólafsson, 2003). The den-

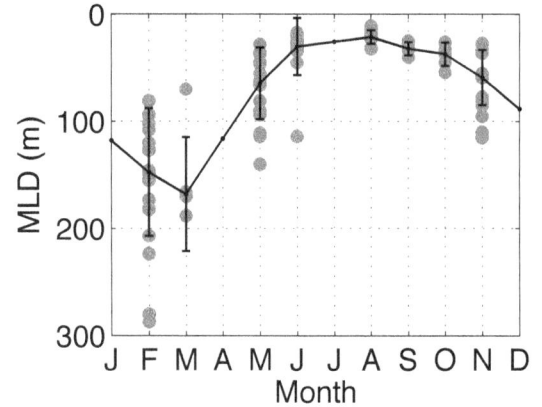

Figure 2. Calculated mixed layer depth (MLD) at the Iceland Sea time series station, using the density difference criteria of $\Delta\sigma_\theta$ $0.05\,kg\,m^{-3}$. The grey dots show the MLD for each year, and the line is the median of the values for each month, and the error bars show the standard deviation (SD). The values for the months without shown data are interpolated.

sity difference criteria $\Delta\sigma_\theta = 0.05\,kg\,m^{-3}$ showed the highest agreement with Ólafsson (2003) and was also used by Zhai et al. (2012), which is why we adopted this criteria in the present study. However, the seasonal drawdown in nutrients and DIC (see Fig. 3) is largely confined to the upper 100 m. Based on this we define the upper 100 m as the surface layer, and calculate the climatological fluxes in and out of this layer. The approach is described in detail below.

3.1 Calculation of deficits

We apply a box-model approach, which was developed for idealised annual plankton cycles (Evans and Parslow, 1985), and has been applied in, for example, the Greenland and the Norwegian seas (Anderson et al., 2000; Skjelvan et al., 2001; Falck and Anderson, 2005). Here we compute deficits (DEF) of nutrients and DIC in the surface layer relative to a defined sub-surface layer:

$$\mathrm{DEF}_X = \int_{100}^{0} ([X_{\mathrm{SSL}}] - [X_{\mathrm{SL}}]),\qquad(1)$$

where X is the concentration of the constituent of interest (here nutrients and DIC), SSL is the sub-surface layer, and SL is the surface layer. Thus the deficit increases when there is a decrease in carbon or nutrients in the surface layer. While the surface layer is chosen to be the upper 100 m, the sub-surface layer is defined as the layer between 100 and 200 m, for which monthly mean concentrations are calculated and applied in Eq. (1). Applying this on the monthly mean profiles, the deficits are calculated for every 10 m interval in the upper 100 m, relative to the monthly mean concentration in the sub-surface layer, multiplied with 10, and summed up for each month (Anderson et al., 2000).

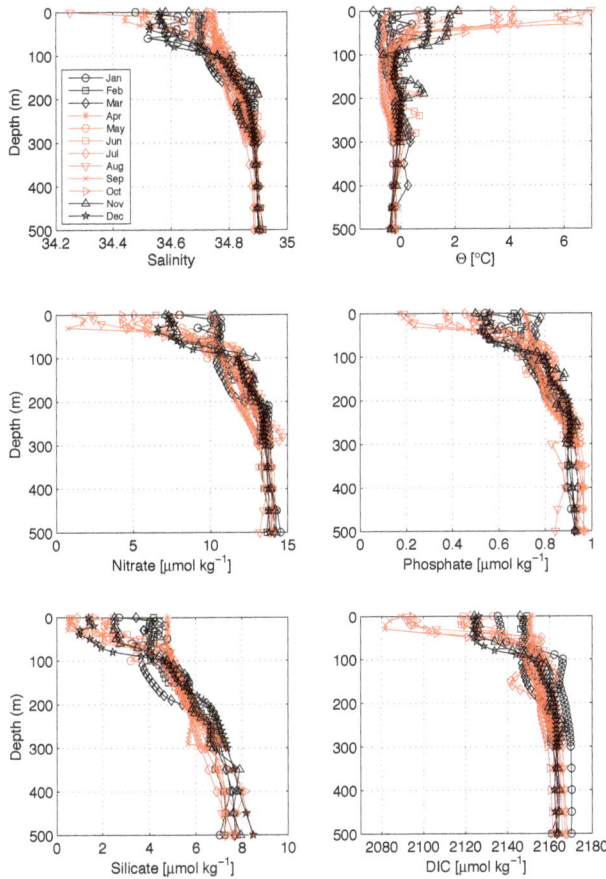

Figure 3. Mean monthly concentration profiles (upper 500 m) in the Iceland Sea, of salinity (upper left), potential temperature (upper right), nitrate (middle left), phosphate (middle right), silicate (lower left), and DIC (lower right). The black profiles indicate months with an increase in MLD (compared to previous month) and the red profiles depict months with a decreased or very shallow (< 40 m) MLD (see Fig. 2).

3.2 Flux calculations

The change in the deficit ($\Delta\mathrm{DEF}^X$) of constituent X are explained by the sum of the fluxes into and out of the surface layer; the vertical exchange with the deeper layers (F_{vert}), the horizontal fluxes (F_{hor}), the biological production (F_{bio}), and the air–sea exchange (F_{atm}):

$$\Delta\mathrm{DEF}^X = F_{\mathrm{vert}}^X + F_{\mathrm{hor}}^X + F_{\mathrm{bio}}^X + F_{\mathrm{atm}}^X. \tag{2}$$

Positive fluxes indicate a transport out of the surface layer. Regarding the time series station as a very thin section the horizontal fluxes will balance, and F_{hor} could then be set to zero. We also assume no atmospheric input of nutrients, and thus F_{atm} is only of importance for the calculations of the DIC fluxes. The uncertainty in the different fluxes is estimated from error propagation of the standard deviations of the different terms in the flux calculations. The uncertainties are discussed in Sect. 6.

The vertical flux to the surface layer can be calculated from Eq. (3) (Anderson et al., 2000; Skjelvan et al., 2001; Falck and Anderson, 2005):

$$F_{\mathrm{vert}}^X = \frac{v_{\mathrm{mix}}}{H}\mathrm{DEF}^X, \tag{3}$$

where v_{mix} is the vertical entrainment velocity, and H is the thickness of the surface layer. We estimate v_{mix} through changes in the calculated mixed layer depth (following, for example, Skjelvan et al., 2001), and apply this for the periods with a deepening of the mixed layer, which is the period from September to March seen from the development of the MLD (Fig. 2). During the period from April to August there is a decrease in the MLD, and for this period we apply a background mixing through the base of the mixed layer of 0.1 m d^{-1} (Anderson et al., 2000; Skjelvan et al., 2001), which corresponds to a shallowing of 3.0 m month^{-1}. The applied entrainment velocities are shown in Table 1. We here define v_{mix} as negative to get a negative flux when directed into the surface layer.

The flux due to biological activity is given by Eq. (4):

$$F_{\mathrm{bio}}^X = \Delta\mathrm{DEF}^X - F_{\mathrm{vert}}^X - F_{\mathrm{atm}}^X. \tag{4}$$

For the nutrients we assume a negligible atmospheric source, but when calculating the biological production from DIC, F_{bio} needs to be corrected for the air–sea flux (see below). The resulting fluxes are positive as long as the production is greater than the decay of organic matter, as is the case when there is a net biological uptake, removing DIC and nutrients from the surface layer.

The air–sea flux of carbon can be calculated from the difference in partial pressure of CO_2 between seawater and air, the gas transfer velocity k, and the solubility of CO_2 in seawater, K_0:

$$F_{\mathrm{atm}} = k K_0 \Delta p CO_2, \tag{5}$$

where

$$\Delta p CO_2 = p CO_2^{\mathrm{sea}} - p CO_2^{\mathrm{air}}. \tag{6}$$

The solubility of CO_2 in the Iceland Sea surface water was calculated after Weiss (1974), using long-term monthly mean values of salinity and temperature in the upper 30 m. For the dependence of wind speed on the transfer velocity k we used the parameterisation of Sweeney et al. (2007) after Wanninkhof (1992):

$$k = 0.27u^2\sqrt{\frac{660}{Sc}}, \tag{7}$$

where u is the long-term surface wind speed (m s^{-1}), and Sc is the Schmidt number. The transfer coefficient was then converted to m month^{-1} by multiplying with $(365.25/12) \times (24/100)$.

Table 2. Summary of annual fluxes ($mol\,m^{-2}\,yr^{-1}$) of carbon, nitrate, phosphate, and silicate to the surface layer (upper 100 m) of the Iceland Sea; vertical flux (F_{vert}), air–sea flux (F_{atm}), and biological production (F_{bio}). Negative values indicate a flux into the surface layer. The horizontal fluxes are assumed to balance over the year and were set to zero.

	F_{vert} ($mol\,m^{-2}\,yr^{-1}$)	F_{atm} ($mol\,m^{-2}\,yr^{-1}$)	F_{bio} ($mol\,m^{-2}\,yr^{-1}$)
Carbon	-2.9 ± 0.5	-4.4 ± 1.1	7.3 ± 1.0[a]
Nitrate	-0.45 ± 0.09	–	0.45 ± 0.14[b]
Phosphate	-0.026 ± 0.005	–	0.026 ± 0.010
Silicate	-0.26 ± 0.06	–	0.26 ± 0.16

[a] corresponds to NCP
[b] corresponds to new production

To calculate the partial pressure in the atmosphere from the molar fractions obtained from GLOBALVIEW we used the formulation:

$$pCO_{2,atm} = XCO_2 \left(P_b - P_w \right), \qquad (8)$$

where P_b is the barometric pressure (in atmospheres), and P_w is the saturation water vapour pressure calculated from temperature and salinity in the sea surface layer, according to Cooper et al. (1998). Monthly mean seawater pCO_2 values were calculated from observational data over the 13-year time period in the upper 30 m.

4 Results

The deficits of nutrients and DIC in the upper 100 m decrease from January to March (Fig. 4), as a result of the deepened mixed layer depth (Fig. 2). The increase in the deficits after March, related to biological production, continues until a maximum in September, after which the deficits decrease again. There is a small decrease in deficit in phosphate from May to June, which coincides with an almost unchanged deficit in silicate and a slower rate of change of DIC. At the same time the change in the nitrate deficit continues largely as before (Fig. 4). There is a significant uptake of nutrients from winter to late summer (Fig. 3), but on average the system, never gets fully depleted. The calculated fluxes deduced from a change in the deficits, related to vertical mixing, air–sea exchange, and biological production, are presented in the following section and are summarised in Table 2 and Fig. 5.

4.1 Vertical fluxes

The calculated vertical fluxes add carbon and nutrients to the mixed layer all year around, even though the fluxes during the period of shallow MLD are small. The annual vertical fluxes of DIC and nutrients to the mixed layer was estimated to be $2.9 \pm 0.5\,mol\,C\,m^{-2}\,yr^{-1}$, $0.45 \pm 0.09\,mol\,N\,m^{-2}\,yr^{-1}$, $0.026 \pm 0.005\,mol\,P\,m^{-2}\,yr^{-1}$,

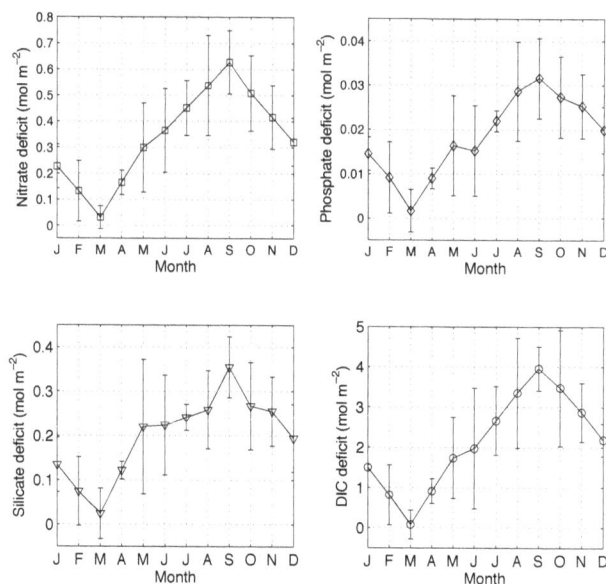

Figure 4. Calculated monthly-mean deficits of nitrate, phosphate, silicate, and carbon, in the upper 100 m in the Iceland Sea. For the calculations we used mean monthly values for the 100–200 m depth range as reference. The error bars show the propagated error (uncertainty) from the standard deviation of the respective reference concentrations and the average monthly standard deviation in the surface layer. As for the MLD calculations, for the months sampled less than three times in the time series we have used interpolated values. See text for details.

and $0.26 \pm 0.06\,mol\,Si\,m^{-2}\,yr^{-1}$, for DIC, nitrate, phosphate, and silicate, respectively. The flux of DIC equals $35\,g\,C\,m^{-2}\,yr^{-1}$. The presented uncertainties are calculated from error propagation of the terms in Eq. (3) (see details in Sect. 6.2).

4.2 Air–sea flux of CO_2

The air–sea flux is directed into the surface layer all year around, as the region is permanently undersaturated with respect to atmospheric CO_2 (Fig. 5). The calculated annual flux was $4.4 \pm 1.1\,mol\,C\,m^{-2}\,yr^{-1}$, which is consistent with the estimate of Ólafsson et al. (2009) of $4.5\,mol\,C\,m^{-2}\,yr^{-1}$. When converted, the calculated flux into the Iceland Sea is $53\,g\,C\,m^{-2}\,yr^{-1}$.

4.3 Biological production

The biologically related fluxes of carbon and nutrients all show a two-peak seasonality, with the first maximum in April–May, and a second, larger peak in September. Phosphate shows a slightly different evolution, with no flux in June, and a broader peak in late summer, with a small maximum in August. The nutrients also show a negative flux in October, when there is still a net uptake of carbon.

Table 3. Stoichiometric (median) ratios of computed monthly vertical fluxes and of biological production during the period of seasonal drawdown (net community uptake).

	Vertical flux[a] (annual)	Net uptake[a] (Apr–Sep)
N : P	18.4	18.2[b]
C : N	6.20	9.00
C : P	112	159[b]
C : Si	11.1	25.9
N : Si	1.67	2.72
Si : P	10.5	13.3[b]

[a] We use the median of the monthly values since some months show large deviations.
[b] Since the biologically related flux of phosphate is zero in June these numbers are only based on April–May, and July–September.

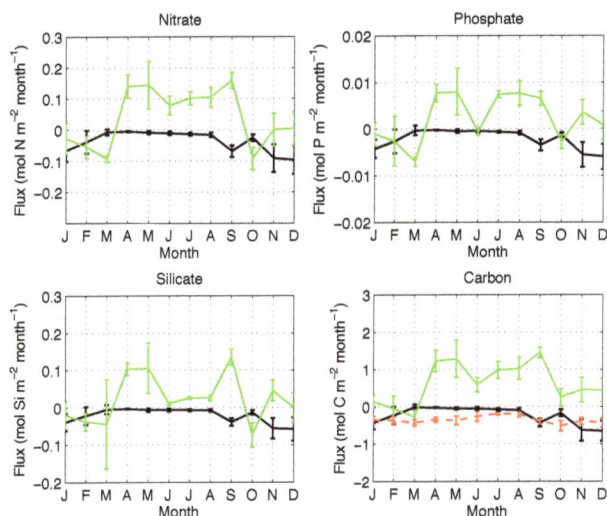

Figure 5. Calculated seasonal fluxes to the upper 100 m in the Iceland Sea, for nitrate, phosphate, silicate and DIC. All fluxes are in $\mathrm{mol\,m^{-2}\,month^{-1}}$. The figures show the vertical flux (F_{vert}; solid black line), the biological production (F_{bio}; green solid line), and the air–sea flux of CO_2 (F_{atm}; red dashed line for carbon). The error bars show the propagated errors (see Sect. 6). Note that the scale on the y axis is different for all constituents.

The change in the deficit (ΔDEF) equals zero over the course of the year, and hence there is a balance between the calculated fluxes (Eq. 2). For the nutrients, with the assumption of negligible horizontal and air–sea fluxes, there is a balance between the net vertical fluxes and the net biological fluxes, and the latter amounts to $0.45 \pm 0.14\,\mathrm{mol\,N\,m^{-2}\,yr^{-1}}$, $0.026 \pm 0.010\,\mathrm{mol\,P\,m^{-2}\,yr^{-1}}$, and $0.26 \pm 0.16\,\mathrm{mol\,Si\,m^{-2}\,yr^{-1}}$, respectively (Table 2). Following the definition of new production (Dugdale and Goering, 1967), and our assumptions of negligible horizontal and air–sea flux of nitrate, the addition of nitrate from vertical mixing must equal new production. In the Iceland Sea this amounts to $0.45 \pm 0.09\,\mathrm{mol\,N\,m^{-2}\,yr^{-1}}$.

The biologically driven change in DIC, corrected for vertical flux and air–sea exchange, corresponds to NCP, with positive numbers illustrating net autotrophy, and negative values net heterotrophy. There is a very small or negative NCP in the first part of the year, but from March to October there is a net autotrophic production (Fig. 5). There is also a small positive NCP in December, but this could be due to the fact that the values have been interpolated because there is less data available in December and January. This will not be discussed further.

The net annual NCP corresponds to the export production, when assuming steady state. In the Iceland Sea this sums up to $7.3 \pm 1.0\,\mathrm{mol\,C\,m^{-2}\,yr^{-1}}$, or $88 \pm 12\,\mathrm{g\,C\,m^{-2}\,yr^{-1}}$.

The seasonal drawdown of nitrate, corresponding to the period of net community uptake (i.e. increasing deficit; April to September; see Fig. 4), relates to the total production. This period shows positive biological fluxes, and the sum of these amounts to $0.72 \pm 0.10\,\mathrm{mol\,N\,m^{-2}\,yr^{-1}}$. The difference between the new and total production ($0.27 \pm 0.15\,\mathrm{mol\,N\,m^{-2}\,yr^{-1}}$) gives the regenerated production, which represents 37 % of the total production. Then we get an f ratio (i.e. the ratio between new and total production) of 0.63 in the Arctic domain of the Iceland Sea. Performing the same calculations for phosphate and silicate

gives a total production of $0.036 \pm 0.006\,\mathrm{mol\,P\,m^{-2}\,yr^{-1}}$ and $0.40 \pm 0.07\,\mathrm{mol\,Si\,m^{-2}\,yr^{-1}}$.

4.4 Stoichiometry of the calculated fluxes

An evaluation of the stoichiometric relationships between carbon and nutrients show varying values during the year, as well as for the different fluxes (Table 3).

Evaluating the stoichiometry for the biological production is not straightforward since the flux of carbon and nitrate do not show the same direction for all months. The change in deficits of DIC and nitrate (Fig. 4), however, both show a net uptake from April to September, so we will use this period to evaluate the biologically related stoichiometry. The C : N ratios of the monthly biological production (Fig. 7), during the period of seasonal drawdown of DIC and nitrate, differ between the early and the late part of the season, with C : N ratios of 8.8–8.9 in April and May, and 9.1–9.8 between July and September, while the value in June is 7.4.

5 Discussion

5.1 Primary production in the central Iceland Sea

The main aim of this study is to investigate primary production and related stoichiometry in the central Iceland Sea. This domain is dominated by Arctic waters, and is the least productive of the waters around Iceland (e.g. Gudmundsson, 1998; Assthorsson et al., 2007). However, it could be repre-

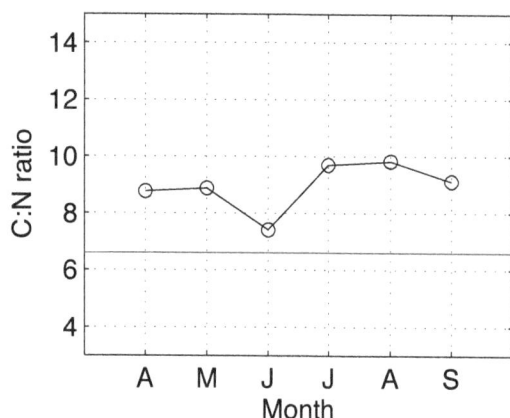

Figure 6. Average monthly C : N ratios for biological production (see Fig. 5) during the period of seasonal drawdown (April–September) of DIC and nitrate in the Iceland Sea. Then red line show the Redfield C : N ratio of 6.6.

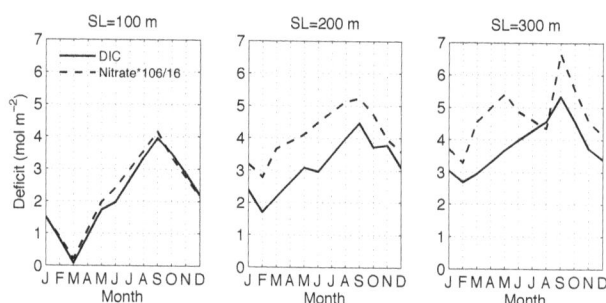

Figure 7. Comparison of calculated monthly-mean deficits of DIC and nitrate in the Iceland Sea, for different thickness of the surface layer (SL). The nitrate deficits are multiplied with the Redfield C : N ratio of 6.6.

sentative of the whole Arctic domain in the Nordic Seas, with similar hydro-chemical properties.

How realistic is our estimated annual net production (NCP) of $88 \pm 12 \, \mathrm{g \, C \, m^{-2} \, yr^{-1}}$ in the Iceland Sea? Gudfinnsson (2012) found, from his data of daily productivity, an average annual phytoplankton productivity of $65 \, \mathrm{g \, C \, m^{-2} \, yr^{-1}}$, and Thordardottir (1984) presented an average annual primary production (1958–1982) in the Arctic domain, in the vicinity of the time series station, of $75 \, \mathrm{g \, C \, m^{-2} \, yr^{-1}}$, based on measured ^{14}C uptake at light saturation. A modelling study (Skogen et al., 2007), suggests a mean annual production in the Iceland Sea at $70 \, \mathrm{g \, C \, m^{-2} \, yr^{-1}}$, with an f ratio of ~ 0.7. These estimates show a large agreement with the estimates in our study, giving more trust in our results, and the approach. The uncertainty in our presented fluxes, and the approach in general, are discussed in Sect. 6.

From remote sensing data, Zhai et al. (2012) gave a production estimate in the Arctic domain of $179 \pm 36 \, \mathrm{g \, C \, m^{-2} \, yr^{-1}}$. This is more than twice as high as the estimates based on in situ data. This has also been seen in other comparisons between production estimates based on in situ and remote sensing data (e.g. Richardson et al., 2005; Körtzinger et al., 2008; Frigstad et al., 2015).

The negative nutrient flux in October, when there is still a net uptake of carbon (Fig. 5), is similar to what have been observed in the Norwegian Sea (Falck and Anderson, 2005), which were explained largely by a build-up of dissolved organic matter (DOM), which is relatively low in nutrients. We will discuss this further below, in relation to the stoichiometry of the production.

5.2 Variable stoichiometry

The evaluation of the C : N ratios during seasonal drawdown (April to September) of DIC and nitrate (Fig. 6) showed a clear deviation from the Redfield C : N ratio of 6.6, except in June, when the production was lower. The consumption of carbon relative to nitrate in excess of Redfield, a phenomena termed "carbon overconsumption" (Toggweiler, 1993), was higher during the late summer production (C : N ratio > 9) compared to the early production peak (C : N ratio < 9). Similar increases in carbon overconsumption during the later part of the productive season have been described in several studies from different ocean regions, and have been explained by the build up of low-nitrogen DOM (e.g. Toggweiler, 1993; Williams, 1995; Kähler and Koeve, 2001; Körtzinger et al., 2001). Without any data of DOM in the central Iceland Sea we cannot find direct evidence supporting this mechanism in our study, but the similarity to the Atlantic-dominated Norwegian Sea (Falck and Anderson, 2005) suggest that this may be a general feature also in the Nordic Seas. This should be evaluated further in the future. Nonetheless, different mechanisms seem to affect the flux of carbon and nitrogen during the season, as shown for different regions (e.g. Banse, 1994; Kähler and Koeve, 2001; Frigstad et al., 2011).

If we compare the total new production and NCP during the year, from the values in Table 2, we get a net C : N ratio of 16.2. This means that, if we were to convert the computed new production into export production, using the ratios of Redfield et al. (1963) (6.6), or Takahashi et al. (1993) (7.3), we would underestimate the export production by 55–60 %, assuming our estimated export production is reasonable. This confirms the findings of, Sambrotto et al. (1993), who found that the actual carbon production exceeds any estimate based on nitrogen consumption, converted by the Redfield C : N ratio, by 36–81 %.

This illustrates the problem in converting new production into NCP, or export production, using constant stoichiometric ratios. As discussed by Laws (1991) these terms may not be related, and would assume that nitrate and carbon are assimilated by autotrophs during new production, in the same ratio as carbon and nitrate are recycled by heterotrophs. Furthermore, C : N ratios have been observed to differ both between seasons (e.g. Körtzinger et al., 2001; Frigstad et al., 2011) and between regions (e.g. Koeve, 2006; Tamelander et al., 2013; Frigstad et al., 2014), with values as high as ~ 15.

An evaluation of the relationship between DIC and nitrate in the surface water using the time series data (not shown) gives a high agreement with the estimated stoichiometry in the region by Takahashi et al. (1993). However, this value represents the relationship between measured properties in the surface waters over the year, which includes the net effect of air–sea exchange, biological activities, and mixing. Due to this, Banse (1994) cautioned against using observed in situ DIC : nitrate relationships to make statements about elemental ratios during biological production, and respiration, and recommended smaller closed, controllable systems to find mechanistic explanations of uptake ratios in the surface layer.

5.3 Comparison to production estimates for other parts of the Nordic Seas

How representative of the Nordic Seas are our estimated production terms in the Iceland Sea? The average NCP in the Nordic Seas, based on an oxygen budget, have been estimated to $\sim 36\,\mathrm{g\,C\,m^{-2}\,yr^{-1}}$ (Falck and Gade, 1999). This is roughly half of the annual NCP we find in the central Iceland Sea. However, to evaluate regional differences we compare with estimates for the different basins in the area.

For the Greenland Sea, Richardson et al. (2005) estimated the annual primary production to $81\,\mathrm{g\,C\,m^{-2}\,yr^{-1}}$, or $70\,\mathrm{g\,C\,m^{-2}\,yr^{-1}}$, if excluding observations within the ice or at the ice edge. Anderson et al. (2000) estimated the annual new production, in the upper 150 m, of $34\,\mathrm{g\,C\,m^{-2}\,yr^{-1}}$, based on a box model similar to ours, and nitrate data (using a C : N ratio of 7.5). With an f ratio of 0.56 (Smith, 1993) this corresponds to a total production of $61\,\mathrm{g\,C\,m^{-2}\,yr^{-1}}$ (Richardson et al., 2005). The likely range of annual primary production in the Greenland Sea is in the range 60–$100\,\mathrm{m^{-2}\,yr^{-1}}$ (Richardson et al., 2005), which is in agreement with the range of estimates for the Iceland Sea.

In the Norwegian Sea, the primary production has been estimated to $80\,\mathrm{g\,C\,m^{-2}\,yr^{-1}}$ (Rey, 2004) and that the new production is 60 % of that. It has also been pointed out that where zooplankton grazing is high, as in the Norwegian Sea, new production may be underestimated (Bathmann et al., 1990) and could be as high as 80 %. Results from a modelling study (Skogen et al., 2007), suggests a mean annual production in the Norwegian Sea at $65\,\mathrm{g\,C\,m^{-2}\,yr^{-1}}$, with an f ratio of ~ 0.75.

Falck and Anderson (2005) used a box model approach similar to the present study, and for the Norwegian Sea, they assumed the export production to correspond to the vertical flux of nutrients to the surface layer (upper 100 m),

which equalled $0.23\,\mathrm{N\,m^{-2}\,yr^{-1}}$, or $18\,\mathrm{g\,C\,m^{-2}\,yr^{-1}}$; when using the traditional Redfield C : N ratio (6.6). Their new production estimate amounted to $0.51\,\mathrm{mol\,N\,m^{-2}\,yr^{-1}}$, or $41\,\mathrm{g\,C\,m^{-2}\,yr^{-1}}$, using the same ratio. If equating their vertical flux of nitrate with new production, and their total production with the sum of all positive biological fluxes during the year, we get an f ratio of 0.43. This is clearly lower than the earlier estimates mentioned above (Rey, 2004; Skogen et al., 2007).

Earlier estimates of new production in the Norwegian Sea (70° N, 0° E) are in the range 21–$29\,\mathrm{g\,C\,m^{-2}\,yr^{-1}}$ (Bodungen et al., 1995). These values agree with estimates of NCP, based on oxygen fluxes in the Norwegian Sea, of ~ 24–$32\,\mathrm{g\,C\,m^{-2}\,yr^{-1}}$ (Skjelvan et al., 2001). The new production estimate is in reasonable agreement with what we estimate for the Iceland Sea, but it is clear that previous NCP estimates based on oxygen budgets are significantly lower than what we get in the Iceland Sea. This could partly be due to the oxygen-to-carbon conversion applied, mostly based on the traditional Redfield ratio, but the only way to unravel real or artificial differences is to analyse the whole region with the same method. This should be pursued in the near future to investigate regional differences, but also to evaluate trends and changes in the system. Nevertheless, the range of methods and approaches, both based on observations and models, and different assumptions, including ours, still seems to reach some consensus of annual primary production in the Nordic Seas of ~ 60–$100\,\mathrm{g\,C\,m^{-2}\,yr^{-1}}$. More work is needed to evaluate regional similarities and differences in stoichiometry and any temporal trends in primary production. This will aid understanding of the variability drivers in biological production, both natural and anthropogenic, and how the increasing levels of atmospheric CO_2 will affect the biological carbon pump.

6 Uncertainties

One obvious source of error is the fact that our approach only makes long-term averages for all months, so any trends in the observed properties will cause some uncertainty in the resulting values. With this in mind we proceed to evaluate the uncertainty of the approach and the individual fluxes.

6.1 Deficit calculations

The uncertainties in the deficit calculations are related to the interannual variability in the observed concentrations in the surface layer and in the sub-surface reference concentrations, and the uncertainties arising from the averaging procedures of the monthly profiles. The uncertainty in the monthly surface layer concentrations (seen from the average monthly standard deviation) is largest for silicate (values up to 40–50 %), but for nitrate and phosphate there is a maximum in late summer/early autumn, when the concentrations are lowest by 20–30 %. Due to the high concentrations of DIC the uncertainty in these numbers is insignificant. If we propagate the uncertainties in the surface concentrations and the reference concentrations and use this as the overall uncertainty in the monthly deficits we get the values depicted in Fig. 4, which are quite substantial for some of the months, with a relative error of up to 60–75 % at or just after the early peak in production, but lower (10–40 %) during the later part of the year. The uncertainty in the values from the first part of the year, during the period of deepened mixed layer, is rather low in an absolute sense, compared to later in the year, but due to the low deficits in this period the relative errors get very large (see Fig. 4).

There is a potential error in assessing the production, and related terms, in the upper 100 m, when the MLD apparently reaches deeper in winter. However, the vertical distribution of nutrients and DIC do show a homogeneous upper 100 m in winter, followed by a gradient down to stable concentration at depths below ~ 300 m. Profiles of salinity show the same feature (Fig. 3). Deficits were also calculated for the upper 200 m (referenced to the monthly means between 100 and 200 m), and the upper 300 m (referenced to the monthly means between 300 and 400 m). The resulting deficits of carbon and nutrients showed an increasing degree of decoupling with increasing depth of the surface layer, as shown in Fig. 7. The C:N ratio during the period of net biological uptake also varies considerably more with thicker surface layer (not shown) compared to the upper 100 m. With a surface layer down to 200 m the C:N uptake ratio is 20 during the spring peak, below 4 in June, and shows values between 13 and 19 from July to September. A surface layer of 300 m gives C:N uptake ratios of 10 during the spring peak, followed by negative values during summer, and a value of 4 in September. This suggests that processes other than biological assimilation contributed much more to the distribution of nutrients and carbon at these depths Since we mainly want to evaluate the fluxes of importance for the production, and these seem to be confined to the upper 100 m, we argue that the applied method best captures the biological production with the relatively shallow surface layer we use. This may also be connected to the different water masses present in the Iceland Sea, so it is important to evaluate different surface layer thickness in different regions.

6.2 Vertical flux

The uncertainty in the vertical fluxes could be significant. With the assumption that the air–sea fluxes, as well as the horizontal fluxes of nutrients could be neglected, the increase in nutrient concentration during periods of deepened mixed layer depths should equal the vertical fluxes. Since we estimate the vertical entrainment velocity from the observed changes in MLD, there is both an uncertainty related to the chosen method to calculate MLD, and the variability in the monthly MLD during the time series. The variability-driven uncertainty in the mean monthly MLD is on average

~ 30 % (Fig. 2). The calculated uncertainty in the vertical fluxes of DIC, and nutrients are all in the range 17–22 % (see Table 2).

6.3 Air–sea exchange

From the propagation of the errors due to spread in mean $p\mathrm{CO}_2$ values for atmosphere and sea surface, and putting this error estimate in the flux calculation for each month, we get an annual uncertainty of 1.1 mol C m^{-2}, which is 25 % of the estimated annual flux. This agrees with previous findings from the North Atlantic and the Nordic Seas (Körtzinger et al., 2001; Olsen et al., 2003). Körtzinger et al. (2008) have estimated a maximum error in calculated CO$_2$ fluxes of 40 %.

6.4 Biological production

Since the biological production is calculated as the residual of all other terms (Eq. 4) it also carries the uncertainty of each of these terms. Some of the uncertainty could be connected to interannual variability in the timing of the peak in the productive events, something that should be evaluated further in later studies. To estimate the uncertainty in the ΔDEF term we use the relative error in the calculated deficits, and multiply these with the ΔDEF values for each month, for each constituent. The relative error in the deficit for the months with very low values (February–March) is unrealistically large. For these months we instead use the uncertainty in MLD as the minimum error. For February this is ~ 50 %, and for March ~ 30 %. The total estimated errors in the biologically related fluxes are in the range 31–61 % for the nutrients (highest for silicate), but only 14 % for carbon (Table 2).

7 Conclusions

The computed monthly fluxes of dissolved inorganic carbon, nitrate, phosphate and silicate in the Iceland Sea show similarities in the seasonality, but also a decoupling during the year, illustrating different mechanisms effecting the uptake and remineralisation of the different constituents. We estimate an Iceland Sea new production of 0.45 ± 0.09 mol N m^{-2} yr^{-1}, based on nitrate added to the surface layer via vertical mixing, and an annual net community production (NCP) of 7.3 ± 1.0 mol C m^{-2} yr^{-1} (or 88 ± 12 g C m^{-2} yr^{-1}). The presented NCP shows a high agreement with earlier estimates of primary production in the Iceland Sea, and to other parts of the Nordic Seas. The estimated C:N ratios during net biological uptake are in the range 7.4–9.8, and thus indicate that a conversion of the nitrate-based new production to carbon using traditional Redfield C:N would markedly underestimate the primary production in the Iceland Sea.

Acknowledgements. We thank the two anonymous reviewers for helpful comments. This research was supported from the European Union FP7 projects GreenSeas (265294), EURO-BASIN (264933), and, CarboChange (264879). Siv Lauvset is acknowledged for valuable help with netctd files.

Edited by: E. Marañón

References

Anderson, L. A. and Sarmiento, J. L.: Redfield ratios of remineralization determined by nutrient data analysis, Global Biogeochem. Cy., 8, 65–80, 1994.

Anderson, L. G., Drange, H., Chierici, M., Fransson, A., Johannessen, T., Skjelvan, I., and Rey, F.: Annual carbon fluxes in the upper Greenland Sea based on measurements and a box-model approach, Tellus B, 52, 1013–1024, 2000.

Astthorsson, O. S., Gislason, A., and Jonsson, S.: Climate variability and the Icelandic marine ecosystem, Deep-Sea Res. PT II, 54, 2456–2477, 2007.

Banse, K.: Uptake of inorganic carbon and nitrate by marine plankton and the Redfield Ratio, Global Biogeochem. Cy., 8, 81–84, 1994.

Bathmann, U. V., Peinert, R., Noji, T. T., and Bodungen, B. V.: Pelagic origin and fate of sedimenting particles in the Norwegian Sea, Prog. Oceanogr., 24, 117–125, 1990.

Bodungen, B. V., Anita, A., Bauerfeind, E., Haupt, O., Koeve, W., Machado, E., Peeken, I., Peinert, R., Reitmeier, S., Thomsen, C., Voss, M., Wunsch, M., Zeller, U., and Zeitzschel, B.: Pelagic processes and vertical flux of particles: an overview of a long-term comparative study in the Norwegian Sea and Greenland Sea, Geol. Rundsch., 84, 11–27, 1995.

Church, M. J., Lomas, M. W., and Muller-Karger, F.: Sea change: Charting the course for biogeochemical ocean time-series research in a new millennium, Deep-Sea Res. PT II, 93, 2–15, 2013.

Cooper, D. J., Watson, A. J., and Ling, R. D.: Variations of P_{CO_2} along a North Atlantic shipping route (UK to the Caribbean): A year of automated observations, Mar. Chem., 60, 147–164, 1998.

Daly, K. L., Wallace, D. W. R., Smith, W. O., Jr., Skoog, A., Lara, R., Gosselin, M., Falck, E., and Yager, P. L.: Non-Redfield carbon and nitrogen cycling in the Arctic: Effects of ecosystem structure and dynamics, J. Geophys. Res., 104, 3185–3199, 1999.

de Boyer Montégut, C., Madec, G., Fischer, A. S., Lazar, A., and Iudicone, D.: Mixed layer depth over the global ocean: An examination of profile data and a profile-based climatology, J. Geophys. Res., 109, C12003, doi:10.1029/2004JC002378, 2004.

Dugdale, R. C. and Goering, J. J.: Uptake of new and regenerated forms of nitrogen in primary productivity, Limnol. Oceanogr., 23, 196–206, 1967.

Eppley, R. W. and Peterson, B. J.: Particulate organic matter flux and planktonic new production in the deep ocean, Nature, 282, 677–680, 1979.

Evans, G. T. and Parslow, J. S.: A model of annual plankton cycles, Biol. Oceanogr., 3, 327–347, 1985.

Falck, E. and Anderson, L. G.: The dynamics of the carbon cycle in the surface water of the Norwegian Sea, Mar. Chem., 94, 43–53, 2005.

Falck, E. and Gade, H. G.: Net community production and oxygen fluxes in the Nordic Seas based on O_2 budget calculations, Global Biogeochem. Cy., 13, 1117–1126, 1999.

Falkowski, P. G., Barber, R. T., and Smetacek, V.: Biogeochemical Controls and Feedbacks on Ocean Primary Production, Science, 281, 200–206, 1998.

Frigstad, H., Andersen, T., Hessen, D. O., Naustvoll, L.-J., Johnsen, T. M., and Bellerby, R. G. J.: Seasonal variation in marine C : N : P stoichiometry: can the composition of seston explain stable Redfield ratios?, Biogeosciences, 8, 2917–2933, doi:10.5194/bg-8-2917-2011, 2011.

Frigstad, H., Andersen, T., Bellerby, R. G. J., Silyakova, A., and Hessen, D. O.: Variation in the seston C : N ratio of the Arctic Ocean and pan-Arctic shelves, J. Marine Syst., 129, 214–223, 2014.

Frigstad, H., Henson, S. A., Hartman, S. E., Cole, H., Omar, A. M., Jeansson, E., Pebody, C., and Lampitt, R. S.: Links between surface productivity and deep ocean particle flux at the Porcupine Abyssal Plain (PAP) sustained observatory, in preparation, 2015.

Fritsch, F. N. and Carlson, R. E.: Monotone Piecewise Cubic Interpolation, SIAM J. Numer. Anal., 17, 238–246, 1980.

Gaarder, T. and Gran, H. H.: Investigation of the production of phytoplankton in the Oslo Fjord, Rapp. P.V. Cons. Int. Explor. Mer, 42, 1–48, 1927.

GLOBALVIEW-CO2: Cooperative Atmospheric Data Integration Project – Carbon Dioxide, NOAA, GMD, Boulder, Colorado, available via anonymous FTP to ftp://ftp.cmdl.noaa.gov, Path: ccg/co2/GLOBALVIEW, 2012.

Guðfinnsson, H. G.: Breytingar á blaðgrænumagni, frumframleiðni og tegundasamsetningu svifþörunga í Íslandhafi/Changes in chlorophyll a, primary production and species composition in the Iceland Sea. Technical Report Hafrannsóknastofnunin (Marine Research Institute), Reykjavík, 164, 45–67, 2012 (in Icelandic).

Gudmundsson, K.: Long-term variation in phytoplankton productivity during spring in Icelandic waters, ICES J. Mar. Sci., 55, 635–643, 1998.

Kähler, P. and Koeve, W.: Marine dissolved organic matter: can its C : N ratio explain carbon overconsumption?, Deep-Sea Res. PT I, 4, 49–62, 2001.

Kalnay, E., Kanamitsu, M., Kistler, R., Collins, W., Deaven, D., Gandin, L., Iredell, M., Saha, S., White, G., Woollen, J., Zhu, Y., Leetmaa, A., Reynolds, R., Chelliah, M., Ebisuzaki, W., Higgins, W., Janowiak, J., Mo, K. C., Ropelewski, C., Wang, J., Jenne, R., and Joseph, D.: The NCEP/NCAR 40-Year Reanalysis Project, B. Am. Meteorol. Soc., 77, 437–471, 1996.

Koeve, W.: C : N stoichiometry of the biological pump in the North Atlantic: Constraints from climatological data, Global Biogeochem. Cy., 20, GB3018, doi:10.1029/2004GB002407, 2006.

Körtzinger, A., Koeve, W., Kähler, P., and Mintrop, L.: C : N ratios in the mixed layer during the productive season in the northeast Atlantic Ocean, Deep-Sea Res. PT I, 48, 661–688, 2001.

Körtzinger, A., Send, U., Lampitt, R. S., Hartman, S., Wallace, D. W. R., Karstensen, J., Villagarcia, M. G., Llinás, O., and DeGrandpre, M. D.: The seasonal pCO_2 cycle at 49° N/16.5° W in the northeastern Atlantic Ocean and what it tells us about biological productivity, J. Geophys. Res., 113, C04020, doi:10.1029/2007JC004347, 2008.

Laws, E. A.: Photosynthetic quotient, new production and net community production in the open ocean, Deep-Sea Res., 38, 143–167, 1991.

Ólafsson, J.: Winter mixed layer nutrients in the Irminger and Iceland Seas, 1990–2000, ICES Mar. Sc., 219, 329–332, 2003.

Ólafsson, J. and Olsen, A.: Nordic Seas nutrients data in CARINA, Earth Syst. Sci. Data, 2, 205–213, doi:10.5194/essd-2-205-2010, 2010.

Ólafsson, J., Ólafsdottir, S. R., Benoit-Cattin, A., Danielsen, M., Arnarson, T. S., and Takahashi, T.: Rate of Iceland Sea acidification from time series measurements, Biogeosciences, 6, 2661–2668, doi:10.5194/bg-6-2661-2009, 2009.

Ólafsson, J., Ólafsdottir, S. R., Benoit-Cattin, A., and Takahashi, T.: The Irminger Sea and the Iceland Sea time series measurements of sea water carbon and nutrient chemistry 1983–2008, Earth Syst. Sci. Data, 2, 99–104, doi:10.5194/essd-2-99-2010, 2010.

Olsen, A., Bellerby, R. G. J., Johannessen, T., Omar, A. M., and Skjelvan, I.: Interannual variability in the wintertime air-sea flux of carbon dioxide in the northern North Atlantic, 1981–2001, Deep-Sea Res. PT I, 50, 1323–1338, 2003.

Olsen, A.: Nordic Seas total dissolved inorganic carbon data in CARINA, Earth Syst. Sci. Data, 1, 35–43, doi:10.5194/essd-1-35-2009, 2009.

Pálsson, Ó. K., Gislason, A., Gudfinnsson, H. G., Gunnarsson, B., Ólafsdóttir, S. R., Petursdottir, H., Sveinbjörnsson, S., Thorisson, K., and Valdimarsson, H.: Ecosystem structure in the Iceland Sea and recent changes to the capelin (Mallotus villosus) population, ICES J. Mar. Sci., 69, 1242–1252, 2012.

Passow, U. and Carlson, C.A.: The biological pump in a high CO_2 world, Mar. Ecol.-Prog. Ser., 470, 249–271, 2012.

Platt, T., Harrison, W. G., Lewis, M. R., Li, W. K. W., Sathyendranath, S., Smith, R. E., and Vezina, A. F.: Biological production of the oceans: the case for a consensus, Mar. Ecol.-Prog. Ser., 52, 77–88, 1989.

Quay, P., Stutsman, J., and Steinhoff, T.: Primary production and carbon export rates across the subpolar N. Atlantic Ocean basin based on triple oxygen isotope and dissolved O_2 and Ar gas measurements, Global Biogeochem. Cy., 26, GB2003, doi:10.1029/2010GB004003, 2012.

Redfield, A. C., Ketchum, B. H., and Richards, F.A.: The influence of organisms on the composition of seawater, in: The Sea, edited by: Hill, M. N., John Wiley, New York, 26–77, 1963.

Reuer, M. K., Barnett, B. A., Bender, M. L., Falkowski, P. G., and Hendricks, M. B.: New estimates of southern ocean biological production rates from $O_2\,Ar^{-1}$ ratios and the triple isotope composition of O_2, Deep-Sea Res. PT I, 54, 951–974, 2007.

Rey, F.: Phytoplankton: the grass of the sea, in: The Norwegian Sea Ecosystem, edited by: Skjoldal, H. R., Tapir Academic Press, Trondheim, Norway, 97–136, 2004.

Richardson, K., Markager, S., Buch, E., Lassen, M. F., and Kristensen, A. S.: Seasonal distribution of primary production, phytoplankton biomass and size distribution in the Greenland Sea, Deep-Sea Res. PT I, 52, 979–999, 2005.

Sabine, C. L., Feely, R. A., Gruber, N., Key, R. M., Lee, K., Bullister, J. L., Wanninkhof, R., Wong, C. S., Wallace, D. W. R., Tilbrook, B., Millero, F. J., Peng, T.-H., Kozyr, A., Ono, T., and Rios, A. F.: The oceanic sink for anthropogenic CO_2, Science, 305, 367–371, 2004.

Sambrotto, R. N., Savidge, G., Robinson, C., Boyd, P., Takahashi, T., Karl, D. M., Langdon, C., Chipman, D., Marra, J., and Codispoti, L.: Elevated consumption of carbon relative to nitrogen in the surface ocean, Nature, 363, 248–250, 1993.

Skjelvan, I., Falck, E., Anderson, L. G., and Rey, F.: Oxygen fluxes in the Norwegian Atlantic Current, Mar. Chem., 73, 291–303, 2001.

Skjelvan, I., Falck, E., Rey, F., and Kringstad, S. B.: Inorganic carbon time series at Ocean Weather Station M in the Norwegian Sea, Biogeosciences, 5, 549–560, doi:10.5194/bg-5-549-2008, 2008.

Skogen, M. Budgell, W. P., and Rey, F.: Interannual variability in Nordic Seas primary production, ICES J. Mar. Sci., 64, 889–898, 2007.

Smith, W. O.: Nitrogen uptake and new production in the Greenland Sea: The spring Phaeocystis bloom, J. Geophys. Res., 98, 4681–4688, 1993.

Sweeney, C., Gloor, E., Jacobson, A. R., Key, R. M., McKinley, G., Sarmiento, J. L., and Wanninkhof, R.: Constraining global air-sea gas exchange for CO_2 with recent bomb ^{14}C measurements, Global Biogeochem. Cy., 21, GB2015, doi:10.1029/2006GB002784, 2007.

Swift, J. H. and Aagaard, K.: Seasonal transitions and water mass formation in the Iceland and Greenland seas, Deep-Sea Res., 28, 1107–1129, 1981.

Tamelander, T., Reigstad, M., Olli, K., Slagstad, D., and Wassmann, P.: New production regulates export stoichiometry in the ocean, PLoS ONE, 8, e54027, doi:54010.51371/journal.pone.0054027, 2013.

Takahashi, T., Ólafsson, J., Goddard, J. G., Chipman, D. W., and Sutherland, S. C.: Seasonal variations of CO_2 and nutrients in the high-latitude surface oceans: A comparative study, Global Biogeochem. Cy., 7, 843–878, 1993.

Thordardottir, T.: Primary production north of Iceland in relation to water masses in May–June 1970–1989, International Council for the Exploration of the Sea, CM 1984/L:20, 17 pp., 1984.

Toggweiler, J. R.: Carbon overconsumption, Nature, 363, 210–211, 1993.

Wanninkhof, R.: Relationship Between Wind Speed and Gas Exchange Over the Ocean, J. Geophys. Res., 97, 7373–7382, 1992.

Weiss, R. F.: Carbon dioxide in water and seawater: The solubility of a non-ideal gas, Mar. Chem., 2, 203–215, 1974.

Williams, P. J. I.: Evidence for the seasonal accumulation of carbon-rich dissolved organic material, its scale in comparison with changes in particulate material and the consequential effect on net C/N assimilation ratios, Mar. Chem., 51, 17–29, 1995.

Zhai, L., Gudmundsson, K., Miller, P., Peng, W., Guðfinnsson, H., Debes, H., Hátún, H., White Iii, G. N., Hernández Walls, R., Sathyendranath, S., and Platt, T.: Phytoplankton phenology and production around Iceland and Faroes, Cont. Shelf Res., 37, 15–25, doi:10.1016/j.csr.2012.01.013, 2012.

Permissions

All chapters in this book were first published in Biogeosciences, by Copernicus Publications; hereby published with permission under the Creative Commons Attribution License or equivalent. Every chapter published in this book has been scrutinized by our experts. Their significance has been extensively debated. The topics covered herein carry significant findings which will fuel the growth of the discipline. They may even be implemented as practical applications or may be referred to as a beginning point for another development.

The contributors of this book come from diverse backgrounds, making this book a truly international effort. This book will bring forth new frontiers with its revolutionizing research information and detailed analysis of the nascent developments around the world.

We would like to thank all the contributing authors for lending their expertise to make the book truly unique. They have played a crucial role in the development of this book. Without their invaluable contributions this book wouldn't have been possible. They have made vital efforts to compile up to date information on the varied aspects of this subject to make this book a valuable addition to the collection of many professionals and students.

This book was conceptualized with the vision of imparting up-to-date information and advanced data in this field. To ensure the same, a matchless editorial board was set up. Every individual on the board went through rigorous rounds of assessment to prove their worth. After which they invested a large part of their time researching and compiling the most relevant data for our readers.

The editorial board has been involved in producing this book since its inception. They have spent rigorous hours researching and exploring the diverse topics which have resulted in the successful publishing of this book. They have passed on their knowledge of decades through this book. To expedite this challenging task, the publisher supported the team at every step. A small team of assistant editors was also appointed to further simplify the editing procedure and attain best results for the readers.

Apart from the editorial board, the designing team has also invested a significant amount of their time in understanding the subject and creating the most relevant covers. They scrutinized every image to scout for the most suitable representation of the subject and create an appropriate cover for the book.

The publishing team has been an ardent support to the editorial, designing and production team. Their endless efforts to recruit the best for this project, has resulted in the accomplishment of this book. They are a veteran in the field of academics and their pool of knowledge is as vast as their experience in printing. Their expertise and guidance has proved useful at every step. Their uncompromising quality standards have made this book an exceptional effort. Their encouragement from time to time has been an inspiration for everyone.

The publisher and the editorial board hope that this book will prove to be a valuable piece of knowledge for researchers, students, practitioners and scholars across the globe.

List of Contributors

S.-J. Kao
State Key Laboratory of Marine Environmental Science, Xiamen University, Xiamen, China

B.-Y. Wang
State Key Laboratory of Marine Environmental Science, Xiamen University, Xiamen, China

L.-W. Zheng
State Key Laboratory of Marine Environmental Science, Xiamen University, Xiamen, China

K. Selvaraj
State Key Laboratory of Marine Environmental Science, Xiamen University, Xiamen, China

S.-C. Hsu
Research Center for Environmental Changes, Academia Sinica, Taipei, Taiwan

X. H. Sean Wan
State Key Laboratory of Marine Environmental Science, Xiamen University, Xiamen, China

M. Xu
State Key Laboratory of Marine Environmental Science, Xiamen University, Xiamen, China

C.-T. Arthur Chen
Department of Oceanography, National Sun Yat-sen University, Kaohsiung, Taiwan

F. S. Pacheco
Earth System Science Center, National Institute for Space Research, São José dos Campos, 12227-010, São Paulo, Brazil

M. C. S. Soares
Laboratory of Aquatic Ecology, Federal University of Juiz de Fora, Juiz de Fora, 36036-900, Minas Gerais, Brazil

A. T. Assireu
Institute of Natural Resources, Federal University of Itajubá, Itajubá, 37500-903, Minas Gerais, Brazil

M. P. Curtarelli
Remote Sense Division, National Institute for Space Research, São José dos Campos, 12227-010, São Paulo, Brazil

F. Roland
Laboratory of Aquatic Ecology, Federal University of Juiz de Fora, Juiz de Fora, 36036-900, Minas Gerais, Brazil

G. Abril
Laboratoire Environnements et Paléoenvironnements Océaniques et Continentaux (EPOC), CNRS, Université Bordeaux 1, Avenue des Facultés, 33405 Talence, France

J. L. Stech
Remote Sense Division, National Institute for Space Research, São José dos Campos, 12227-010, São Paulo, Brazil

P. C. Alvalá
Earth System Science Center, National Institute for Space Research, São José dos Campos, 12227-010, São Paulo, Brazil

J. P. Ometto
Earth System Science Center, National Institute for Space Research, São José dos Campos, 12227-010, São Paulo, Brazil

K. H. Salmon
Environment, Earth and Ecosystems, The Open University, UK

P. Anand
Environment, Earth and Ecosystems, The Open University, UK

P. F. Sexton
Environment, Earth and Ecosystems, The Open University, UK

M. Conte
Bermuda Institute of Ocean Sciences, St George's GE01, Bermuda

G. Abril
Laboratoire EPOC, Environnements et Paléoenvironnements Océaniques et Continentaux, CNRS, Université de Bordeaux, France
Programa de Geoquímica, Universidade Federal Fluminense, Niterói, Rio de Janeiro, Brazil

S. Bouillon
Katholieke Universiteit Leuven, Department of Earth & Environmental Sciences, Leuven, Belgium

F. Darchambeau
Unité d'Océanographie Chimique, Université de Liège, Belgium

C. R. Teodoru
Katholieke Universiteit Leuven, Department of Earth & Environmental Sciences, Leuven, Belgium

T. R. Marwick
Katholieke Universiteit Leuven, Department of Earth & Environmental Sciences, Leuven, Belgium

F. Tamooh
Katholieke Universiteit Leuven, Department of Earth & Environmental Sciences, Leuven, Belgium

F. Ochieng Omengo
Katholieke Universiteit Leuven, Department of Earth & Environmental Sciences, Leuven, Belgium

N. Geeraert
Katholieke Universiteit Leuven, Department of Earth & Environmental Sciences, Leuven, Belgium

L. Deirmendjian
Laboratoire EPOC, Environnements et Paléoenvironnements Océaniques et Continentaux, CNRS, Université de Bordeaux, France

P. Polsenaere
Laboratoire EPOC, Environnements et Paléoenvironnements Océaniques et Continentaux, CNRS, Université de Bordeaux, France

A. V. Borges
Unité d'Océanographie Chimique, Université de Liège, Belgium

C. D. MacLeod
Department of Zoology, University of Otago, Dunedin, New Zealand

H. L. Doyle
Department of Chemistry, University of Otago, Dunedin, New Zealand

K. I. Currie
Department of Chemistry, University of Otago, Dunedin, New Zealand
National Institute of Water and Atmospheric Research (NIWA), Dunedin, New Zealand

I. Fedorova
Arctic and Antarctic Research Institute, St. Petersburg, Russia
Hydrology Department, Institute of Earth Science, Saint Petersburg State University, St. Petersburg, Russia

A. Chetverova
Arctic and Antarctic Research Institute, St. Petersburg, Russia
Hydrology Department, Institute of Earth Science, Saint Petersburg State University, St. Petersburg, Russia

D. Bolshiyanov
Arctic and Antarctic Research Institute, St. Petersburg, Russia
Hydrology Department, Institute of Earth Science, Saint Petersburg State University, St. Petersburg, Russia

A. Makarov
Arctic and Antarctic Research Institute, St. Petersburg, Russia

J. Boike
The Alfred Wegener Institute, Helmholtz Centre for Polar and Marine Research, Potsdam, Germany

B. Heim
The Alfred Wegener Institute, Helmholtz Centre for Polar and Marine Research, Potsdam, Germany

A. Morgenstern
The Alfred Wegener Institute, Helmholtz Centre for Polar and Marine Research, Potsdam, Germany

P. P. Overduin
The Alfred Wegener Institute, Helmholtz Centre for Polar and Marine Research, Potsdam, Germany

C. Wegner
Helmholtz Centre for Ocean Research, Kiel, Germany

V. Kashina
Hydrology Department, Institute of Earth Science, Saint Petersburg State University, St. Petersburg, Russia

A. Eulenburg
The Alfred Wegener Institute, Helmholtz Centre for Polar and Marine Research, Potsdam, Germany

E. Dobrotina
Arctic and Antarctic Research Institute, St. Petersburg, Russia

I. Sidorina
Hydrology Department, Institute of Earth Science, Saint Petersburg State University, St. Petersburg, Russia

P. Bragée
Quaternary Sciences, Department of Geology, Lund University, Sweden

F. Mazier
GEODE, UMR5602, Jean Jaures University, Toulouse-Le Mirail, France

A. B. Nielsen
Department of Physical Geography and Ecosystem Science, Lund University, Sweden
Quaternary Sciences, Department of Geology, Lund University, Sweden
Department of Biology and Environmental Science, Linnæus University, Sweden

P. Rosén
Department of Ecology and Environmental Science, Umeå University, Sweden

D. Fred
Quaternary Sciences, Department of Geology, Lund University, Sweden

D. Hammarlund
Quaternary Sciences, Department of Geology, Lund University, Sweden

A. Broström
Quaternary Sciences, Department of Geology, Lund University, Sweden

W. Granéli
Department of Biology, Aquatic Ecology, Lund University, Sweden

A. Kubo
Department of Ocean Sciences, Tokyo University of Marine Science and Technology, 4-5-7 Konan, Minato-ku, Tokyo, 108-8477, Japan

M. Yamamoto-Kawai
Center for Advanced Science and Technology, Tokyo University of Marine Science and Technology, 4-5-7 Konan, Minato-ku, Tokyo, 108-8477, Japan

J. Kanda
Department of Ocean Sciences, Tokyo University of Marine Science and Technology, 4-5-7 Konan, Minato-ku, Tokyo, 108-8477, Japan

J. F. Oxmann
GEOMAR Helmholtz Centre for Ocean Research Kiel, Marine Biogeochemistry, 24148 Kiel, Germany

L. Schwendenmann
School of Environment, The University of Auckland, Auckland 1010, New Zealand

N. J. Silbiger
University of Hawaii, at Manoa, Hawaii Institute of Marine Biology, PO Box 1346, Kaneohe, Hawaii

M. J. Donahue
University of Hawaii, at Manoa, Hawaii Institute of Marine Biology, PO Box 1346, Kaneohe, Hawaii

R. T. Letscher
Earth System Science, University of California, Irvine, CA, USA

J. K. Moore
Earth System Science, University of California, Irvine, CA, USA

Y.-C. Teng
Earth System Science, University of California, Irvine, CA, USA

F. Primeau
Earth System Science, University of California, Irvine, CA, USA

L. Ran
Department of Geography, National University of Singapore, 117570, Singapore

X. X. Lu
Department of Geography, National University of Singapore, 117570, Singapore
College of Environment & Resources, Inner Mongolia University, Hohhot, 010021, China

J. E. Richey
School of Oceanography, University of Washington, Box 355351, Seattle, WA 98195-5351, USA

H. Sun
Institute of Geology and Geophysics, Chinese Academy of Sciences, Beijing, 100029, China

J. Han
Institute of Geology and Geophysics, Chinese Academy of Sciences, Beijing, 100029, China

R. Yu
College of Environment & Resources, Inner Mongolia University, Hohhot, 010021, China

S. Liao
Tongguan Hydrographic Station, Yellow River Conservancy Commission, Tongguan, 714399, China

Q. Yi
Toudaoguai Hydrographic Station, Yellow River Conservancy Commission, Baotou, 014014, China

E. Jeansson
Uni Research Climate, Bergen, Norway

R. G. J. Bellerby
Uni Research Climate, Bergen, Norway
Norwegian Institute for Water Research (NIVA), Bergen, Norway

Skjelvan
ophysical Institute, University of Bergen, Bergen,
orway

. Frigstad
orwegian Environment Agency, Oslo, Norway

R. Ólafsdóttir
arine Research Institute, Reykjavik, Iceland

J. Olafsson
Marine Research Institute, Reykjavik, Iceland
Institute of Earth Sciences, University of Iceland,
Reykjavik, Iceland